Cultures without Culturalism

Cultures without Culturalism

THE MAKING OF SCIENTIFIC KNOWLEDGE

Karine Chemla and Evelyn Fox Keller, *editors*

Duke University Press · Durham and London · 2017

Printed and bound by CPI Group (UK) Ltd, Croydon, CR0 4YY
Designed by Courtney Leigh Baker
Typeset in Arno Pro by Westchester Publishing Services

Library of Congress Cataloging-in-Publication Data
Names: Chemla, Karine, editor. | Keller, Evelyn Fox, [date] editor.
Title: Cultures without culturalism : the making of scientific
knowledge / Karine Chemla and Evelyn Fox Keller, editors.
Description: Durham : Duke University Press, 2017. |
Includes bibliographical references and index.
Identifiers: LCCN 2016042485 (print)
LCCN 2016044201 (ebook)
ISBN 9780822363569 (hardcover : alk. paper)
ISBN 9780822363729 (pbk. : alk. paper)
ISBN 9780822373094 (e-book)
Subjects: LCSH: Science—Social aspects. | Culture. |
Social epistemology.
Classification: LCC Q175.55.C85 2017 (print) |
LCC Q175.55 (ebook) | DDC 303.48/3—dc23
LC record available at https://lccn.loc.gov/2016042485

COVER ART: Karen Margolis, *Ichor*, 2010. Mixed media on paper, 14 × 11 inches. Courtesy of
the artist.

For Cale, Chloe, and Joachim:
the generation for whom we launch this foray
into the dynamic multiculturalism
of the future

CONTENTS

This volume is the product of two workshops the editors organized and convened (in 2008 and 2011) at Les Treilles (France). These workshops were based on research the editors carried out, when Evelyn Fox Keller was awarded a Chaire Blaise Pascal by the Fondation de l'Ecole Normale Supérieure (2005–2007). The first workshop convinced the participants of the value of publishing a book based on the papers that had been given. A selection of papers were then revised for greater coherence and discussed during the second workshop, with recommendations for further revisions. Finally, Karine Chemla's residency during the summer 2013 at Les Treilles allowed her to work on the completion of the book. We thank the Fondation des Treilles for its hospitality in surroundings that were inspirational and for its support in the preparation of the book. Many thanks to Karen Margolis for her indispensable contribution to the completion of the volume. Part of the research presented in the book led to the European Research Council project SAW (European Union's Seventh Framework Programme [FP7/2007-2013] / ERC grant agreement 269804), for which Chemla is principal investigator, together with Agathe Keller and Christine Proust. Chemla acknowledges funding from the project in support of her work on the introduction to the book. It is a pleasure to express our gratitude to Duke University Press and, more specifically, to Kenneth Wissoker and his colleagues, Jade Brooks, Nicole Campbell, Heather Hensley, Danielle Houtz, Christine Riggio, and Trish Watson, for their warm and rigorous support, and also to the two referees whose thoughtful suggestions have been decisive in improving the book.

Introduction

A Phenomenon and a Multiplicity of Concepts

Scientific knowledge was once perceived to be universal and hence unified. The practice of science was similarly regarded as unitary (at least ideally), which is another way of stating a second thesis, that of the univocality of the scientific method. Yet, even a superficial observation of different groups of scientists at work shows diversity among local and collectively shared ways of doing science. Decades of work in science studies have produced incontrovertible evidence establishing the need to acknowledge this diversity and to explore its significance. The question is, how?

Many efforts have been made to respond to this need, with Thomas S. Kuhn's (1962, 1970) concept of paradigm perhaps the best known. A key feature of the concept of paradigm is that it allows us to distinguish among approaches to what might appear to be the same range of problems by different groups of practitioners, often working in radically disparate time periods. Ludwik Fleck had developed kindred arguments in his publications on *Denkkollektiv* (thought collectives) (see Fleck 1935, 1979; see also Cohen and Schnelle 1986). And more recently, scholars in science studies focusing on a variety of sociological, historical, and philosophical projects have developed several other concepts to address such concerns. Alistair Crombie introduced the concept of styles of thought to distinguish among six modes of "scientific inquiry, argument and explanation," all within what he regarded as a "specific style of rationality created within European culture."[1] Ian Hacking embraces this anthropological project but speaks instead of styles of reasoning, or of styles of scientific thinking and doing. His concern is to emphasize the dimension of doing so crucial to ways of finding out but missing from Crombie's concept. The shift in language is also important to Hacking's philosophical project of tracking different ways of introducing new objects, new sentences about these objects, and new criteria both for determining what sentences are subject to the judgment of true or false and for adjudicating their truth or falsity (see Hacking 1982, 1992, 2012).

Both Crombie's and Hacking's concepts seek to describe ways of doing science as *longue durée* phenomena. Other concepts have emerged in the context of projects aimed at grasping similar but shorter-term phenomena. With the introduction of "epistemic cultures," sociologist Karin Knorr Cetina seeks to understand knowledge societies in terms that are not only economic (seeing knowledge as a productive force) but also cultural. In this regard, epistemic cultures (alternatively, cultures of knowledge settings or different machineries of knowing) are for her constitutive units of knowledge societies. Her approach to epistemic cultures is mainly through ethnographic studies (carried out, e.g., in laboratories) to identify the entities an epistemic culture brings into play and the relations between these entities such a culture establishes (Knorr Cetina 2005, 65, 67–68; see also Knorr Cetina 1999). By contrast, Evelyn Fox Keller reviews the history of twentieth-century approaches to the problem of embryonic development, focusing on the failure of communication among different collectives working on the same phenomena. She introduces the notion of epistemological culture to account for the reluctance, voiced by one group, to accept the types of explanation, or even the questions, put forward by another as meaningful (see Keller 2002). For her, epistemological factors (e.g., types of explanations sought, or modes of reasoning employed) appear as features of a scientific culture that are essential to take into account in characterizing collectively shared ways of doing science and in explaining failures of communication between groups.

Although the set of concepts mentioned above is far from exhaustive, it suffices to illustrate the variety of approaches that historians, philosophers, sociologists, and anthropologists of science have been advocating, all in the effort to further identify the characteristics that lend particular ways of making scientific knowledge their specificity and that distinguish among the diverse cultures of scientific practice.

How do these concepts relate to one another? To what extent are they redundant? To what extent incompatible? In what ways might they inform and enrich more concrete, empirically based, historical and philosophical studies? These are obvious theoretical questions to be addressed in attempting to sharpen the methodological tools needed to account for diversity in scientific practice. This book is partly devoted to such an inquiry. However, during this effort we have encountered other questions that also need to be addressed.

Part I: Some Problems Attached to Concepts of Culture,
and the Theses of the Book

Several concepts put forward in the effort to attend to the diversity in ways of doing science invoke the notion of culture—a notion that brings with it problems long debated in the humanities, although less so in science studies. We suggest that the time has come to fill this gap.

An overriding problem that has been much emphasized, especially in the anthropological literature, derives from the enormously difficult task of defining what one means by "culture" (see Geertz 2000, 11–13; Knorr Cetina 2005, 71). One of the tasks this book addresses is clarifying what appears important to the meaning of this term in the particular context of scientific practice.

Closely associated with this problem of definition are two risks that the recent interest in local cultures of scientific practice brings with it—risks to which the notion of culture is itself prone and to which our own intellectual histories surely make us especially sensitive: the temptation posed by concepts of culture to slight the dynamic and interactive character of the formation of any kind of cultural identity, and the pitfall of cultural essentialism (what we call culturalism), the view of cultures as essentially homogeneous, static, and fixed by prior constraints or race, gender, or nationality.[2]

In this introduction, we first query the necessity of a notion of culture for our project and then go on to examine these two risks in greater detail. This enables us to then present our own approach to the disunities of scientific practice, which we claim can avoid such risks.

In chapter 1, Donald MacKenzie begins with a reflection on the general difficulties of the concept of culture in the social sciences, and in science studies in particular. Would it be better, he asks, to restrict ourselves to such notions as clusters of practices? In relation to his own case study of securities and ratings in the last financial crisis, and more specifically of the practice of evaluation, he decides not, for doing so would deprive us of a crucial explanatory resource; it would also lose some key features of the situations under study.

More specifically, in MacKenzie's own case study, the concept of culture brings into focus certain key features that prove of general validity. First, it draws attention to the different ontologies associated with different practices (in this case, different assumptions about "what economic value consists of and [about] the nature of the economic processes that create it"). Second, culture captures the different processes of socialization associated with these ontologies, as well as the mechanisms for interaction among participants. Finally, the notion of culture focuses attention on how patterns of change in current practice depend on local histories of past practice. Indeed, these suggestions, along with others to which we return later, signal features of scientific cultures that recur throughout this book.

MacKenzie concludes that, in this case at least, culture remains a useful and even indispensable analytical concept, despite the problems it raises.[3] Accepting this conclusion, however, still leaves us with the task of addressing concerns about the risks associated with the notion of culture. But first, do these risks actually manifest in the history and philosophy of science? This question is addressed from a variety of perspectives in the remaining chapters of part I.

Kenji Ito's contribution in chapter 2 of this volume, focusing on histories of science in Japan, bears witness to the fact that such problems are quite real. He identifies (and deplores) essentialist tendencies in a number of prominent publications on the history of physics in Japan. These historical writings, he notes, give differences between physics in Japan and elsewhere pride of place and often account for them in culturalist terms (i.e., by reference to the actors' Japaneseness). Ito's primary concern with the latter view is its tacit view of scientific activities in Asia as exotic and as shaped by stereotypic national identities of "the East." In other words, his focus is on the particular form of essentialism described by Edward Said as orientalism (see Said 1978). Note, however, that in this case orientalism is as much a tacit assumption as it is a product of the history of science. Differences in scientific practices are interpreted as manifesting a Japanese culture and identity, while the interpretation in turn helps shape or reinforce the stereotype on which it is based. This orientalizing stance, Ito stresses, is not just common in historical discussions of science in Japan but is even encouraged by the expectations of the profession. Of particular importance here is his analysis of the forces at play in contemporary history of science that contribute to the promotion of essentialist historiography.[4]

Ito's main criticisms of such orientalist tendencies in the history of science are fourfold. First, they misrepresent the de facto diversity of science in Japan. Second, they overlook the transnational dimensions of current practices.[5] Third, in their preoccupation with difference, these historiographies obscure the extent to which the appearance of sameness (or uniformity) between twentieth-century physics in Japan and elsewhere is in fact a puzzle and not a natural or spontaneous outcome of the universality of science. Sameness in scientific practice or knowledge actually appears as an achievement deriving from actors' intentional effort to overcome differences. Indeed, the issue of shaping sameness, and the work through which actors achieve it, recurs throughout this book and is clearly a phenomenon in critical need of historical analysis.

Finally, Ito criticizes the empirical support provided by these historiographies. This point requires clarification. Indeed, if one can expose problems in the empirical basis that should support observers' essentialist conclusions with respect to scientific practice and knowledge in Japan, the tricky issue is that such essentialist tendencies at the same time echo many accounts by Japanese actors themselves. How are we to deal with this issue? Ito dubs this phenomenon self-orientalism and challenges us to deal with it, at the very least reminding us to exercise caution in distinguishing between actors' and observers' categories. The introduction of culture as an actors' category and actors' use of it are phenomena that observers must examine in a critical way and not absorb automatically into their analytic toolbox. Whether or not actors' conceptions and uses of culture are those relevant and useful for an observer's project must be treated as an open question.

Chapter 3 provides a good example of the need for such an approach. Guillaume Lachenal analyzes the circumstances and contexts in which actors might find it useful to claim authority for resources they perceive as indigenous. His case study illustrates the use of cultural essentialist categories by scientific actors in contexts extending far beyond scientific communities. His particular focus is on the controversy following the announcement in 2001 of the discovery of a vaccine against AIDS by a well-known Cameroon professor of medicine and former minister of health. Of relevance to us here is the fact that the vaccine was defined by local actors as Cameroonian and hence as opposed to the culture of transnational biomedical research. Lachenal also relates the case to other episodes in which actors claim an inherent "Africanity" for their cures in the expectation that such claims might enhance their value.

The risks attached to strategies of enhancing value of scientific research through appeals to native culture are manifest and would only be exacerbated were they to be embraced by anthropologists, sociologists, historians, and philosophers of science. In turn, the use of culturalist categories by observers can also provide support for their deployment by actors themselves, escalating the risks yet further. As Robert Kowalenko (2011) warns us, the story of AIDS in South Africa might well illustrate how disastrous the consequences can be.[6]

Like Ito, Lachenal insists that observers need to take actors' culturalist statements as objects of research and not embrace them uncritically, and he clearly illustrates the value of such research with his own analysis of how the combination of the organization of worldwide biomedical research and local politics influenced the development of these episodes of self-orientalism. Noteworthy for our argument is the fact that, in Ito's case, actors' culturalism relates to the resources available for the practice of science—the Japanese would have access to specific resources helpful in the advancement of a theory—while in Lachenal's case study, Africanity is also claimed for the results of the inquiry, that is, for the knowledge produced. The distinction between these two levels will prove of interest in what follows.[7]

Worries with a History

Problems associated with cultural essentialism with respect to science have worried historians of science at least since World War II. Sinologist Joseph Needham, for example, repeatedly voiced concern that perceiving science in such ways ran the risk of denying what he called the "continuity" of both mankind and science (Chemla 2014). Relatedly, he also described the dangers he attached to the conception that mankind could be decomposed into distinct cultures, separate from one another, each developing its own science, valid only for the originating culture, specific to it, and incommensurate with the other sciences (see Chemla 2014 for references to specific quotations). By contrast, the historiographies of science and culture that Needham and historian Lucien Febvre developed from 1947 onward, in the context of the newly established UNESCO, all emphasized that mankind could not be meaningfully cut into pieces (see Petitjean and Domingues 2007). Interestingly, when in the 1970s Needham became explicit about risks specific to the history of science, he had in mind not only cultures of the type discussed by Ito and Lachenal in this book, which actors of that time were loudly re-claiming as specific, but also the ways of pinpointing diversity in collectively shared ways of doing science, which observers, like Kuhn with his notion of paradigms, were already beginning to advocate.

Needham's strategy in reaction to the dilemma resembles that advocated by Ito: he focused on "sameness." He did so in a specific way. He systemati-cally read similarity between scientific results found in two distinct parts of the planet as proof of a circulation of knowledge from one to the other, and hence as proof that knowledge could be shared worldwide. However, this strategy has been criticized for methodological flaws, and rightly so. State-ments of similarity overshadowed significant differences, and the hypothesis of independent occurrence was almost systematically ruled out. Moreover, Needham's strategy actively denies the manifest diversity in cultures of scientific practice that we want to address. Is there no way of accounting for such diversity without running into the problems associated with cultural essentialism that both Ito's and Lachenal's chapters demonstrate?

The question is not confined to science studies. Evelyn Fox Keller in chap-ter 4 examines what historians and philosophers of science might learn from decades of debate in feminist theory about similar issues. Recognition of the force of gender categories in organizing our conceptual and social landscapes marked a critical milestone in the emergence of feminist theory, but discus-sion quickly became enmired in worries about the lure of cultural essential-ism. The question for feminist scholars—how can one recognize the force of gender categories while avoiding the pitfalls of gender essentialism?—bears an obvious parallel to the one we are concerned with here. At the same time, the history of debates in feminist studies also reveals another pitfall: an anxi-ety over essentialism so great as to threaten the very effort to understand the significance of cultural differences. Exactly the same questions arise here. How do we avoid throwing out the baby with the bath water? How can one write about scientific cultures (or gender) as both recognizable and consequential, without inviting the assumption that such cultures (or categories) are fixed and closed to external influence?

To summarize the points made so far, we can readily recognize the impor-tance of highlighting differences between cultures of scientific practice in the history and philosophy of science, both in efforts to combat tacit assump-tions of uniformity/spontaneous universality (and hence the hegemony implied by such assumptions) and in the effort to do justice to the actual life of the sciences. But doing so is fraught with dangers. Up to now we have focused on the dangers of culturalism, both in science studies and in feminist theory. These risks relate to the shaping of collectives as separate using either the way they practice science (against the assumption of uniformity) or the kind of knowledge they produce (against the assumption of universality of science). But in this latter respect, Keller emphasizes, highlighting such differences also creates other problems, eliciting other dangers. In particular, it raises the old but still critical problem of relativism—the other problem Needham attempted to avoid by insisting on the universality of science.

Can the conclusions of different scientific cultures be evaluated in relation to each other? Or does the recognition of differences among them imply that they can be judged only from within? Must we assume (as was once common) that these conclusions have equal relevance to the world (or worlds) to which they ostensibly refer? These questions are consubstantial with the very project of considering differences among cultures. Here again, it proves vital for the analysis to distinguish between actors' claims, with respect to the specificity of their knowledge, and observers' assessments. Actors' claims are important objects of research, much in the way Lachenal advocates in his chapter. Keller reminds us, as observers, of pitfalls of ignoring how much of the natural world we inhabit is in fact shared. She also insists on the questions compelled by the recognition of diversity among cultures of scientific practice about how these different cultures connect to one another and how a wider and more critical consensus can be—and often is—reached. She argues that, especially in the light of the global problems that now loom, these questions too are part of the challenge we must face.

Our Approach to Cultures of Scientific Practice

We can now lay out the agenda of this book more clearly. For us, a key outcome of part I is that it sheds light on critical aspects in notions of culture that are at least sometimes invoked by actors themselves. In these conceptions, cultures are all-encompassing. They are general contexts in which scientific activity takes place. They leave their imprint on the practices and bodies of knowledge achieved, granting them their value in the eyes of the collective sharing the culture in question. Furthermore, these cultures are taken to be impervious to change. Most important, the scientific activity of the actors has no impact on their culture—*that* remains unaffected by external influences. These features of a notion of culture, we claim, are precisely those that allow the emergence of culturalist agenda and give rise to the risks identified. But do they accord with what we observe in scientific practice? We don't believe so.

If, following MacKenzie's analysis, we acknowledge the need for science studies to put into play some concept of culture, this book aims at developing another approach. Its core idea is introduced in MacKenzie's chapter and explored from various viewpoints in the remaining parts of the book. Above all, MacKenzie (along with many of the other contributors) insists that a culture of scientific practice is a product of what actors do—it is an outcome rather than a cause of their activity. In other words, the book places emphasis on the fact that actors shape the immediate context in which they carry out their scientific activity. The term "context" can refer to many different notions, but it is especially its use in relation to the specifics of scientific practice on which this book concentrates.

Furthermore, as the outcome of activity, a culture of scientific practice is subject to constant change—in relation to the problems actors address and the goals they pursue and in the ways they draw on the resources available to them to mold and remold their objects of research, their values, and so forth. Finally, we suggest that establishing bridges between cultures of scientific practice is also part of what actors do. Overcoming differences in knowledge and practice, constructing sameness (or even universality), and achieving consensus are not properties of scientific practice and knowledge that are given a priori but are outcomes of actors' knowledge activities. However, this part of scientific work—whether considered synchronically or diachronically—has as yet not been systematically examined as a general phenomenon in the history and philosophy of science (we return to this issue later).[8]

We will proceed in three main steps. Part II of the book examines specific components of these cultures, whereas part III takes a more global view. Finally, part IV addresses the historiographic implications of our suggestions for investigating cultures of scientific practice while avoiding the pitfalls of culturalism.

Part II: A Toolbox for Investigating Cultures
of Scientific Practice

The essays in part II engage the issues of cultural diversity by adopting an analytic approach, discussing constituent elements and features of cultures that appear essential to characterize a given way of carrying out scientific activities.

The thesis of culture as something that actors do comes out forcefully from Nancy J. Nersessian's case studies in chapter 5, where she examines the constitution of new cultures arising at the interface of biology and engineering. By focusing especially on the material models that are deployed in different laboratories, she demonstrates how these devices not only constitute an organizing center for the laboratory's culture but also come to represent its signature. Here, actors' shaping of the laboratory cultures in each of the cases in question is made manifest by the central role played by a material device, collectively designed and transformed throughout the research process.

Nersessian investigates the processes by which these models both reflect and help shape the laboratory's material, conceptual, and social practices. First, they allow people in the lab to connect their work to one another, serving as hubs through which the collective remains connected. These devices are thus central to the mechanisms of interaction among participants. Second, the devices provide the means through which cognitive operations are carried out. Cognitive operations in the laboratory are hence intimately tied to the laboratory's culture, as indeed are the results they yield. Nersessian thereby introduces a new general phenomenon, echoed in other chapters as well: the knowledge produced in the context of a scientific culture presents specificities that can be correlated with features of the culture itself.

The aspects so far evoked could have been captured through looking at the laboratory cultures as epistemic cultures. However, a third key dimension is precisely what Keller's epistemological cultures bring into focus: these devices, Nersessian argues, come to embody the epistemological values and hypotheses of the lab. Indeed, the study shows that, far from being static, the devices and hence the cultures they help constitute are shaped and reshaped by the actors in the laboratory according both to past results and to the (evolving) dictates of particular experimental agendas. Also, illustrating MacKenzie's comments about the historicity of epistemological cultures, they embody the cognitive history of the lab—a history that is recorded by the actors both as a potential guide to future developments and as a conscious reminder of the evolution of the laboratory's culture. The correlation between the scientific culture, on the one hand, and the questions addressed and the goals pursued in the laboratory, on the other, is here not only established but also clarified.

The cases Nersessian studies also illustrate how specific cultures are shaped as separate from other cultures in which similar problems are addressed. At the same time, these cases highlight the openness and ongoing interactivity of the cultures with respect to their environments—in this context, specifically the other cultures with which they are in communication. The devices are thus hybrid—borrowing from biomedicine and engineering while at the same time feeding back into the cultures from which they derive; the values they embody are equally hybrid and dynamic, collectively shaped through the research process.

In chapter 6 Mary S. Morgan focuses on the ways in which a scientific culture shapes an object of study while also fashioning new resources for research. With her case study of the emergence of an epistemic object out of a wider cultural transformation, she illustrates yet another form of openness of a local scientific culture. Her concern is with the "glass ceiling," a phenomenon that became an object of study for social scientists—indeed, that became visible to the public at large—only in late twentieth-century America, its visibility arising in direct response to the rise of second-wave feminism. Also, and of particular interest to us here, is the extent to which the scientific study of the glass ceiling made use of qualitative personal experience—a mode of analysis that had previously been absent from conventional social science studies but that was now demanded by feminist critiques. Here too, the dynamic and open nature of the way of working and its shaping by researchers are manifest, in this case affecting the type of the data used as resources. The inclusion of these data may have been a response to external pressure, but it clearly led to a radical shift in the internal epistemological culture of those studying the glass ceiling.

Morgan draws from this historical example some penetrating observations about potential differences between scientific cultures in the social and natural sciences, arising not only from the intimate relation between the collectives of scientist-observers and the larger community in which they live but also from the relations among communities of observers, of observed, and of generators of questions and concerns. As her case study clearly illustrates, cultures of social-scientific practice are de facto both open to and dependent upon the environments in which they are embedded. Indeed, it is in good part the larger communities that "raise questions and prompt what is found problematic and thus what is studied." They also influence how it is to be studied. Thus, in addition to providing the context for scientific research, these wider communities may be far more involved in defining matters of content and ontology than they are in the study of the natural sciences.

Chapters 5 and 6 demonstrate the centrality of communication among various types of contiguous collectives to a culture of scientific practice. We see actors borrowing elements that they then recycle to shape tools, topics of research, and even data. Forms of communication forged and adopted to connect laboratories and their environments thus emerge as essential features of the constitution and transformations of (at least some) scientific cultures, and it is precisely this topic that Claude Rosental addresses in chapter 7.

Rosental's case study deals with a very specific form of such communication, one that has become omnipresent in certain present-day scientific and technological practices: the public demonstration (demo). The simple fact that forms of communication vary illustrates their historicity and, accordingly, the malleability of cultures that adopt and adapt to new modes of communicating. In this specific case, demos depend on new techniques of communication, and they emerged in parallel with new modes of funding and management. By concentrating on them, Rosental in fact also suggests the fruitfulness of a cultural approach to funding schemes and management.

The focal point of Rosental's analysis is the use of demos in facilitating communication among managers, policy makers, and the public, on the one hand, and scientific workers on the other, and the impact of this form of communication on the knowledge culture itself. In the fields studied by Rosental, the need to produce such demonstrations for managers of large-scale programs shaped the organization of scientists' work in rather specific ways (a feature brought into focus by Knorr Cetina's notion of epistemic cultures). Further, Rosental argues that such modes of communication have a recognizable impact on the content of the work produced (issues that Keller's notion of epistemological cultures asks us to think about). Finally, because demos can be addressed to and visualized by a wider audience— that is, specific features and modalities of communication—this wider audience's reactions can also contribute to defining the goals that practitioners set themselves.

Such demos have become an essential part of the scientific practices that use them. By virtue of their role in mediating communication between two different cultures (i.e., between managers of science programs and groups engaged in specific projects), the demos are inevitably shaped by the culture of the managers to whom they are directed. With the example of modes of communication, Rosental demonstrates once again the dynamic character of cultural formation as bodies of knowledge and forms of practice evolving in response to interactions among different communities.

Thus far, all of the chapters in part II concentrate on specific scientific cultures, and through a close study of some of their constituent elements and forms of communication they underscore the dynamic and interactive nature of the constitution of these cultures. By focusing on yet another cultural element, David Rabouin in chapter 8 also draws our attention to patterns of cross-fertilization between different cultural formations in the context of a single discipline. Especially noteworthy are the patterns of intercultural circulation prominent in his discussion of style in the context of mathematics. In contrast to other notions of style, notably Hacking's "style of reasoning," Rabouin suggests that a style in mathematics can best be characterized as a way of writing. In the sense that he elaborates, a way of writing is an essential component of the scientific culture to which a mathematician belongs.[9]

For this notion of style, Rabouin draws inspiration from a paper written by the mathematician Claude Chevalley in 1935. In effect, Rabouin, as observer, adopts an actor's category in his analytic toolbox. The historical context in which Chevalley introduced the concept of style, Rabouin argues, might shed light on its usefulness for us today, at the same time as it indicates the variety of views actors formed about this concept at that time. Just shortly before Chevalley's paper, the notion of style had been invoked with another meaning by the German mathematician Ludwig Bieberbach to denigrate Jewish mathematics and praise the German or Aryan style of doing mathematics. The risks attached to these views were unfortunately realized quite dramatically, and historians like Needham never lost sight of them. Rabouin suggests that Chevalley's 1935 article, putting forth a different notion of style that emphasized circulation not just between France and Germany but among mathematicians of all origins, was nothing less than naive, and not unrelated to his historical context.

In any case, Rabouin highlights several interesting features that a notion of style conceived along these lines possesses. First, it helps account for the disunities in a mathematical culture of actors who otherwise share the same practices. In particular, Rabouin suggests that actors can share a way of writing but can differ in their interpretations of this writing. In short, they need not have a completely uniform way of working for their culture to be fruitful and dynamic. This remark points to a general and crucial issue of intracultural variation that awaits systematic research. Second, in contrast to Hacking's style of reasoning, in which all of mathematics falls under a single style (or, at best, under two styles), Rabouin's notion of style enables us to distinguish among mathematical collectives. It also seems to better describe how actors actually work. Particular styles, understood as ways of writing, are cultural elements that readily circulate among local cultures. How actors might specify them, in local contexts, and which changes styles undergo in the process of circulation are likewise issues worth pondering. Perhaps more important, we can infer from Rabouin's argument that styles are thus able both to connect local cultures and to mediate between local and more global features. We return below to this type of phenomenon.

Such circulation not only illustrates cultural interaction but also helps account for the stability that mathematical knowledge so often displays. Indeed, insofar as the phenomenon of the stabilization of mathematical knowledge resembles the process toward, and establishment of, sameness with which Ito is concerned, it would be interesting to see if the circulation of ways of writing physics might also prove of explanatory value.

Part III: Toward a More Global Approach—
Elements of a History of Culture Making

Taken together, the chapters in part II demonstrate how actors shape constituent elements specific to the culture in which they work. Importantly for our purpose, the chapters highlight how the formation of these elements draws on both internal and external resources and, more generally, how they are shaped by their larger political and social contexts. Actors design their working tools in response, for instance, to the directions of research and goals chosen, or to requirements from outside cultural formations with which they need to comply. In short, the making of scientific cultures has a history, and this history is as meaningful as the history of concepts and theories. What is more, these histories are intimately related to one another. Indeed, if these cultural elements provide analytic tools for understanding how collectives work, they also have explanatory power with regard to the types of knowledge produced in these contexts. We learn that knowledge and culture are dynamically correlated, with elements of both constantly circulating, taken up, and reworked by actors through the same type of processes.

Similar conclusions arise from the more global approaches to scientific cultures presented in part III. These studies illustrate the variation (according to the cases and issues dealt with) in scale at which analysis must be conducted. Albeit from quite different perspectives, the first two chapters focus on the resources individual actors bring into play in their shaping of their culture.

Koen Vermeir in chapter 9 explores the immediate context in which the term "culture" (as an actor's category, borrowed from agriculture) was first systematically employed to refer to the cultivation of a group of humans, in discussions of educational reform in Jesuit institutions in early modern Europe. These early uses of the term "culture" bring to the fore an issue of general importance to our concerns: the specific type of activity required for cultivating oneself or one's talent—in other words, what MacKenzie calls the process of socialization required to reach the state of culture valued in a particular social context. Vermeir's study also sheds light on how the actors themselves contribute to designing the process of acculturation.

Vermeir identifies the successive editions of Antonio Possevino's *Cultura Ingeniorum* (starting in 1593) as the main sites attesting to the shaping of this Jesuit epistemic culture. Through his analysis, he is able to show the relevance of both goals (e.g., Possevino's political goals, dictated by the agenda of the Jesuit company) and social context (notably, that of the Counterreformation) in the shaping of that culture. As is often the case with actors' categories of culture, the crafting of an identity that could be opposed to others was at the core of the project. But, for Possevino, culture was an activity through which this identity could be achieved, in contrast to other actors' conceptions of culture as a state deriving either from nature or from belonging to a group. Symptomatically, to define this new epistemic culture, Vermeir shows Possevino nevertheless borrowing elements from various sources, selecting and remolding them for his own uses—indeed, much as Georges Cuvier later did in shaping a new scientific culture in paleontology, as Bruno Belhoste describes in chapter 10.

For Possevino, the emphasis was placed on the making of cultivated men, able to fulfill tasks useful for the Jesuit company. For Cuvier, by contrast, the point was to create an environment, modes of reasoning, and practices that would enable him to draw conclusions about new epistemic objects. To approach the constitution of this culture, Belhoste first focuses on the working spaces developed by Cuvier at the end of the eighteenth century and beginning of nineteenth century in his effort to identify and restore extinct species. However, the task Belhoste sets himself leads him to suggest extending the meaning of the notion of working space, to designate less a physical location than a *dispositif*, the system of social, material, and epistemological resources that Cuvier assembled and mobilized (all of which could be found in the urban setting of Paris) to pursue his goals.

Cuvier's achievements depended on the particular way of carrying out research in paleontology that he forged out of this dispositif. He drew crucially from three distinct cultures, from which he created a unique synthesis: the traditions of museum culture, with the knowledge machineries and epistemological values attached to them; the culture of late Newtonianism then dominating the physical sciences in France, and especially the value it placed on specific styles of reasoning and types of explanation; and, finally, the epistemological resources of contemporary historical and antiquarian scholarship. Cuvier also put into play the many social and administrative resources provided and/or facilitated by his position at the National Museum of Natural History, as well as, of course, the wealth of analytic, observational, and graphical skills he personally had honed from years of experience. Belhoste's study shows us how Cuvier was able to fashion from these disparate resources a new and phenomenally productive way of carrying out research in paleontology, even while recycling features from other cultural formations. Cuvier's example (like Possevino's) illustrates the ability of a single individual, while clearly drawing from the social, institutional, and professional world around him, to forge a practice that could evolve from one man's way of doing science into a disciplinary culture. In fact, the forging of this culture represents one of the most important outcomes of Cuvier's work.

So far, we have mainly focused on the local shaping of a culture at a time, examining how this process involved taking up elements from other cultures. However, such a perspective, attentive to a single culture, or at most the cultures from which the latter drew, does not allow us to consider what happens through borrowing and reshaping at higher levels. The question that presents itself now is, does every culture of scientific practice follow a specific path, or are there phenomena that connect and relate these processes with one another? In brief, what does a more global perspective reveal about how these local cultures exchange with one another? Hans-Jörg Rheinberger in chapter 11 offers a crucial contribution to precisely these questions.

In his own effort to account for the generation of meaning in the context of scientific activity, Rheinberger finds it necessary to shift focus to what he calls a meso-level of analysis, intermediate between the micro-level of most laboratory studies and macro-level histories of disciplines. To this end, he invokes the concept of cultures of experiment—networks of experimental systems that are not only held together by material forms of sharing (i.e., by the sharing of technologies, of human agents with scientific know-how, and of biological substrates, objects, and/or experimental environments) but that actually unfold from these material interactions (see Rheinberger 1997, esp. chap. 9, for a development of these ideas).

Sharing, a form of communication that establishes bridges between a local culture and other cultural formations, appears here as a crucial operation, which incidentally evokes Rabouin's suggestion with respect to how styles as ways of writing work in mathematics. However, it would be a mistake to take sharing as a transparent operation. What is shared across a network, how it is shared, and what the effects of sharing are—these are crucial issues that remain in need of further analysis. Indeed, they are the elements by which another form of culture is shaped. What Rheinberger in fact suggests is that here, too, it is part of actors' activity to develop and shape these modes of sharing (indeed, in shaping the entire meso-level of cultural formation), and it behooves us to attend to the work thereby involved.

Rheinberger's particular case study focuses on the culture of in vivo experimentation, and he traces the historical evolution of this culture through the interactions between experimental systems developed in physiology and biochemistry. Despite operating on a different scale, experimental systems play much the same role for Rheinberger in the generation of scientific cultures as do Nancy Nersessian's devices: both serve simultaneously in the formation of a scientific community and as generators and embodiments of its current state of knowledge. In a sense, this view offers a suggestion with respect to both how phenomena like Kuhn's paradigms can take shape, and how the collective in the context of which they are adopted is dynamically formed.

Finally, as to the larger question of what, if anything, is special about scientific/experimental cultures, Rheinberger calls our attention to the inadequacies of the conventional opposition between culture and nature—an opposition that typically echoes the equally canonical division between mind and matter. In opposition to such dichotomies, Rheinberger insists on grounding culture itself in material interactions, while at the same time emphasizing that these interactions constitute the nodes through which meaning/culture is collectively engendered. In so doing, he points to the particular modes of circulation between mind and matter, nature and culture, as a way of addressing the question of what is special about experimental cultures. Rheinberger's approach thus further contributes to elucidating the dynamic and collective construction of sameness.

Through the clear-cut contrast it presents with Rheinberger's work in chapter 11, Fa-Ti Fan's study of an experimental culture in chapter 12 highlights the role actors play in shaping such cultures. It shows different modes of data shaping and data sharing, other types of actors involved, other processes of socialization, and other kinds of networks and forms of communication. Finally, it also brings into focus the impact of other cultural formations in the process of shaping these cultures.

By comparison with Rheinberger's case study, where the scale was smaller than that of a discipline, Fan's analysis of earthquake prediction as practiced in communist China during the Cultural Revolution in the 1960s and 1970s requires more of a macro perspective. It deals not with an academic culture but with a "people's science" of earthquake monitoring. The broader political and cultural context in which this scientific culture took shape can scarcely be ignored. Fan's main point is to stress the interaction of scientific ideas and political beliefs in shaping earthquake monitoring practices. From this political environment developed a culture of seismological practice with an agenda of its own. In particular, Fan emphasizes the variety of means (publication, training, material devices, and even a rhetoric of cultural essentialism) deployed to facilitate the inclusion of all citizens. This new knowledge culture further employed specific knowledge machinery: particular instruments, modes of observation, criteria for the mobilization of personnel and for the collection of data, and modes of communication. Each of these features reflected a value more generally placed on mass participation in the production and consumption of scientific knowledge, and together they account for the marked differences between the culture of seismology that developed in the China of that time and other, more exclusively academic cultures.

Fan's culture of earthquake prediction is further distinguished by methodological assumptions (e.g., an emphasis on the role of macroscopic phenomena in short-term prediction, the relatively greater significance attached to correlations over mathematical modeling, and the importance of a phenomenological approach) that embody prized epistemological values (e.g., precision). Noteworthy for us is the fact that all these cultural features can be correlated with the type of knowledge produced in this context.

Fan's study thus echoes some of the points made by Morgan in her chapter concerning the influence of the wider community on both the knowledge produced and the practices developed by a particular scientific culture. In like manner, Fan's observations of Mao extolling "Chinese science" as a complement to Western science echo Lachenal's and Ito's concerns about the opportunistic uses of self-orientalism. Like these other contributors, Fan too warns against the uncritical adoption of this actors' category in the observers' analytic toolbox.

With this chapter, we conclude our exploration of the approach to culture that is at the core of this book. We began with the theses that cultures are an outcome of actors' knowledge activity, in parallel and, more important, in correlation with other outcomes such as concepts, results, and theories. Exploring these theses led us to notice other equally important phenomena (e.g., the ways in which cultures of scientific practice borrow both from one another and from other cultural formations). Moreover, we can observe the work that actors perform in establishing the networks of sharing that shape higher-level cultures.

Part IV: Historiographic Implications

Yet a final question still remains: what might be the historiographic implications of identifying multiple coexisting scientific cultures? This is the question that part IV addresses, first from a synchronic and then from a diachronic viewpoint. Both of the case studies in part IV are devoted to mathematics, a discipline for which the historiographic implications of the notions of scientific culture, or of a multiplicity of cultures, have only recently begun to be systematically examined. Furthermore, previous discussions of such issues have either taken "culture" in the singular, or have taken an approach cast in conventional culturalist terms. We aim for a discussion more faithful to the actual practices of mathematics.

Caroline Ehrhardt takes a synchronic perspective in chapter 13, illustrating what can be gained by considering the distinct professional mathematical cultures of nineteenth-century Europe. She makes the strategic choice of concentrating on a single work, one that was widely read at the time: Galois's memoir about the resolution of equations. To shed light on the multiplicity of mathematical cultures, she focuses on the variety of ways in which Galois's ideas, as outlined in this memoir, were appropriated and developed by different groups of mathematicians working in different contexts. Her key point is to suggest that the notion of local culture helps us account for the various readings that were made of Galois's writings and the various mathematical theories that were built on the basis of these readings. Of particular relevance in this case is the fact that these local cultures differ in their practices of both proof and definition, in their ways of writing, in the mathematical contexts in which they interpret and develop Galois's writings, in their goals, and in their epistemological choices. At the same time, however, sufficient commonality needed to underlie all of these local cultures—this points to the existence of the meso-level introduced by Rheinberger—for Galois's writings to be meaningful to all of them in spite of their differences. What Ehrhardt shows is that these separate developments were later reworked and integrated into what came to be considered a unified Galois theory (despite interesting variations from one context to another). She thereby highlights the mathematical fruitfulness of having different mathematical cultures reworking, each in its own way, the same ideas. More generally, at several points, Ehrhardt also emphasizes how mathematicians regularly combine results and practices developed in different mathematical cultures. In this respect, textbooks appear as a specifically important site for the construction and circulation of such hybrid theories—in Rheinberger's terms, one of the sites for the fashioning of a knowledge that can be shared at a meso-level.

Recognizing the coexistence of different mathematical cultures thus enables the historian to account for the various ways in which a single work can come to be appropriated and enriched. In addition, it sheds light on the problem of the highly nonlinear processes by means of which knowledge comes to appear universal. At the same time, like Rabouin, Ehrhardt stresses the limits of uniformity, showing that, even within a given mathematical culture, mathematicians can and do follow their own individual trajectories.

Karine Chemla in chapter 14 also focuses on the historiographic implications of identifying distinct mathematical cultures but from a diachronic perspective. Also, because it deals with long-term history, and more specifically with ancient history, her chapter raises a new set of issues. Sociologists and anthropologists have rightly emphasized fieldwork as providing privileged access to the description of cultures of scientific practice. However, were this the only access, historians of science would be seriously limited in their inquiry. Several chapters in the book have argued that sources from the past also shed light on the character of scientific cultures. Ancient history, with the scarcity of documents that usually haunts it, represents a particular challenge.

The first issue Chemla addresses in her description of mathematical cultures of the past is precisely one of method. Sources that were written in relation to specific cultures of mathematical practice, she argues, often contain clues that historians can use to describe these cultures. Employing a set of Chinese mathematical sources selected from writings composed between the first and the thirteenth centuries, she demonstrates the existence of distinct mathematical cultures in ancient China, appearing at different times and displaying both considerable overlap with and significant differences from one other.

She argues that the main historiographic benefit provided by an interest in ancient cultures is that the identification and description of such cultures give us crucial information for interpreting the available sources. She illustrates this claim with the example of quadratic equations, demonstrating how characterizing the cultures to which the various Chinese sources bear witness enables the historian not only to identify distinct kinds of knowledge about these equations, attested to by these sources, but also to capture in a new way the continuities and differences among them. In brief, conceptual history requires a history of culture making.

Chemla's main thesis—the second historiographic implication she points to—is that the description of cultures thereby yields important tools to carry out conceptual history and brings out new phenomena. Indeed, she argues that the development of new concepts of quadratic equations to which her corpus attests can be directly correlated with aspects of the mathematical culture in which they take shape. However, there is no determinism in this correlation. Rather, Chemla suggests, it indicates "that cultures also change partly in relation to the conceptual work done . . . as much as the concepts change in relation to how actors worked." Especially noteworthy for the general project of the book are the interconnections and correlations between these cultural and conceptual changes, similar to those Nersessian identified in her anthropological study of laboratories. The history of concepts thus appears as deeply intertwined with the history of culture making.

Conclusion: Suggestions for Future Research

We began this introduction by inquiring about the relation between and among the various concepts that have been recently introduced to characterize the specificity of different ways of making scientific knowledge. The essays in this book suggest a clear answer to that question: differences in cultures of scientific practice are multidimensional. Far from offering alternative descriptions/conceptions of differences among ways of practicing science, each of the various labels (e.g., styles of thought, styles of reasoning, epistemic cultures, or epistemological cultures) captures certain dimensions of these differences. One can argue over the suitability of these labels for particular contexts and suggest alternatives (as, e.g., Rabouin does in response to Hacking), but it is useful to recognize that which dimensions are focused on likely reflect the particular aims of the author.

Whatever the case may be, it thus appears that these labels are complementary. This is why we have mainly spoken of "culture" in this introduction, emphasizing the features brought into focus by this or that more specific concept. Interestingly enough, these labels each reflect the disciplinary culture of their author and the kinds of source material on which he or she has been working. It is, for instance, not by chance that it was through an ethnographical study devoted to, on the one hand, present-day high-energy physics and, on the other, molecular biology that sociologist Knorr Cetina was struck by the relevance of various types of machineries of knowing in the production of knowledge. Such a dimension would not have appeared so prominently through a research work devoted to ancient Greek mathematics and yet might nonetheless prove fruitful for it. Similarly, it is not by chance that Keller's historical work on different disciplinary approaches to the study of embryonic development drew her attention to the philosophical and even logical dimensions of scientific culture.

The variety of case studies explored from different disciplinary perspectives and that make use of different kinds of sources thus appears to be a clear asset. Not surprisingly, the essays in this book, written by scholars with different backgrounds working on different time periods, different disciplines, and different regions of the planet, not only show how different concepts of cultural difference can inform and enrich concrete studies but also invite us to consider other dimensions that might be relevant in characterizing scientific cultures. In science studies, too, the diversity of scientific cultures is an asset, provided that we strive to achieve the meso-level, where our different practices and bodies of conceptual knowledge can be at least partially integrated.

Another question we raised in the beginning of this introduction is, how can one write about scientific cultures as recognizable and consequential categories, without inviting the assumption that they are fixed and closed to external influence? And here, too, we suggest the essays conjointly support and inform the theses we have put forward. The concept of culture that emerges from these studies is something fluid, dynamic, and porous, all properties that result from the fact that, as we have repeatedly stressed, scientific cultures are de facto

- forged by actors in relation to the questions they address, the goals they set themselves, and the resources they have available and recycle;
- in constant interaction with both other cultures and the external environment; and, for that very reason,
- open rather than closed.

These remarks thus call for the development of a history of culture making, in parallel with the histories that attend to concepts or theories.

One important consequence emerges from the features listed above, and it is duly noted and examined in some of the essays. The bodies of knowledge produced in these cultures present correlations with features of the cultures in which they took shape. However, more than determinism, what we see is a phenomenon of coconstruction, whereby each of the two terms, knowledge and culture, is shaped in intimate relation to the other. Moreover, the elements of knowledge thus produced are de facto taken as resources in other cultural contexts. This phenomenon reveals another dimension of actors' activity: establishing bridges between cultures. One facet of the phenomenon is the establishment of meso-level entities introduced by Rheinberger. Another facet is the creation of syntheses of the kind studied by Ehrhardt. These are some of the processes through which the making of sameness is carried out.

Finally, just a few words about the problem of relativism with respect to the sciences. In our view, this problem is sorely in need of clarification. Indeed, the problem is closely related to issues, addressed by several of the contributors here, concerning the stabilization, unification, and even universalization of scientific knowledge, all of which require further investigation. But even on the basis of these brief forays, it seems evident that the emergence of the very question of relativism depends, at least to some extent, on notions of culture as fixed, closed, and impervious to outside influence. If, by contrast, we recognize the fostering of interaction among cultures of scientific practice (both with one another and with the worlds around them) as a key dimension of actors' work, we can begin to recognize the construction of consensus as an ongoing process—one that persists in the face of differences of interpretation, interests, and the purposes for which knowledge is sought. Consensus, sameness, and universality may be ideals to work toward, but they can never be fully realized. And fortunately so, for variation is essential for the fertilization of new cultures.

Notes

1. See Crombie 1994. See the summary of the project in Crombie 1995, which he presents as a "comparative historical anthropology of thinking" (232), comparing between "different civilizations and societies" (227) as well as between different modes of inquiry actors within Western Europe identified and distinguished. The quotations in the main text are on pp. 225, 232, and 225, respectively.

2. In other words, we use the term "culturalism" as defined by Jens-Martin Eriksen and Frederik Stjernfelt (2009): "Culturalism is the idea that individuals are determined by their culture, that these cultures form closed, organic wholes."

3. Knorr Cetina (2005, 68–71), addressing the same issue, also answers positively, emphasizing what in her view the concept of epistemic culture adds to the consideration of practices.

4. We further suggest that any discourse in terms of "Western science" or "Chinese science" is likely to derive from, and further reinforce, assumptions of this type.

5. Noteworthy is the fact that these two criticisms highlight more general ways in which the historical record can be distorted in historiographies of this kind.

6. Kowalenko (2011, 9) notes that Hacking's concept of style of reasoning is not immune to such risks, since "by Hacking's very unrestrictive characterisation [of] styles of scientific thinking, . . . African magical thinking amounts to a distinct African style of scientific thought." Nothing prevents Hacking's concept from being enrolled as support of the validity of cures for AIDS developed in the context of self-proclaimed African exceptionalism. Hacking (2012, 608) shows his awareness of similar objections and promises to reply to them in a forthcoming publication. He is also aware of the fact that some concepts of style have a bleak history. We return to this below.

7. Incidentally, these remarks can help us further clarify the purpose of this book. It is clear that the problems we have in mind are not with a notion of "culture" in the singular. We are not dealing with culture as a range of phenomena separate from science and whose relation to science should be investigated. The problems we address relate to uses of "cultures" in the plural, as illustrated by both Ito's and Lachenal's chapters. The culturalism we focus on emerges in such contexts, and we are specifically interested in understanding the role science and the history of science play at large in these culturalist developments.

8. In one sense, our project bears a resemblance to that of Peter Galison and David J. Stump (1996). The groups of scholars whose contributions are gathered in that book and this one share a similar profile. The main difference lies in the key theses just outlined, and the concerns about ways of conceptualizing "disunity" from which we proceed (see esp. Galison 1996, 2).

9. Note that, compared with material devices (Nersessian), data (Morgan), and forms of communication (Rosental), "styles" now bring into focus ways of working specific to cultural formations, along with the processes of socialization they require.

References

Chemla, K. 2014. "The Dangers and Promises of Comparative History of Science." *Sartoniana* 27: 13–44. Available at http://www.sartonchair.ugent.be/file/288.

Cohen, R. S., and T. Schnelle. 1986. *Cognition and Fact: Materials on Ludwik Fleck.* Dordrecht: Kluwer.

Crombie, A. C. 1994. *Styles of Scientific Thinking in the European Tradition: The History of Argument and Explanation Especially in the Mathematical and Biomedical Sciences and Arts.* London: Duckworth.

Crombie, A. C. 1995. "Commitments and Styles of European Scientific Thinking." *History of Science* 33: 225–238.

Eriksen, J.-M., and F. Stjernfelt. 2009. "Culturalism: Culture as Political Ideology." *Eurozine,* January 9. Accessed January 25, 2016. http://www.eurozine.com/articles/2009-01-09-eriksenstjernfelt-en.html.

Fleck, L. 1935. *Entstehung und Entwicklung einer wissenschaftlichen Tatsache: Einführung in die Lehre vom Denkstil und Denkkollektiv.* Basel: Benno Schwabe.

Fleck, L. 1979. *The Genesis and Development of a Scientific Fact.* Foreword by Thomas Kuhn. Translated by T. J. Trenn and R. K. Merton. Chicago: University of Chicago Press.

Galison, P. 1996. "Introduction: The Context of Disunity." In *The Disunity of Science: Boundaries, Context, and Power*, ed. P. Galison and D. J. Stump, 1–33. Palo Alto, CA: Stanford University Press.

Galison, P., and D. J. Stump. 1996. *The Disunity of Science: Boundaries, Context, and Power*. Palo Alto, CA: Stanford University Press.

Geertz, C. 2000. *Available Light: Anthropological Reflections on Philosophical Topics*. Princeton, NJ: Princeton University Press.

Hacking, I. 1982. "Language, Truth and Reason." In *Rationality and Relativism*, ed. M. Hollis and S. Lukes, 48–66. Oxford: Blackwell.

Hacking, I. 1992. "'Style' for Historians and Philosophers." *Studies in History and Philosophy of Science* 23(1): 1–20.

Hacking, I. 2012. "'Language, Truth and Reason' Thirty Years Later." *Studies in History and Philosophy of Science* 43: 599–609.

Keller, E. F. 2002. *Making Sense of Life: Explaining Biological Development with Models, Metaphors, and Machines*. Cambridge, MA: Harvard University Press.

Knorr Cetina, K. 1999. *Epistemic Cultures: How the Sciences Make Knowledge*. Cambridge, MA: Harvard University Press.

Knorr Cetina, K. 2005. "Culture in Global Knowledge Societies: Knowledge Cultures and Epistemic Cultures." In *The Blackwell Companion to the Sociology of Culture*, ed. M. D. Jacobs and N. W. Hanrahan, 65–79. Oxford: Blackwell.

Kowalenko, R. 2011. "Styles of Scientific Thinking Can Kill." In *Two-Day Workshop "On Hacking's 'Style(s) of Thinking,'"* ed. J. Ritchie and J. Wanderer, 8–10. Cape Town: University of Cape Town. Accessed July 29, 2015. http://www.cilt.uct.ac.za/sites/default/files/image_tool/images/160/Conference%20abstracts.pdf.

Kuhn, T. 1962. *The Structure of Scientific Revolutions*. Chicago: University of Chicago Press.

Kuhn, T. 1970. *The Structure of Scientific Revolutions*. 2nd ed., with postscript. Chicago: University of Chicago Press.

Petitjean, P., and H. M. B. Domingues. 2007. "1947–1950: Quand l'Unesco a cherché à se démarquer des histoires européocentristes." halshs-00166355. Accessed June 15, 2016. http://halshs.archives-ouvertes.fr/halshs-00166355/document.

Rheinberger, H.-J. 1997. *Toward a History of Epistemic Things: Synthesizing Proteins in the Test Tube*. Stanford, CA: Stanford University Press.

Said, E. W. 1978. *Orientalism*. London: Routledge and Kegan Paul.

I. STATING THE PROBLEM
Cultures without Culturalism

1
———

On Invoking "Culture" in the Analysis of
Behavior in Financial Markets

How does one *represent* other cultures? What is *another* culture? Is the notion of a distinct culture . . . a useful one, or does it always get involved either in self-congratulation (when one discusses one's own) or hostility and aggression (when one discusses the "other")?—EDWARD W. SAID, *Orientalism*

In recent decades, the concept of "culture" has had a strangely bifurcated history. The critique implicit in Edward Said's questions has been deeply influential, especially in social anthropology, for which culture was the single most central concept. By 1996, an encyclopedia of the discipline suggested the possibility of "abandon[ing] talk of different 'cultures' altogether, because of its taint of essentialism" (Barnard and Spencer 1996, 142), in other words, because of its connection to the simplistic idea that a culture was a kind of "package" that was "coherent inside and different from what is elsewhere" (Mol 2002, 80). Simultaneously, however, invocation of culture has increased sharply in areas in which it had not been hugely salient previously, such as in the social studies of science in the form of, for instance, the "local scientific cultures" of Barry Barnes, David Bloor, and John Henry (1996), the "epistemic cultures" of Karin Knorr Cetina (1999), and the "epistemological cultures" of Evelyn Fox Keller (2002). Indeed, "culture" has escaped the boundaries of academia: it has become common to find the term used in everyday language in a sense roughly similar to its social science usages. When teaching students thirty or so years ago, it was necessary to explain that the term did not refer simply to high culture such as opera; now such a warning is hardly necessary.

Of course, suspicion of the concept of culture arises above all in relation to Euro-American representations of non-Euro-American peoples (the ellipsis in the quotation from Said conceals a reference to race, religion, and civilization), and the cultures invoked in the social studies of science are not, for example, national cultures but far more local. Nevertheless, Said's questions are worth asking: is the notion of culture a useful one in other contexts? With the partial exception of Knorr Cetina (1999), invocations of "culture" in the social studies of science have not tended to devote much discussion to the polysemic, politically treacherous aspects of the term.[1] In this chapter, I ask whether a notion of culture broadly inspired by these usages in the social studies of science can productively be applied to a specific aspect of behavior in financial markets: evaluation, that is, efforts to determine the economic worth of financial instruments. (I include efforts to estimate the intrinsic value of financial instruments, either in absolute terms or relative to that of other instruments, as well as efforts simply to judge whether the price of an instrument is likely to rise or to fall.)[2]

There are two reasons for shifting the focus from science to the financial markets. First, evaluation in financial markets is of enormous importance yet very poorly understood. For example, a crucial role of financial markets is to channel investment capital to some activities and not to others, and the amounts of capital involved are huge.[3] The evaluation of financial instruments—of shares, government or corporate bonds, and so forth—is a crucial aspect of this channeling role. Second, shifting attention to the financial markets highlights an issue that tends not to be prominent when "culture" is applied to science: the theory of action associated with the concept. In their ordinary social-science usage, invocations of culture are often associated informally with an implicit theory of action as based on habit, belief, and routine rather than on rational choice. There is nothing inherently necessary in this association, however: culture is still relevant even when one views actors as reflexive and rational. Since rational, reflexive action is to be found in the financial markets if it is to be found anywhere, those markets provide a useful arena in which to explore the usefulness of the concept of culture in contexts in which such action is prevalent.[4]

This chapter proceeds as follows. First, I explain my empirical focus: two similarly structured types of financial instrument important to the 2007–2008 global financial crisis, asset-backed securities (ABSs) and collateralized debt obligations (CDOs). The next two sections then describe how ABSs and CDOs were evaluated. The contrast between how this was done in the two cases shows how clusters of practices can differ consequentially even when similarly structured instruments are being evaluated. I then turn to the question of what would be needed to justify moving beyond the notion of clusters of practices to a stronger notion of evaluation cultures, suggesting four criteria that would need to be met. The conclusion then returns to Said's question of the usefulness of the notion of culture, and I express what are inevitably rather personal views on that.

The empirical research on which this chapter is based is a detailed study of the evaluation and governance of complex financial instruments such as ABSs and CDOs. This research draws upon two main sources. First is a set of 101 interviews conducted with analysts, managers, traders, and such who worked with these instruments.[5] These interviews took a broadly oral-history form, with interviewees being led through their careers in respect to the instruments in question, with a view to examining the main developments in the markets for those instruments and in the ways in which they were evaluated. Second, the trade press and technical literature were searched to assemble a corpus of documents relevant to analysis of those market developments and evaluation practices; such documents include, for example, the technical reports in which the credit rating agencies outlined the models they employed in generating their ratings.

ABSs and CDOs

Let me begin by explaining what ABSs and CDOs are. That is most easily done by explaining how the creator of an ABS or CDO—typically, a major investment bank—goes about setting one up. First, it creates a separate special-purpose legal vehicle, such a trust or special-purpose corporation. The vehicle then buys a set of interest-bearing debt instruments (in the case of ABSs, typically mortgages or other forms of consumer debt; in the original CDOs, loans made to corporations or bonds issued by them). It raises the capital to buy those mortgages, loans, or bonds by selling investors securities

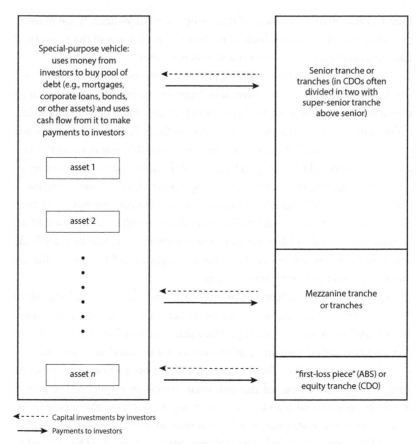

FIGURE 1.1. An asset-backed security (ABS) or collateralized debt obligation (CDO; simplified and not to scale). Investors in lower tranches receive payments only if funds remain after payments due to investors in more senior tranches are made.

that are claims on the cash flow—the interest payments and eventual capital repayments—from the pool of debt instruments.[6]

In all but the simplest ABSs, several different classes (usually called tranches) of securities are sold (see figure 1.1).[7] The highest tranche of securities (called senior or sometimes super-senior) is the safest: the purchasers of these have the first claim on the cash flow from the pool of mortgages and such (at least after management fees have been deducted). Only after these more senior claims are met are the claims of investors in lower tranches of an ABS or CDO met. Tranches at the bottom of the hierarchy are thus riskiest. If the debt instruments in the pool suffer defaults (e.g., if mortgagors stop repaying the loans made to them), then there may be a shortfall in the payments due to investors in the lowest tranche. If there are larger numbers of defaults, investors in that tranche may lose the entirety of their investment, and holders of the next most senior tranche may start to be hit. Because the higher tranches are safest, they can be sold at lower spreads (i.e., with interest rates only slightly above Libor [London interbank offered rate] or other benchmark interest rate), while lower tranches need to offer higher spreads.[8]

The motivations for setting up ABSs or CDOs varied. Originally, the goal was often for a bank or other financial institution to raise money for further lending by selling loans it had already made, or to remove those loans from its balance sheet to reduce the capital that regulators required them to hold in respect to those loans. Increasingly, though, ABSs and CDOs began to be set up simply because it was profitable to do so, for example, because of the management fees that could be earned. Institutional investors bought tranches of ABSs and CDOs because they offered higher rates of return than simpler products such as bonds, with the same credit ratings and thus apparently the same level of risk.[9] We now turn to how these complex products were evaluated.

Evaluating ABSs

Consider the most important class of ABSs, mortgage-backed securities.[10] One particular historical contingency strongly shaped their evaluation: the fact that the securitization of mortgages in the United States (in other words, the packaging of them into pools and the sale of securities that are claims on the income from those pools) began its modern history in 1970 as a government-backed program (for its historical origins, see Quinn 2009), in which investors were made good by government-backed agencies in the event of defaults on the underlying mortgages. Because investors could thus ignore the risk of default, they focused primarily on a different risk: prepayment. Deliberate government intervention in the U.S. mortgage market after the Great Depression (at the peak of which, "nearly 10 percent of homes were in foreclosure" [Green and Wachter 2005, 94–95]) led to the dominance of a specific form of mortgage that Richard Green and Susan Wachter call simply the "American mortgage": a long-term, fixed-interest-rate mortgage with no penalty for prepayment (i.e., for redeeming the mortgage early). The American mortgage thus both protects mortgagors from interest-rate rises and gives them the valuable option of redeeming the mortgage early and refinancing if interest rates fall. The obverse of that benefit to mortgagors, however, is a risk to the investor: that he or she will receive his or her money back early at a point at which (because of low interest rates) it cannot be reinvested as profitably. The evaluation of mortgage-backed securities was thus primarily a matter of determining by how much the borrower's option to prepay reduced the value of those securities.

The focus on prepayment risk in the evaluation of U.S. mortgage-backed securities continued even after private-label (i.e., not government backed) securities with tranched structures such as that shown in figure 1.1 (the early government-backed securities had simpler structures) began to be issued from 1977 onward, and also after mortgage lending moved beyond the prime mortgages that the government-backed agencies would purchase or insure to subprime (e.g., loans to mortgagors with impaired credit histories). Investors in the more senior tranches even of subprime mortgage-backed securities continued largely to ignore the risk of default and still focused primarily on prepayment (see MacKenzie 2011).

As the 2007–2008 crisis showed only too clearly, there is a potential agency problem in securitization (i.e., in the packaging of loans into pools and the selling of securities based on those pools): if the risk of default on these loans is thus passed on to external investors, then the originators of the loans that will go into the pool have a much reduced incentive to monitor the capacity of borrowers to repay. Indeed, that agency problem undermined all the pre-1970 waves of mortgage securitization in the United States (see Snowden 1995). For around twenty-five years after the rebirth of private-label mortgage securitization in the United States in 1977, however, the agency problem was held at bay, in part because of the activities of two sets of gatekeepers.

The first was the credit-rating agencies: Moody's, Standard & Poor's, and Fitch. Ratings were essential to the successful sale of mortgage-backed securities. It was very hard to find buyers for securities without an investment-grade rating (i.e., BBB– or above); indeed, the lowest externally sold tranche of a mortgage-backed security was often the "mezzanine" tranche with a BBB– or BBB rating. By far the largest demand was for AAA-rated securities at the very top of the hierarchy.

This made the evaluation of mortgage-backed securities by ratings agencies a crucial matter. The agencies were concerned exclusively with default: they considered prepayment to lie outside their ambit. The evaluation practices they employed gradually evolved from the analysis of the overall character-istics of mortgage pools (e.g., the average loan-to-value ratio of the mort-gages in the pool) to logistic regression or hazard-rate models of default on individual mortgages (using a wider range of variables, e.g., creditworthiness scores of the borrowers; see Poon 2007, 2009). Crucially, there was no explicit modeling of the phenomenon that the evaluators of CDOs were to call "correlation" (see below). Interdependence among defaults was handled by other means, such as the use of historically based "stress scenarios," above all the mortgage defaults of the Great Depression; at Standard & Poor's, for example, the criterion for a rating of AAA was that the tranche in question could withstand Great Depression default rates or their equivalents for the pool in question. The use of stress scenarios then made it possible mathematically to treat mortgage defaults as independent events, because it could be argued that correlation was already taken into account (at least implicitly) in the adverse macroeconomic circumstances crystallized in the scenarios. Another source of what CDO specialists were later to call "correlation"—the exposure of a geographically limited pool of mortgages to local economic conditions—was also handled procedurally rather than by explicit mathematical modeling: pools of mortgages considered insufficiently diversified geographically were subject to ratings penalties.

The second set of gatekeepers was the investors in the mezzanine tranches of mortgage-backed securities (as noted, those were typically the lowest of the tranches to be sold to external investors).[11] Their role was pivotal, because the mezzanine tranches were the hardest to sell, and the investors in those tranches were those with their capital most immediately at risk. Investing in mezzanine tranches was typically a specialized activity, conducted by institutional investors with considerable experience in the mortgage market. They would frequently ask for the "loan tapes" (the electronic records of the mortgages in the pool), which investors in more senior tranches almost never did, and inspect the tapes in detail, for example, looking for clusters of particularly risky mortgages. If they found such clusters, they would sometimes demand that the composition of the pool be changed before they would invest. The creators of mortgage-backed securities had to take such demands seriously, because failure to sell the mezzanine tranche would typically mean that a mortgage-backed security could not successfully be created.

Evaluating CDOs

CDOs were a later development than mortgage-backed securities. While, as noted, the first modern private-label U.S. mortgage-backed security was issued in 1977, the first CDO was created only in 1987. The structure of CDOs was typically similar to that of mortgage-backed securities and other ABSs (see figure 1.1), but the composition of the pool of debt instruments differed: instead of mortgages or other consumer debt, the pool of a CDO would typically be corporate debt. Originally, the evaluation of CDOs was broadly similar to the evaluation of mortgage-backed securities, with the exception of the fact that prepayment was a relatively minor issue (with no equivalent of the deliberate government action on behalf of mortgagors, loans to corporations are often floating rate or involve substantial prepayment penalties). For instance, the way in which analysts at rating agencies evaluated CDOs was originally quite similar to how they evaluated mortgage-backed securities. Stress scenarios were prominent, and poor diversification (e.g., too much of a CDO's pool coming from one industry) was again penalized procedurally. When evaluating a CDO, analysts at Standard & Poor's, for example, would "notch" (i.e., reduce by one or more ratings grades) the debt instruments issued by corporations in a given industry if that industry formed more than 8 percent of the CDO's pool (MacKenzie 2011).

However, the practices employed in rating CDOs were affected by a historical contingency of a kind quite different from those that shaped the evaluation of mortgage-backed securities. The evaluation of CDOs was influenced deeply by the development of modern mathematical modeling of financial derivatives, in particular the most celebrated of all such models, the Black-Scholes option pricing model (Black and Scholes 1973). An "intellectual descendant" of the Black-Scholes model, the Gaussian copula family of models, came to dominate the evaluation of CDOs, first in the late 1990s at major banks, and then from November 2001 onward at the rating agencies. (November 2001 was the date at which Standard & Poor's first released its Gaussian-copula-style software system CDO Evaluator.)[12]

In a Gaussian copula model (see MacKenzie and Spears 2014a, 2014b), the way in which defaults are not independent events—that is, correlation—is modeled explicitly and mathematically rather than being handled implicitly and procedurally as in the evaluation of mortgage-backed securities.[13] The correlation of two corporations is most usually taken to mean the correlation between the changing market values of their assets, and a Gaussian copula model takes the matrix of the correlations of all the pairs of corporations in the CDO's pool, along with estimates of the probability of default for each of the corporations and of the recovery rate for each corporation (i.e., the extent to which the losses suffered by the holders of a corporation's bonds and its other creditors after it defaults are less than total), and produces an estimate of the risk of default for each of the CDO's tranches. If correlation is low, for example, only the lowest tranche of a typical CDO will be at any substantial risk. If correlation is high, however, more senior tranches are also at risk, because defaults are likely to come in clusters of sufficient size to cause losses to investors in those tranches.

Despite the original similarities between the ways in which ABSS and CDOS were evaluated, the evaluation of the two classes of instrument became largely separate organizationally. In banks, quite separate groups typically handled ABSS, on one hand, and CDOS, on the other, and the same was true in the rating agencies (at least in their main offices in New York). As discussed in MacKenzie (2011), this cognitive and organizational separation was an important aspect of the credit crisis that has not received sufficient attention. Crucial to the crisis (the source, e.g., of the single largest concentration of losses) were ABS CDOS, a new category of instrument developed at the end of the 1990s, in which the assets in the pool of a CDO were no longer corporate debt but tranches of mortgage-backed securities and other ABSS. The ABS CDO was a kind of nested Russian doll: a tranched, structured instrument, each component of which was itself a tranche of a structured instrument.

All three of the rating agencies found a similar organizational solution to the problem of how to evaluate a nested instrument of this kind: the evaluation of the overall structure was handled by CDO groups, which relied on the ratings of the underlying ABS tranches that their colleagues in the separate group who evaluated securities of that kind had produced. This enabled the evaluation of ABS CDOs to proceed in a manner similar to the evaluation of a CDO in which the underlying assets were corporate bonds or loans made to corporations: in effect, a tranche of a mortgage-backed security rated BBB was treated in the analysis of the CDO in a way very similar to how a BBB corporate bond was treated.

Usually, although not always, higher correlation values were employed in the analysis of ABS CDOs than were used for CDOs whose pools consisted of corporate debt,[14] but not sufficiently high to prevent the construction of ABS CDOs being a very profitable activity. In particular, it was possible to create a pool of the mezzanine tranches of mortgage-backed securities, with their typical rating of BBB, and to build a CDO in which around three-quarters of the structure could gain an AAA rating. This might look like magic or alchemy but was in fact the consequence of the assumption of only modest correlation, the analog of the way in which, although an individual toss of a coin can easily produce a tail, twenty independently tossed coins are most unlikely all to turn up tails.

Because of the huge demand from ABS CDOs for the mezzanine tranches of ABSS, the second group of the traditional gatekeepers (investors in those mezzanine tranches) was sidelined completely: while, as noted, such investors would frequently scrutinize the pool of mortgages underlying an ABS in detail, the creators of ABS CDOs had much less incentive to do so, since they were going to pass the risk of default on to investors in the CDO. That left only the ratings agencies in a gatekeeper role, and (as we now know) that was insufficient to prevent the agency problem latent in securitization from becoming manifest.

It was less a question of the ratings agencies lowering their standards (although there is some evidence of this; see Financial Crisis Inquiry Commission 2011) than of widespread "arbitraging" of rating-agency models: the packaging of debts of increasingly poor quality in such a way as still to be evaluated favorably by those models. The process was by no means the only cause of the financial crisis (in several cases, for instance in the United Kingdom and Ireland, banks rendered themselves insolvent by old-fashioned reckless lending, especially in commercial property, rather than via complex structured instruments), but it was a central part of the story of the crisis, pivotal, for example, in the downfall or near-downfall of Citigroup (the world's largest bank), AIG (the world's largest insurer), Merrill Lynch, UBS, and so on. The particular toxicity of ABS CDOs lay not simply in the huge losses incurred on them but in their detrimental effects on standards of mortgage lending and in the way ABS CDOs concentrated losses at the very pinnacle of the global financial system—above all, via the retention or purchase by banks of the apparently safest, super-senior tranches of ABS CDOs, and via the insuring of those tranches against default by AIG and the specialist insurers known as "monolines" (see Tett 2009; MacKenzie 2011).

Evaluation Cultures?

Evaluation practices do thus differ, even when financial instruments with very similar structures such as ABSs and CDOs are being evaluated, and the differences between practices are consequential.[15] Should we conceive of differences among evaluation practices as constituting different evaluation cultures? I posit that four criteria need to be met before we should do so: the presence of (a) different practices associated with distinctive ontologies, (b) processes of socialization, (c) mechanisms of interaction among participants, and (d) path-dependent patterns of change.

First, we would need to find different practices associated with at least somewhat distinctive ontologies, in other words, with nonidentical views of what economic value consists of and of the nature of the economic processes that create it.[16] Very different such views can be found historically, among both political economists and lay people. For example, mercantilists believed that value was created by "capturing the turn in traded goods":[17] the difference between the prices at which a commodity could be bought and be sold. For classical political economists, value was created above all in production; for Marx, for instance, the ultimate source of all value was labor.

Broad-brush differences in ontology can still be found in today's financial markets. Perhaps the most pervasive example is whether financial instruments have an intrinsic value, or whether that notion is meaningless and the value of a financial instrument is nothing other than the price someone else is prepared to pay for it. Generally speaking, the evaluation of both ABSs and CDOs proceeded within an intrinsic-value ontology: those instruments were analyzed as claims on a future income stream, with the price others were prepared to pay for the instruments not taking an especially salient role (in many cases, tranches were bought simply because of the spread they offered, rather than because it was hoped that they could be resold to others at a higher price).

The most prominent difference in ontology between the evaluation of ABSs and CDOs concerns, as suggested above, correlation. The CDO specialists I talked to seemed almost all to consider correlation to be a real phenomenon, albeit one that was often frustratingly difficult to measure, while a specialist in mortgage-backed securities told me that he and his colleagues simply did not think that way: that is "not how we do it in mortgages; it's never been done that way." To them, an ABS tranche was simply not analogous to a corporate bond. As already noted, that analogy was the ontological implication of the way in which CDO correlation models such as the Gaussian copula were applied to ABSs in the evaluation of ABS CDOs. To specialists in mortgage-backed securities, however, this analogy was a misleading oversimplification of those securities, which were "among the most complicated financial instruments to understand," as another of them put it to me.

The difference between the ontologies of the worlds of ABSs and of CDOs was consequential. The key justification of awarding higher ratings to the tranches of a structured security than to the debt instruments forming its pool is diversification of the pool. Implicitly, and as far as I can tell almost entirely unnoticed by participants, that argument was appealed to *twice* in the rating of ABS CDOs: first as the justification for awarding investment-grade ratings to tranches of subprime mortgage-backed securities, and then as the justification for awarding higher ratings to the tranches of an ABS CDO than to the component mortgage-backed securities. However, ratings penalties for geographical concentration meant that mortgage-backed securities were already quite well diversified, at least to the extent that the geographical distribution of lending in the United States permitted (subprime lending was heavily concentrated in a number of states, e.g., California and Florida). In consequence, the packaging of ABSs into CDOs seems often not to have added a great deal of extra diversification.

The difference in ontologies meant that lack of diversification was conceptualized as correlation only in the evaluation of the overall CDO and was treated procedurally and implicitly in the evaluation of the underlying mortgage-backed securities. So the way in which credit was being given for diversification twice over was less than obvious. In my 101 interviews, I found only two interviewees who had clearly identified this issue prior to the crisis. One had tried to persuade the senior management of his agency that ratings practices in relation to ABS CDOs were therefore flawed but had been unable to do so.

A second criterion I would propose for judging whether an evaluation culture exists is the existence of processes of socialization by which newcomers learn the ontology and practices in question. Surprisingly—given the importance of socialization—this is an area where data with respect to finance are weak, although some traces of socialization mechanisms can certainly be found in ethnographies such as Ho (2009). (Unfortunately, the research underpinning this chapter did not encompass socialization: the issue of evaluation culture did not occur to me until too late in the research process.) Historically, it seems that evaluation practices in finance were mostly learned by apprenticeship. More recently, evaluation practices are taught formally by industry bodies, most prominently the CFA Institute (which offers the qualification of Chartered Financial Analyst), and by universities via MBA programs and master's courses in financial mathematics. Very likely, learning evaluation practices either via apprenticeships or via these more formal routes is not simply a matter of acquiring narrow skills but also of becoming acquainted with wider ontologies and priorities. There is an intriguing piece of quantitative evidence that is consistent with this: Oguzhan C. Dincer, Russell B. Gregory-Allen, and Hany A. Shawky (2010) found that investment managers with MBAs construct riskier portfolios than do managers with a CFA qualification.

A third criterion of the existence of an evaluation culture must surely be the presence of mechanisms of interaction among participants, including those who work for different firms. One might imagine that competitors would not speak to one another, but they often do, if only because people in finance frequently move from firm to firm. Those in different firms with esoteric expertise (e.g., in the mathematical models used to evaluate CDOs) will often attend the same industry meetings, may well know each other, and indeed will quite often have worked together. They can often talk to each other about the value of financial instruments by employing specialist language that makes full sense only to those schooled in the underlying ontology, language that can sometimes greatly facilitate communication about esoteric instruments. Particularly prominent, for example, when specialists discuss CDOs—sometimes even when they bargain over the spreads that tranches need to offer—is the term "implied correlation": the level of correlation consistent with the interest rate offered by a tranche, which can be determined only by using a mathematical model of correlation. Being someone's competitor does not prohibit—indeed, may even require—communication with that individual. Jan Simon (2010), for instance, demonstrates the frequency and importance of communication between hedge fund managers who were apparent competitors in that they were pursuing the same strategy in the same domain of the financial markets.[18]

A fourth indicator of the existence of an evaluation culture would be path-dependent patterns of change. It is hard to imagine it being justifiable to invoke the concept of culture when actors approach every new situation entirely afresh, but in the cases of ABSs and CDOs we have found plentiful evidence that this is not the case: past practices are resources for current evaluation activities. In part, this is because of the shared understandings that a common history brings; in part, it is simply because it is usually easier to modify an existing practice than to invent an entirely new one. One example of the influence of past practices is the way in which the government backing of mortgage securitization and the historic dominance of "the American mortgage" led to evaluation practices that focused on prepayment risk. As noted, those practices continued in the evaluation of mortgage-backed securities that were not government backed and did not consist just of American mortgages. Another example of path dependence is the way in which ABS CDOs were not analyzed afresh in the rating agencies as an entirely new class of instrument; rather, existing CDO practices were applied to them with only relatively minor modifications. Those practices formed a rich and readily available set of resources, in many cases already crystallized in software packages available off the shelf, making their employment in the evaluation of ABS CDOs the easiest course of action. In a context of heavy workloads, time pressure, and sometimes senior management unwilling to provide the wherewithal to explore different approaches, it is unsurprising that this was the path followed.

Conclusion

There is therefore evidence that a notion of evaluation culture, conceptualized as above, might be applicable to the analysis of behavior in the financial markets. One reason that this is an attractive possibility is that the political valence of this kind of application of the concept of culture differs from the situation that gave rise to suspicion of the concept, in which the researcher's society is more powerful than the society studied. Research on financial markets is "studying up" (Nader 1974): on any ordinary criterion of power, those who are studied are more powerful than the researcher. Anthropologists have rightly become intensely self-conscious regarding the task of representing other cultures, but I feel no such political difficulty with respect to finance: it can, and does, speak for itself powerfully. Although I am not an anthropologist, the question Laura Nader asked four decades ago still resonates with me: "What if, in reinventing anthropology, anthropologists were to study the colonizers rather than the colonized, the culture of power rather than the culture of the powerless, the culture of affluence rather than the culture of poverty?" (Nader 1974, 289).

There is, however, an important nuance here that needs to be acknowledged. Successfully applying the concept of culture requires fieldwork. Although for practical reasons my research has involved almost no participant observation (access for this in finance is hard to negotiate, and the most relevant sites are far from where I live and work), extensive interviewing has been necessary. Very busy people have given me their time,[19] and have taken some risks in doing so. To take but one example, during the research reported here I had formed the impression that one of the rating agencies had been more stringent in its evaluation of ABS CDOs than the others. After I had asked him questions predicated on this assumption, one of my interviewees suggested I speak to his boss. I phoned him immediately after the interview, and he agreed to see me the very next day. He gently corrected my misapprehension, enabling me to see that my favorable view of his agency rested on my failure fully to understand the implications of the particular way in which it implemented its Gaussian-copula-style model. He put me right, at the cost of his time and at the expense of what might have been minor favorable publicity for his agency.

Interactions such as this make the "othering" of bankers and others who work in the finance sector emotionally difficult. In public discourse (e.g., among politicians) that othering has been strong since the 2007–2008 crisis, and after giving talks I sometimes find that audiences are faintly disappointed by my lack of condemnation of those I have been studying. Certainly, I *have* found error and unscrupulous behavior in finance, but I find it difficult to be sure that bankers, for example, are in general any worse people than we academics are—it is just that their errors and lack of scruples are more consequential than ours normally are.

Furthermore, just because one might feel comfortable politically with an invocation of culture and with representing that culture does not mean that the invocation is intellectually unproblematic. Said (1978, 324) notes that "it is sobering to find, for instance, that while there are dozens of organizations in the United States for studying the Arab and Islamic Orient, there are none in the Orient itself for studying the United States, by far the greatest economic and political influence in the region." Were that to be corrected— were institutes for the study of the United States to flourish in Tunis, Baghdad, or Cairo—it would be just as much a mistake for them to essentialize American culture as for Western orientalists to posit universal moral traits or belief systems characteristic of "the East." If Harold Garfinkel's (1967) "cultural dopes," who simply act out the scripts of a preexisting culture, are not to be found in the East or the South (and they are not), there is no reason to expect to find them in the West or North.

To be applicable to financial markets, the concept of culture must be de-essentialized. Culture is something people do; it is made and unmade in action. It provides intellectual resources—often borrowed from elsewhere and not necessarily mutually consistent—and while the availability of resources for action may help shape action, resources are not determinants of action. There are no cultural dopes—not in finance, nor indeed anywhere else. I have yet to meet a market participant whom I would not class as skeptical and reflexive, and indeed these characteristics are surely necessary for success in the market, for example, because of the importance of being aware of how others will evaluate the financial instruments in question—it is surely not a coincidence that the world's most celebrated hedge fund manager is a particular proponent of reflexivity (Soros 1994).[20]

Let me return, then, to Said's question: "Is the notion of a distinct culture . . . a useful one?" To the extent that the answer is no, I suspect that the problem lies in too strong a notion of distinctness, the associated temptation to essentialize culture, and the trap of treating culture simply as the cause of action rather than as an aspect of action. Such difficulties, however, are not reason to abandon "culture." While it is tempting, as Alan Barnard and Jonathan Spencer note, to discard the noun and keep merely the adjective "cultural," abandoning the noun implies also abandoning its plural, "cultures," and thus discarding "the very important pluralizing element, the element which marked off modern anthropological usage in the first place" (Barnard and Spencer 1996, 142).

To abandon "culture" would also be to abandon a word that is useful precisely because, as noted above, it has entered popular discourse. In one of the more recent of my interviews, quite unprompted by any question from me, a mortgage specialist told me that in two of the banks in which he had worked he had observed what he called a cultural clash between mortgage experts and specialists in the corporate debt underpinning traditional CDOs. I was both pleased by the confirmation of my hypothesis and momentarily discomfited (because I am enough of a traditional social scientist to enjoy the conceit that my findings are discoveries rather than phenomena already known to those I am studying). It wasn't right to be discomfited: it is a virtue of culture that it is comprehensible to nonacademics (too often, work in the humanities and social sciences that rightly aspires to be politically progressive is written in such a way that it becomes inaccessible to outsiders). To find ways of writing and thinking about culture without becoming culturalist (i.e., without positing cultures as essences, as mutually disjoint organic wholes determining action) is thus a task of no small importance, precisely because we are dealing with a loaded and evocative word, and one that has escaped the confines of academia. It is this chapter's postulate that evaluation cultures in financial markets are a useful site for this writing and thinking.

Notes

The research reported here was funded primarily by grants from the U.K. Economic and Social Research Council (RES-062-23-1958) and the European Research Council (FP7 291733: EPIFM); for more detail on the episodes discussed here, see MacKenzie 2011 and MacKenzie and Spears 2014a, 2014b, on which this chapter draws. I am grateful to Jonathan Spencer for a helpful discussion of the career of the concept of culture in anthropology, and to Mary Morgan, Bruno Belhoste, Jonathan Regier, and two anonymous readers for insightful comments on the first draft of this chapter, although I must apologize to them for the fact that space constraints have made it impossible to take up all their suggestions. Of course, all errors remain my responsibility.

1. I intend no critique here, or at least no critique that I would not apply to my own usage of the term "culture" (see MacKenzie 2001, chap. 9). Other relevant invocations include the "cultures of economic calculation" of Kalthoff 2006 and "calculative cultures" of Mikes 2009.

2. For a useful discussion of the distinction between evaluation in the sense of the assessment of value and valorizing (creating or adding to value), see Vatin 2013. Of course,

even in financial markets evaluation can have nonmonetary aspects, such as in ethical investing or Islamic finance (see, e.g., Maurer 2005). I concentrate in this chapter on monetary evaluation because I want to investigate the invocation of the term "culture" in contexts in which its applicability is not obvious. There is a rapidly growing wider literature on evaluation and valorizing (see, e.g., Berthoin Antal, Hutter, and Stark 2015), which unfortunately cannot be reviewed here for reasons of space.

3. For instance, in 2010 a single asset management firm, BlackRock—admittedly the world's largest—controlled assets totaling $3.45 trillion, which was more than the GDP of Germany (Kolhatkar and Bhaktavatsalam 2010); by 2015, that figure had risen to $4.6 trillion.

4. I should make clear that in referring to "rational" action, I apply the term narrowly to mean action that is understandable in terms of actors choosing the course of action they view as most likely to achieve their goals, given the pattern of incentives they face, whether or not the resultant actions are rational in a broader sense (e.g., beneficial to the wider economy or society). It is clear that while the behavior that generated the credit crisis may often have been rational in the narrow sense, it was not rational in that broader sense.

5. Of these interviews, 63 took place in London, 27 in New York, and 11 elsewhere; 29 of the interviews, which focused on CDOs and similar instruments, were conducted prior to the eruption of the credit crisis in June 2007, and 72 were conducted after the onset of the crisis.

6. Such CDOs with this structure would be described as cash CDOs; another important category of CDO is the synthetic CDO, which instead of buying a pool of debt assets sells protection on them (i.e., insures them against default) and uses the income from the sales of protection to pay investors.

7. In many ABSs, the first-loss piece is eliminated by overcollateralization (i.e., by issuing securities with an aggregate face value less than that of the total assets in the pool), thus leaving the mezzanine tranche(s) the lowest in the hierarchy of seniority.

8. In CDOs, holders of the equity tranche usually have no fixed claims at all: they receive whatever cash is left after all other claims have been met. If things have gone well, this residual can represent a high rate of return; if there have been many defaults, holders of the equity tranche may lose the entirety of their investment.

9. At later stages in the run-up to the crisis, CDOs themselves started buying up tranches of other CDOs, often via implicit quid pro quo arrangements.

10. Other ABSs' pools include debt such as auto loans, student loans, and credit card receivables. These were peripheral to the credit crisis and are not discussed here.

11. My attention was first drawn to the importance of mezzanine investors by Adelson and Jacob 2008.

12. By "Gaussian-copula-style" I mean systems, such as the original version of CDO Evaluator, that are one-period models (modeling whether, not when, the assets in the CDO's pool will default) and are thus not fully fledged copula models of the kind introduced by Li 2000.

13. A copula function—a formalism introduced to mathematical statistics by Sklar 1959—joins together the distribution functions of uniformly distributed variables to

yield a specific multivariate joint distribution function. A Gaussian copula yields the distribution function of a multivariate normal distribution.

14. For a detailed discussion of correlation assumptions, see MacKenzie 2011.

15. For evidence of differences among evaluation practices in other contexts, see Smith 1989 and Lépinay 2011.

16. The recent vogue in the social sciences for the notion of ontology (especially following the work of Eduardo Viveiros de Castro, e.g., Viveiros de Castro 2004) has given rise to the suspicion that it is "just another word for culture," to quote a motion to that effect debated at the 2008 Meeting of the Group for Debates in Anthropological Theory, University of Manchester (the debate is recorded in the June 2010 issue of *Critique of Anthropology*). As I use the word here, "ontology" is an aspect of culture, not a synonym of it. Nor in invoking distinctive ontologies do I wish to imply that they are wholly disjoint. For reasons of brevity and clarity, here I focus on differences in ontology, but there were of course commonalities as well.

17. I owe the phrase (and the suggestion of these examples of ontologies) to Mary Morgan.

18. For less systematic evidence consistent with Simon's conclusion, see Hardie and MacKenzie 2007.

19. See Ho 2009 on the long working hours in investment banking.

20. For example, a common form of othering in critiques of behavior in the financial markets is the suggestion that participants are what one might call "model dopes," uncritically accepting the output of mathematical evaluation models. Again, I have encountered no such dopes, nor have the others who have addressed this question; see esp. Svetlova 2009 and Beunza and Stark 2012.

References

Adelson, M., and D. Jacob. 2008. "The Subprime Problem: Causes and Lessons." *Journal of Structured Finance* 14(1): 12–17.

Barnard, A., and J. Spencer. 1996. "Culture." In *Encyclopedia of Social and Cultural Anthropology*, ed. A. Barnard and J. Spencer, 136–143. London: Routledge.

Barnes, B., D. Bloor, and J. Henry. 1996. *Scientific Knowledge: A Sociological Analysis*. London: Athlone.

Berthoin Antal, A., M. Hutter, and D. Stark, eds. 2015. *Moments of Valuation: Exploring Sites of Dissonance*. Oxford: Oxford University Press.

Beunza, D., and D. Stark. 2012. "From Dissonance to Resonance: Cognitive Interdependence in Quantitative Finance." *Economy and Society* 41: 335–359.

Black, F., and M. Scholes. 1973. "The Pricing of Options and Corporate Liabilities." *Journal of Political Economy* 81: 637–654.

Dincer, O., R. B. Gregory-Allen, and H. A. Shawky. 2010. "Are You Smarter than a CFA'er?" Social Science Research Network. Accessed November 22, 2010. ssrn.com /abstract=1458219.

Financial Crisis Inquiry Commission. 2011. *The Financial Crisis Inquiry Report*. Washington, DC: Government Printing Office.

Garfinkel, H. 1967. *Studies in Ethnomethodology.* Englewood Cliffs, NJ: Prentice Hall.

Green, R. K., and S. M. Wachter. 2005. "The American Mortgage in Historical and International Context." *Journal of Economic Perspectives* 19(4): 93–114.

Hardie, I., and D. MacKenzie. 2007. "Assembling an Economic Actor: The *Agencement* of a Hedge Fund." *Sociological Review* 55: 57–80.

Ho, K. 2009. *Liquidated: An Ethnography of Wall Street.* Durham, NC: Duke University Press.

Kalthoff, H. 2006. "Cultures of Economic Calculation." In *Towards a Cognitive Mode in Global Finance,* ed. T. Strulik and H. Willke, 156–179. Frankfurt am Main: Campus.

Keller, E. F. 2002. *Making Sense of Life: Explaining Biological Development with Models, Metaphors, and Machines.* Cambridge, MA: Harvard University Press.

Knorr Cetina, K. 1999. *Epistemic Cultures: How the Sciences Make Knowledge.* Cambridge, MA: Harvard University Press.

Kolhatkar, S., and S. V. Bhaktavatsalam. 2010. "The Colossus of Wall Street." *Bloomberg Businessweek,* December 13–19, 60–67.

Lépinay, V. 2011. *Codes of Finance: Deriving Value in a Global Bank.* Princeton, NJ: Princeton University Press.

Li, D. X. 2000. "On Default Correlation: A Copula Function Approach." *Journal of Fixed Income* 9(4): 43–54.

MacKenzie, D. 2001. *Mechanizing Proof: Computing Risk, and Trust.* Cambridge, MA: MIT Press.

MacKenzie, D. 2011. "The Credit Crisis as a Problem in the Sociology of Knowledge." *American Journal of Sociology* 116: 1778–1841.

MacKenzie, D., and T. Spears. 2014a. "A Device for Being Able to Book P&L: The Organizational Embedding of the Gaussian Copula." *Social Studies of Science* 44: 418–440.

MacKenzie, D., and T. Spears, 2014b. "The Formula That Killed Wall Street: The Gaussian Copula and Modelling Practices in Investment Banking." *Social Studies of Science* 44: 393–417.

Maurer, B. 2005. *Mutual Life, Limited: Islamic Banking, Alternative Currencies, Lateral Reason.* Princeton, NJ: Princeton University Press.

Mikes, A. 2009. "Risk Management and Calculative Cultures." *Management Accounting Research* 20: 18–40.

Mol, A. 2002. *The Body Multiple: Ontology in Medical Practice.* Durham, NC: Duke University Press.

Nader, L. 1974. "Up the Anthropologist: Perspectives Gained from Studying Up." In *Reinventing Anthropology,* ed. D. Hymes, 284–311. New York: Vintage.

Poon, M. 2007. "Scorecards as Devices for Consumer Credit: The Case of Fair, Isaac & Company Incorporated." In *Market Devices,* ed. M. Callon, Y. Millo, and F. Muniesa, 284–306. Oxford: Blackwell.

Poon, M. 2009. "From New Deal Institutions to Capital Markets: Commercial Consumer Risk Scores and the Making of Subprime Mortgage Finance." *Accounting, Organizations and Society* 34: 654–674.

Quinn, S. 2009. "Things of Shreds and Patches: Credit Aid, the Budget, and Securitization in America." Working paper, University of California, Berkeley, Department of Sociology.

Said, E. W. 1978. *Orientalism*. London: Routledge.

Simon, J. 2010. "Essays in Hedge Funds: Performance, Persistence and the Decision-Making Process." PhD diss., University of Essex.

Sklar, A. 1959. "Fonctions de répartition à *n* dimensions et leurs marges." *Publications de l'Institut de Statistique de l'Université de Paris* 8: 229–231.

Smith, C. W. 1989. *Auctions: The Social Construction of Value*. New York: Free Press.

Snowden, K. 1995. "Mortgage Securitization in the United States: Twentieth Century Developments in Historical Perspective." In *Anglo-American Financial Systems: Institutions and Markets in the Twentieth Century*, ed. M. D. Bordo and R. E. Sylla, 261–298. New York: Irwin.

Soros, G. 1994. *The Alchemy of Finance: Reading the Mind of the Market*. New York: Wiley.

Svetlova, E. 2009. "Theoretical Models as Creative Resources in Financial Markets." In *Rationalität der Kreativität? Multidiziplinäre Beiträge zur Analyse der Produktion, Organisation und Bildung von Kreativität*, ed. S. A. Jansen, E. Schröter, and N. Stehr, 121–135. Wiesbaden: Springer.

Tett, G. 2009. *Fool's Gold: How Unrestrained Greed Corrupted a Dream, Shattered Global Markets and Unleashed a Catastrophe*. London: Little, Brown.

Vatin, F. 2013. "Valuation as Evaluating and Valorizing." *Valuation Studies* 1(1): 31–50.

Viveiros de Castro, E. 2004. "Exchanging Perspectives: The Transformation of Objects into Subjects in Amerindian Ontologies." *Common Knowledge* 10: 463–484.

2
___

Cultural Difference and Sameness
Historiographic Reflections on Histories
of Physics in Modern Japan

In this chapter, I address historical studies of culture and science in Japan by discussing some case studies of the history of science in Japan that reveal the kinds of culturalism that sometimes affect the way historical studies of science in the non-Western world are written and evaluated. Qualifying certain studies as interesting or not depends strongly on the judgment of the readers, audience, or referees. Such judgments cannot always be explicitly articulated, and they might have various sources. One of the assumptions in this chapter is that such judgments are at least partially based on ideas about what stories historical studies should reveal and how.

My contention is that, when non-Western science is involved, some historians are uncritical about their own preconceptions of what counts as interesting history. In this chapter, I clarify and criticize some of these conceptions and propose alternatives. By "culturalism" I refer to perspectives that view scientific cultures and practices in terms of the essential cultures of the country or region in question and that seek to understand their developments through macroscopic considerations of such cultures. Criticizing these, I argue that national culture is not a useful category for discussing scientific practices. Not only is the notion of national culture too coarse to capture the heterogeneity in scientific cultures, but also the concept of national boundaries falls short of constructing meaningful cultural boundaries. Thus, I call for more careful attention to the local contexts of the times and transnational connections. Furthermore, I argue that in studies of non-Western science, similarities can be as interesting as differences, albeit not for the reasons assumed in the universalist perspective. Both differences and similarities require explanation and provide opportunities for further analyses.

As examples, I discuss two case studies on the history of physics in Japan: the philosophical discussion on quantum mechanics that took place in Japan around the 1930s and the transmission of Feynman diagrams in the 1950s. Japanese debates over the conceptual foundations of quantum mechanics had certain similarities to those in Europe. This is surprising because similarities happened in a totally different location. Yet, such similarities do not necessarily attract much attention because some believe in universality of science, from which similarities in science are natural consequences, whereas differences are unexpected and require explanation. Too often, readers want to know what the differences are and how these differences in the conception of quantum mechanics originated from the "traditional culture" of Japan. Similarly, the transmission of Feynman diagrams to Japan raises the historiographical issue of how to deal with differences and sameness. By way of a self-criticism of my own previously coauthored work on Feynman diagrams that emphasized differences (Kaiser, Ito, and Hall 2004, secs. 1, 9; see also Kaiser 2005), I propose an alternative approach to the same material, emphasizing sameness.

Culturalism and Science in Japan

The history of science written in the English language is becoming more international and inclusive of the non-Western world. The history of science and science studies have investigated the different styles of scientific practices in those regions and have explored the difficulties of communicating across such differences. Certainly attention to different scientific cultures and practices has enriched the scholarship. There is, however, a pitfall: the historiography of science and technology in East Asia is sometimes impaired by a form of culturalism.

Culturalism in this context displays the following tendencies. First, it assumes a form of cultural essentialism, by which I mean the assumption that there are cultural essences of a nation or even of a large geographical region that condition the behaviors and thoughts of its constituents. The process of conditioning can be a passive influence or the active exploitation of cultural resources. According to this assumption, one can study and interpret the behaviors and thoughts in a country or a geographical region in terms of essential cultures. Such essential cultures are often depicted as unique to the area and fundamentally different from the cultures in other countries or regions.

Second, culturalism presupposes a certain form of cultural determinism about science, in particular the assumption that cultural differences can explain the existence of different scientific practices in different places. According to this position, culture is the primary explanatory resource, as if all other explanations were unnecessary. This mode of thinking seems to be adopted when other evidence to support alternative explanations is unavailable (a deficit that can result from various reasons, including, e.g., the analyst's lack of research skills or interest). This position might even be sophisticated enough to recognize the locality of culture, seeking to explain the production of different kinds of knowledge or different scientific practices in terms of such local differences in cultures. However, since the main explanatory resource of cultural essentialism is cultural explanation, cultural essentialism often accompanies cultural determinism. From the culturalist perspective, interesting research is that which finds different scientific practices and cultures, identifies the cultural essence of a country or a region, and tries to explain various social and intellectual phenomena there in terms of this cultural essence.

One root of cultural essentialism in relation to non-Western cultures is orientalism, which has received frequent criticism. In his 1978 book *Orientalism*, Edward Said defines this term as an institutionalized discourse, that is, an imagined or constructed representation of the Orient by the Occident in a way that is fundamentally different from the Occident: it is a way of dealing with the Orient "by making statements about it, authorizing views of it, describing it, by teaching it, settling it, ruling over it" (3). Said's diagnosis points to some of the hidden appeals of culturalism, as a way of dealing with others by labeling them according to imagined cultural essences.

Said's original usage of the term "orientalism" concerned the image of the Middle East constructed by Westerners, but this image need not be limited to that region. This idea of images of the non-Western world constructed as being fundamentally different from the West might no longer be of any analytical utility in today's mainstream East Asian studies. Yet, I argue that orientalism persists not only in popular culture, in which we all too commonly see various forms of orientalism, but also in the history of science in East Asia.

Samurai Science Thesis

One manifestation of the culturalist approach is what I call the "samurai science" thesis, referring to the argument that Japanese science after the Meiji Restoration had something to do with the cultural, social, or ethical traits of Japan's samurai class. There are different versions of the samurai science thesis. Some try to characterize all scientific activities in Japan as under the influence of samurai culture and ethics. Others limit their claims to specific cases. Some claim causal connections between samurai culture and science in Japan, while others consider samurai legacies as cultural resources through which scientists fashioned themselves. Here, I focus on a relatively sophisticated version of the samurai science thesis.

In his book *Science and the Building of a New Japan* (2005), Morris Low presents one kind of samurai science thesis. He explores various aspects of physics and its relation to society and politics in post-WWII Japan, focusing on key figures, including Nishina Yoshio, Sakata Shōichi, Taketani Mituo, Yukawa Hideki, Tomonaga Sin-itiro, Sagane Ryōkichi, and Hayakawa Satio (Low 2005). It is one of the best-researched studies available in English on Japanese physics during this period. The book, however, contains some dubious claims. Low argues, "Qualities associated with the samurai served as a cultural resource, a construct, which helped shape the postwar public persona of physicists" (2). According to him, these qualities shaped the personae of Japan's leading physicists, particularly in the way that they assumed leadership roles among physicists or in society as "public men."

For each physicist, Low's argument follows a similar pattern: physicist X was a "public man," in the sense of having a strong sense of social responsibility, serving the government or the public good, discussing social issues, or doing something good for the scientific community. Low then states that such public service is in accordance with the qualities of the ideal samurai. If physicist X is from a "privileged family," his being a public man is equated with what his ancestors used to do. If physicist X is not from a privileged family, he is still considered to construct his identity in accordance with the qualities of the ideal samurai. For example, with regard to Tomonaga and Yukawa, who had samurai ancestry, Low (2005, 141) writes, "The propensity of Tomonaga and Yukawa to become public men can be traced to the public-spiritedness of the samurai who were their ancestors and the notions of public service that were implicit in the minds of their parents." With respect to Taketani, who was not from the samurai class, he writes, "Taketani lacked the privileged background of Sakata, but, nevertheless, adopted this feeling of public vocation like other intellectuals in postwar Japan" (101). Thus, either way, physicists being public men can be explained in terms of samurai ideals.

Low, however, does not provide any direct evidence regarding how these scientists used the cultural resources of the samurai ideals, fashioned themselves, and constructed their identity as public men through these cultural resources. Low's depiction of a public man is vague and applicable to many scientists who conducted public activities and shouldered responsibilities when they reached a certain senior or celebrity status. Some of the circumstantial evidence that Low provides is also highly problematic. For example, he points out that adoption was common among samurai, and some scientists were adopted (e.g., Yukawa and Sagane). He seems to interpret these facts as suggesting that a connection exists between these scientists and the samurai culture. In reality, adoption was not limited to samurai, because in early modern Japan social functions and occupations were closely tied with family. There is no reason to believe adoption was in any sense samurai-like practice. Another crucial weakness of Low's argument is his assumption that the ideal samurai was the only possible role model for public men, as if people from other social classes would not provide similar norms. Furthermore, Low gives little consideration to other factors that contributed to the public status of some scientists. For example, in the case of Yukawa, it seems much more natural that his Nobel Prize was responsible for his position as a public figure.

To understand how other class cultures could provide leadership ideals, let us take the example of Nishina Yoshio, the focus of chapter 2 of Low's book. Nishina was one of the most important Japanese physicists of the twentieth century. He was principally responsible for creating a strong tradition of atomic physics in Japan. In his book, Low mentions that Nishina was from a "privileged family," and his grandfather served as a *daikan* (magistrate). However, Low fails to mention that the Nishina clan was not part of the samurai class but an agricultural family. Nishina's grandfather, Nishina Arimoto, was accorded samurai status for his extraordinary contribution to the local feudal lord (*daimyo*), the Aoki clan. In addition, one of his sons also had samurai status, but he gave up his samurai status after the Meiji Restoration (Oka 1927, 117). Hence, for example, in the directory of students of the Imperial University of Tokyo, Nishina Yoshio is listed as a commoner (*heimin*), rather than ex-samurai (*shizoku*). Since Low does not mention that

Nishina was from a samurai family, he is probably aware of this fact. It is somewhat troubling to see that he makes this point ambiguous by using the expression "privileged family" (Low 2005, 19).

More important, although the Nishina family was not of the samurai class, the family had an influential and politically important place in the village of Hamanaka, located in today's Okayama prefecture. Hamanaka used to be a very small village, along the coast of Kasaoka Bay. As is the case with many other places in the region, Hamanaka expanded its land through land reclamation that continued for several centuries. In this half-natural and half-artificial environment, the Nishina family owned a large expansion of land, including rice paddies and salt fields. Nishina's grandfather Arimoto excelled in civil engineering and played a leading role in water management, land reclamation, and the creation of salt fields. The Nishinas owned a salt-making business, and the family's salt circulated widely before the Meiji government nationalized this industry. Thus, salt making was responsible for the accumulation of the Nishina family fortune. At the same time, through these enterprises, the Nishinas contributed to local welfare and gained the recognition of the daimyo of the area. Thus, public or community service was certainly the tradition of the Nishina family, and I argue that Nishina's activities in science reflected to a great degree this family tradition. However, this had nothing to do with samurai morality. Rather, the Nishinas were entrepreneurial farmers, whose forte was civil engineering and commerce (Ito 2002, chap. 6).

As far as Nishina was concerned, the idea of a leader in science had little to do with the samurai tradition. For example, the word that his students used to address him was *oyakata* (boss), which was probably not the term for a samurai leader but for the boss of townspeople (Tomonaga et al. 1982). Moreover, in his effort to create a new research tradition for atomic physics, Nishina's scientific leadership had much more to do with what he learned from Niels Bohr than any Japanese cultural resources (Ito 2002, chap. 5). In introducing new science from elsewhere, this was a natural thing to do, and not limited to Nishina.

Nishina was not a singular example. There are other physicists from non-samurai classes, whose behavioral patterns and languages did not display any stereotypical characteristics of samurai. Honda Kōtarō, for example, one of the most influential scientific leaders in prewar Japan, was from a farmer's family.[1] He could be rather stereotypically conceived as a farmer because of his strong accent of the Mikawa region and his work style marked by extreme diligence and patience.

Adopting leadership personae from European scientists with whom Japanese scientists worked when they studied overseas was probably even more common. Role models of European or American scientists were cultural resources that Japanese leading scientists used when they self-fashioned to be leaders of science. One of the first Japanese physicists to study abroad, Tanakadate Aikitu, studied in Glasgow and worked with William Thomson, Lord Kelvin (see Koizumi 1975; Yoshida and Sugiyama 1997). He was so heavily influenced during his stay in the United Kingdom that he continued to adopt not only British manners and attire but also Thomson's research and teaching styles (Nakamura 1943, 80–84). Similarly, Sagane had a long professional relationship with Ernest Lawrence at Berkeley and came to adopt a similar research style (Sagane Ryōkichi Kinen Bunshū Shuppankai 1981).

Hence, Low's argument is highly problematic as far as Nishina and some other physicists are concerned. This is not simply a matter of factual errors and empirical problems. Certainly, using a class category to analyze science can be fruitful, as in the case of "gentleman science" in Steven Shapin and Simon Schaffer's *Leviathan and the Air-Pump* (1989). In addition, as we will see in the next section, analyzing how scientists fashioned themselves is in itself a legitimate approach and can be useful for understanding those scientists. However, Low connects too hastily stereotypical and orientalist images of samurai without confirming that those scientists indeed maintained and projected such images.

It should be noted that Low is not so naive as to assume the samurai culture to be the cultural essence of Japanese physicists. In fact, he is critical of cultural essentialism, as indicated by his thoughtful review on the social studies of science in Japan, where he critically writes about *Nihonjinron*, the theory of Japaneseness (Low 1989; more on Nihonjinron below). Low uses the idea of the samurai to capture the self-fashioning and identity construction of Japanese physicists. Such an image of samurai-like leadership might indeed be applicable to a few Japanese physicists to some degree. However, to make such a claim, one would need evidence that these scientists had such images and used them to construct their identities. Making such a connection without convincing evidence seems motivated by an orientalist agenda to make historical research more interesting, along the lines of culturalism.

The orientalist tendency is not limited to Low. Don-Wong Kim, in the introduction of his 2007 biography of Nishina, writes that Nishina "was typically Japanese. . . . He had behaved exactly as the typical Japanese did when he met and dealt with senior Japanese scientists or nonscientists outside his laboratory" (n.p.), as if there were such a thing as "typical Japanese," which Kim does not specify. Kim further asks, "How much did being Japanese influence Nishina's scientific work and contribute to his success?" Apparently, Kim assumes that "being Japanese" is such an attribute that one can discuss its influence on someone's scientific work and even its contribution to that person's success. Both examples seem to be clear indications of Kim's commitment to cultural essentialism.

Yukawa Hideki's Self-Orientalism and Complementarity

This kind of constructed representation of one "national culture," labeled as orientalism by Said, can be put forward not only from outsiders but also from within. It is possible that natives have certain essentialist cultural theories about themselves and behave in conformity with this theory, particularly with some intention to appear in a certain stereotypical way to outsiders. Such self-fashioning can sometimes be useful in self-presentation. Unfortunately, however, the case of Yukawa suggests that it is doing real harm in the history of Japanese science.

The background of Yukawa's self-fashioning is Japan's fairly visible tradition of native orientalism. Nihonjinron, the theory of Japaneseness, became a popular genre of writing in Japanese in the 1970s, but the genealogy of this trend goes back to the 1930s and Kuki Shūzō's 1930 *Iki no kōzō* (*The Structure of "Iki"*) and Watsuji Tetsurō's 1935 *Fūdo* (*Climate and Culture*). Ruth Benedict's *The Chrysanthemum and the Sword* (1946) can also be included in this genre, in the sense that those works were marketed and consumed in Japan as a form of Nihonjinron (Ogasawara 2006; Saeki and Haga 1987). According to one survey, 698 books belonging to this genre were published from 1946 to 1978, and 58 percent of them were published from 1970 onward (Nomura Sōgō Kenkyūjo 1978). These writings discuss the uniqueness of the Japanese society, culture, and people. Most of them are written in Japanese for Japanese readers, but there are also examples of Nihonjinron written by Japanese in foreign languages for foreigners to present a supposedly unique Japanese culture.

Some notable Japanese intellectuals advocate cultural essentialism that is similar to Nihonjinron and characterize their science as such. Yukawa is a good example of such self-assumed cultural theorists. In his *Creativity and Intuition* (1973), Yukawa writes that Japanese and Chinese cultures have subconsciously influenced his way of doing physics. Yukawa also discusses the relations between Oriental cultures and science. For example, with respect to the abilities that are important for science, Yukawa argues that the Chinese and Japanese are not as "rational" as Westerners: "A thoroughgoing rationalism eludes them" (56). He also argues, however, that they are good at other aspects that are also important to science. One is intuition, at which, according to Yukawa, the Chinese and Japanese excel. Thus, connecting science and culture, Yukawa fashions himself as an Oriental physicist, whose strength is intuition rather than logic.

Niels Bohr invented the notion of complementarity for the conceptual understanding of quantum mechanics. This notion is, along with Bohr himself, notorious for its vagueness. Apparently, many quantum physicists in his time had difficulty understanding it. Leon Rosenfeld, a close collaborator of Bohr, writes that he once asked Yukawa whether the Japanese physicists had experienced the same difficulty as their Western colleagues in assimilating the idea of complementarity and in adapting to it. Yukawa answered, "No, Bohr's argumentation has always appeared quite evident to us." To Rosenfeld, who was naturally surprised, Yukawa continued, "You see, we in Japan have not been corrupted by Aristotle" (Rosenfeld 1963, 47).[2]

If we can assume that Rosenfeld's description is reasonably reliable, Yukawa probably meant that complementarity was incompatible with Aristotelian logic, by which Japanese physicists were not affected. In reality, however, no evidence supports Yukawa's claim (except, perhaps, that provided by himself). In a chapter of my dissertation, I try to understand the historical process of how complementarity was introduced and discussed in Japan from the late 1920s to the early 1940s (Ito 2002, chap. 7).[3] I used an empirical approach to study how complementarity was discussed, based on semipopular articles published in major journals and magazines in Japan at that time. A close look at what actually happened in the discussion of complementarity in Japan clearly indicates that debates and controversies concerning complementarity in Japan could not simply be understood in terms of "Japanese culture" or, even less, "Aristotelian logic." Within the Japanese context, there was diversity, and different views often conflicted with one another. Moreover, those various views were not necessarily of Japanese origin. The Japanese intellectual environment in the 1930s already incorporated ideas from European thinkers, and Bohr's complementarity was understood and discussed in conceptual frameworks incorporating European intellectual resources, such as German idealism and Marxism (Ito 2002, chap. 7; see also below).

One reader expressed disappointment in response to this work. He stated that he had an interest in "Eastern," especially Japanese, thought, and considered complementarity to be important; hence, he felt that he would be sympathetic to the topic of this work. However, he found that "not enough new history of science has emerged" and that I had trivialized Yukawa's philosophical views. Apparently, my work did not satisfy this reader's expectations about connections between "Eastern" thought and complementarity. I suspect that this reaction was motivated by culturalist preconceptions that there were certain unique Eastern philosophical thoughts that mostly determined the reception of complementarity among Japanese physicists, and that the history of non-Western science is interesting when such a story is told. What interests me is the fact that this reader cited Yukawa's *Creativity and Intuition* as counterevidence to my argument and suggested that I should consult this book. There, however, Yukawa was actually very explicit that he was different from the majority of the Japanese with regard to his exposure to Chinese classics (Yukawa 1973, 16). Among the ancient Chinese thinkers, Laozi's and Zhuangzi's ideas left the greatest impression on him. He, however, writes that Taoism is much less known and influential in Japan than are Buddhism and Confucianism, and most Japanese intellectuals of Yukawa's generation were not familiar with Taoism. Even if Yukawa's assertion about himself is more than self-fashioning, he admits that he was not the norm but an exception. The suggestion to view Yukawa as a typical Japanese physicist in relation to complementarity is thus not only a misinterpretation of the history of complementarity in Japan but also a misinterpretation of Yukawa's writing, in both cases projecting an orientalist illusion.

Much critical analysis has been carried out concerning what we call here culturalism in general. Nihonjinron, in particular, has been severely criticized in Japanese studies in North America. Those who study non-Western subjects are likely to encounter Edward Said's work in their undergraduate years. As for the history of science in China, Roger Hart (1999) criticizes the tendency to see an imagined cultural community in order to construct "Chinese science." In spite of these critical studies, culturalism still persists in the history of science in Japan. This is partially because too many readers, especially nonspecialist readers, still find it "interesting." If writers try to accommodate such readers, they can find sources from Japanese scientists like Yukawa.

Complementarity in Japan: Heterogeneity
and Transregionality

The notion of culture is thus a cause of problems. One might simply reject its explanatory value. But despite my reservations, I still consider it as offering some advantage, especially in explaining behaviors that evade social or economic explanations. Even scientific practices, conducted generally by the most highly educated people, do not always derive from calculated, rational decision making that is free of social or economic considerations. Rather than throwing out the baby with the bath water, it would be more productive to discuss the best forms of cultural analysis. In this section, I first discuss what exactly is wrong with culturalism and suggest alternatives to that approach. Then, I apply these considerations to two case studies concerning the history of science in Japan: complementarity and Feynman diagrams.

There are undeniably problems with culturalism. The first is cultural determinism. This problem is avoidable by not resorting to a cultural explanation too easily and by limiting the use of cultural explanations to cases where other explanations are either inappropriate or incomplete. The second is cultural essentialism. To avoid this, we need to abandon the assumption that each geographical group has essential cultural traits. Instead, we should recognize that any region has within it cultural diversity, and that cultural boundaries do not coincide with geographical boundaries. For example, within Japan, there is a vast heterogeneity of cultures, and no single cultural description would capture the whole picture of science in Japan. A finer-grained analysis of cultures would recognize heterogeneous or even conflicting cultural elements within the region in question. At the same time, these different cultures might be shared by other geographical regions. People like Yukawa and Tomonaga might have had more in common culturally with contemporary European or American scientists than with their Japanese ancestors. Hence, I call for attention to the transregional aspects of culture. To illustrate this point, I revisit the case of the introduction of complementarity in Japan and outline my work on this topic.

Heterogeneity and Transregionality of Cultures: The Introduction of Complementarity to Japan

I set up a framework for analyzing how complementarity was understood in Japan in three steps: defining the groups, understanding the culture of each group, and examining the historical process of how complementarity was introduced and discussed in Japan within these different cultures.

I first define five groups of Japanese intellectuals who wrote about complementarity. These groups may overlap; in other words, one person can belong to two or more groups. The groups are defined by their practices, what they do in relation to science, or in terms of their professional identity. The first is the group of physicists who did research in physics or who taught physics at a university. The second is the group of journalists in science, whose job was to write news articles on science. The third group, scientist-literati, refers to a small but extremely visible group of scientists in Japan who wrote literary essays based on science, represented by people like Terada Torahiko and Nakaya Ukichirō. The fourth group is Marxist scientists, who published their writings on science along the lines of the Marxist philosophy of science. This group became extremely influential in postwar Japan but was already active in the 1920s and 1930s. The last group is a school of philosophers who received or delivered education at the University of Kyoto under the influence of Nishida Kitarō.

The intention of this list is not to cover all the participants in the debate. Its main purpose is to clarify the diversity of people who were involved in the discussion of complementarity at that time. One can have multiple identities and belong to more than one group. For example, a research physicist can also be a physics teacher. At the same time, people with different occupations might conduct practices that appear similar. For example, both research physicists and science journalists might have published articles in popular science magazines, yet this same action meant something different to each of them.

The second step is to understand the culture of each group. There is an assumption that, because they shared similar professional identities and common practices, people in the same group shared similar values and worldviews, constituting a reasonably homogeneous culture, within which new ideas were interpreted and evaluated. Research physicists principally evaluated a new concept in terms of its usefulness to their research. Physics teachers understood a new concept in the context of instruction, that is, how to teach physics. Journalists valued a new concept when they could write a news article about it. The interest for scientist-literati was in the literary essays they might write. Marxists and Kyoto school philosophers would try to understand and write about a new concept within their philosophical frameworks.

The third step is to understand the historical process of how complementarity was introduced and discussed in Japan within these different cultures. People from different cultures reacted to complementarity differently. Research physicists were generally indifferent, because of its apparent uselessness for producing scientific papers. A notable exception was Nishina Yoshio, who was a student of Niels Bohr and was involved in the translation of Bohr's first presentation of complementarity in 1927. Some physics teachers used it in textbooks to explain conceptual aspects of complementarity. The interest of journalists in complementarity was aroused by Albert Einstein, Boris

Podolsky, and Nathan Rosen's 1935 paper on the so-called EPR thougtht experiment[4] and appreciation of the news value of the debate that inevitably ensued. Scientist-literati borrowed the notion of complementarity in their literary essays to justify their romanticist notion of science. Kyoto school philosophers also borrowed and used the notion of complementarity to supplement their philosophical doctrines. The reactions from Marxists were mixed, however. Some attacked Bohr for what they saw as idealism in complementarity, while others took the side of Bohr against Einstein, whose idea they regarded as bourgeois and Machian idealism (Ito 2002, chap. 7).

Such a close look at what actually happened in the discussion of complementarity in Japan clearly indicates diversity within Japan. It is thus not helpful to understand debates and controversies in Japan in terms of an overarching Japanese culture. Within the Japanese context, there were different, sometimes conflicting cultures, and these different cultures do necessarily originate from Japan's traditional cultures. Certainly, the Kyoto school philosophers owed much to the Buddhist philosophical tradition (although Buddhism, too, is of foreign origin), and scientist-literati incorporated Japanese classical and modern literary traditions in their essays. However, most of the intellectual resources available to those who struggled to understand complementarity and other conceptual problems of quantum mechanics originated from Europe, which had already occupied important places in Japanese academic life. At the same time, those intellectual traditions that came from the West, or other cultures that had close counterparts in the West (e.g., science journalism), played different roles in different circumstances and gained different cultural meanings within the Japanese context.

Thus, a closer look at what actually happened reveals not only the heterogeneity of cultures among those who discussed the philosophical issues of quantum physics in prewar Japan but also the transnational nature of these cultures. A culturalist approach would be inadequate to understand the historical debates over quantum mechanics in Japan. Rather, one needs to approach cultures by recognizing their heterogeneity and transregionality.

Sameness and Differences: Transmission of the Feynman
Diagrams to Japan

Interests in or overemphasis of differences in scientific practices and cultures can originate from implicit universalism of science. Universalism assumes that there is only one universal form of knowledge about nature, and it is often identified with so-called Western science or what it will become in the future.

Since science is assumed to be universal in this sense, it is the differences, not the similarities, of science in different places that are noteworthy and require explanations. Thus differences are considered as interesting.

We no longer have much trust in universality of science. Instead of assuming universality of knowledge as the norm, we can see that knowledge in different places naturally develops in diverse ways. If we start from diversity of knowledge, what need to be explained about science are the surprising commonalities that emerged through time. Assuming that local contexts matter, it is a natural question to ask how scientific practices have become so compatible among different people in different places. Science studies have investigated how initially diverse scientific or technological possibilities stabilize and converge into a standardized form (see, e.g., Bijker 1997). We can ask similar questions in relation to science and technology in non-Western countries.

In the nineteenth century, for example, one obvious force behind stabilization and standardization in science and technology in the global context was Western expansion and colonialism. Europeans' form of knowledge about nature was so closely tied to the military and industrial powers of nations that non-Western countries hardly had any other choice than to adopt the same forms of knowledge and compete in the same arena. Japan's decision to westernize educational and scientific institutions and introduce Western technology in the late nineteenth century was largely, if not completely, motivated by the military strength of the Western naval powers, symbolized by Commodore Matthew Perry's flotilla and demonstrated by the battles in Chōshū and Satsuma against the Western powers (Ito 2015; see also Hōya 2010; Ishii 1993). Japan thus became, willingly or unwillingly, a participant in the international competition, in which the rules were set by European and American countries.

To discuss the issue of sameness and difference, I focus on the case of the transmission of the Feynman diagrams to Japan. As I mentioned earlier, it was in some sense a self-critique of the work that I coauthored with David Kaiser and Karl Hall on the dissemination of the Feynman diagrams within the United States to Japan and to Russia (Kaiser et al. 2004). The Feynman diagram is a graphical technique for keeping track of calculations in elementary particle physics, accompanied by a set of rules for carrying out theoretical calculations. First introduced by Richard Feynman in 1948, these diagrams greatly facilitated calculations in elemental particle physics. At the same time, since it is a graphical method, explaining it is not entirely straightforward. In discussing the transmission of this skill-like theoretical knowledge, our article demonstrates that even in the transmission of theoretical knowledge there can be issues similar to those that arise in the transmission of skill-like experimental knowledge, or so-called tacit knowledge (see, e.g., Collins [1985] 1992, chap. 3).

In our article, we argue that the transmission of the Feynman diagrams to Japan was difficult due to cultural differences and a lack of personal contacts. It shows how Japanese physicists misinterpreted the Feynman diagrams. In other words, our article depicts physics in Japan as different, and that difference was a hindrance to the introduction of the Feynman diagrams in Japan. One problem with this evaluation is the question of how difficult is "difficult." The transmission of knowledge is obviously not always easy. Our framing of the case in terms of the difficulty of transmission resembles a framing in terms of cultural differences. It is a Kuhnian framework, in which there are incommensurable differences between different paradigms, and communications between such different paradigms often encounter difficulties. Our article thus measures the success and failure, the ease and difficulty of transmitting the Feynman diagrams in terms of differences in conceptual schemes. It emphasizes how difficult it was in the Japanese case and how local differences prevented a smooth transmission. In the conventional models of transmitting science, differences can have only a negative effect on the transmission. Differences are obstacles and sources of noise, prejudice, and misconceptions. The more similar we are to the center of transmission, the easier it will be for us to receive the transmitted message.

I propose framing the questions in the opposite direction: How did Japanese physicists eventually manage to have the same scientific practices as American physicists? What kinds of conditions in Japan facilitated the transmission of the Feynman diagrams? The idea of "the same scientific practices" can be problematic, but in this case it is very clear. American and Japanese physicists drew Feynman diagrams in a way that made these diagrams compatible with each other.

This reversal of the question is partially motivated by the above-mentioned issue of how to measure difficulty. On the empirical level, there is at least empirical evidence that understanding and using Feynman diagrams were perceived as not so difficult. In an interview, Nambu Yōichirō, one of the Japanese theoretical physicists who were central to the new developments in Japan at that time, stated that it was easy for Japanese physicists to learn the technique of the Feynman diagrams when he first encountered it (KEK Shiryōshitsu 2005, 33). This statement is understandable because the Japanese physicists were already developing similar ideas. As mentioned in our article, before the introduction of Feynman diagrams, Japanese physicist Koba Zirō developed his own graphical techniques, called Koba diagrams or Koba transition diagrams. Koba diagrams were not the same as Feynman diagrams. Whether the difference between Koba and Feynman diagrams hindered the introduction of the latter, or the similarities between them helped it, might still be debatable, but at least subjectively, Japanese physicists felt that it helped. This is likely among the conditions that facilitated the introduction of the Feynman diagrams in Japan.

Thus, an alternative narrative would discuss how prior conditions in Japan such as the development of the Koba diagrams led to the creation of the environment in which the same or similar activities—in other words, research on diagrams compatible with that of American scientists—were possible in Japan. Such a study would focus on similarities rather than differences and find the former more interesting than the latter.

Conclusion

Various motivations might encourage one to see differences as interesting, among both Japanese and non-Japanese scholars. Such studies, for example, can be important in showing the existence of kinds of knowledge that differ from Western science. But if such an attitude functions as an unquestionable norm, it might do more harm than good. In particular, it can obscure the potential interest and importance of studies of sameness, for example, of how diverse possibilities of knowledge have stabilized into a reasonably standardized unity. Indeed, understanding what caused this process to occur on the global scale in various disciplines is probably one of the most fundamental questions concerning the form of knowledge that we call science.

How such a process occurs is beyond the scope of this chapter. The cases that I outlined above suggest at least two things. One is the role of preexisting conditions. Rather than pouring tea into an empty cup, the transmission of science to the non-Western world occurred as developments resulting from the interactions between the influx of knowledge from the outside and the preexisting conditions that either helped or hindered the introduction of new knowledge. I previously proposed understanding the replication of a form of knowledge in a different place through a model based on an analogy of resonance. The phenomenon of resonance in physics is best exemplified by the transfer of sound between two sound forks with the same natural frequencies. The transfer of the vibration from one sound fork to the other occurs not only through the sound of the initially vibrating fork but also due to the other sound fork, which has properties capable of replicating the vibration. Similarly, the replication of scientific practices occurs when there are preconditions that allow it, triggered by additional causes (Ito 2002, 14; 2005).

The other question is how replicated scientific practices are stabilized and come to take standardized compatible forms. The model of resonance does not completely explain the sameness of science. One explanation might be the same roots of preconditions in different places. For example, in the case of the Feynman diagrams, American and Japanese physicists shared the same body of physical knowledge before the war; hence, they had the same starting point. Realist philosophers of science might also claim that we all share the same reality; therefore, researchers might be seen as dealing with the same objects, or at least different parts of the same objects. Another explanation might be standardizing the process through interactions or competition. If the rules of the game were set, one might need to adopt the same rules in order to survive.

These questions seem to point to various research directions. But for such investigations to be accepted, both authors and readers would need to overcome culturalist biases. In the "Message from the Chairman" of the English summary version of the 2012 official report by the National Diet of Japan's Fukushima Nuclear Accident Independent Investigation Commission (NAIIC), the commission chair, Kurokawa Kiyoshi, an eminent physician and former president of the Science Council of Japan, writes that the nuclear accident in Fukushima "was a disaster 'Made in Japan'." According to him, "Its fundamental causes are to be found in the ingrained conventions of Japanese culture" (National Diet of Japan 2012, 9). Curiously, such a culturalist explanation of the disaster only appears in the English version of the report; the official Japanese report gives a significantly finer-grained analysis of the institutional cultures of the relevant organizations, without indulging in culturalism. Thus, Kurokawa, following the self-orientalist tradition of Yukawa, presents Japan's nuclear issue in culturalist terms. The non-Japanese audience might have found Kurokawa's conclusion curious, but the NAIIC's report could have been much more helpful to them if it had examined the aspects of the accident that were not unique to the Japanese case. In other words, sameness, rather than difference, would have been more interesting. Perhaps historians of non-Western science should avoid committing similar mistakes.

Notes

1. For Honda's biography see, e.g., Ishikawa 1964.

2. For the arguments on the relation between complementarity and "Eastern thought," see Katz 1986 and Kothari 1985; see also Holton 1973.

3. I am examining this question in detail in a paper currently in preparation.

4. The EPR thought experiment is a thought experiment that Einstein, Podolsky, and Rosen proposed to show that quantum mechanics was not a complete description of the reality. They claimed that in a certain experimental setup it was possible to measure the position and the momentum of a particle more accurately than the limits set by Heisenberg's uncertatinty relations in quantum mechanics. See, for example, Fine 1996.

References

Benedict, R. 1946. *The Chrysanthemum and the Sword: Patterns of Japanese Culture.* Boston: Houghton Mifflin.

Bijker, W. E. 1997. *Of Bicycles, Bakelites, and Bulbs: Toward a Theory of Sociotechnical Change.* Cambridge, MA: MIT Press.

Collins, H. 1992. *Changing Order: Replication and Induction in Scientific Practice.* 2nd ed., with a new afterword. Chicago: University of Chicago Press.

Einstein, A., B. Podolsky, and N. Rosen. 1935. "Can Quantum-Mechanical Description of Physical Reality Be Considered Complete?" *Physical Review* 47: 777–780.

Fine, A. 1996. *Shaky Game.* 2nd ed. Chicago: University of Chicago Press.

Hart, R. 1999. "On the Problem of Chinese Science." In *The Science Studies Reader,* ed. Mario Biagioli, 189–200. New York: Routledge.

Holton, G. 1988. "The Roots of Complementarity." In *Thematic Origins of Scientific Thought: Kepler to Einstein,* rev. ed., 99–146. Cambridge, MA: Harvard University Press.

Hōya, T. 2010. *Bakumatsu Nihon to taigai sensō no kiki: Shimonoseki sensō no butaiura (Late Edo Japan and the Crisis of Foreign War: Behind the Scenes of the Shimonoseki War).* Tokyo: Yoshikawa Kōbunkan.

Ishii, T. 1993. *Meiji Ishin to gaiatsu (The Meiji Restoration and Foreign Pressures).* Tokyo: Yoshikawa Kōbunkan.

Ishikawa, T. 1964. *Honda Kōtarō den (Biography of Honda Kōtarō).* Tokyo: Nikkan Kōgyō Shimbunsha.

Ito, K. 2002. "Making Sense of Ryōshiron (Quantum Theory): Introduction of Quantum Physics into Japan, 1920–1940." PhD diss., Harvard University.

Ito, K. 2005. "The Geist in the Institute: Production of Quantum Theorists in Prewar Japan." In *Pedagogy and the Practice of Science: Historical and Contemporary Perspectives,* ed. David Kaiser, 151–184. Cambridge, MA: MIT Press.

Ito, K. 2015. "La science 'occidentale' sous la restauration Meiji: Mimétisme ou appropriation intelligente?" In *Modernité et globalisation,* vol. 2 of *L'Histoire des sciences et des savoirs,* ed. Kapil Raj and Otto Sibum. Paris: Editions du Seuil.

Kaiser, D. 2005. *Drawing Theories Apart: The Dispersion of Feynman Diagrams in Postwar Physics*. Chicago: University of Chicago Press.

Kaiser, D., K. Ito, and K. Hall. 2004. "Spreading the Tools of Theory: Feynman Diagrams in the United States, Japan, and the Soviet Union." *Social Studies of Science* 34: 879–922.

Katz, E. L. 1986. "Niels Bohr: Philosopher-Physicist." PhD diss., New York University.

KEK Shiryōshitsu. 2005. *Oral History Interview 2005 Yoichiro Nambu* (in Japanese). Tsukuba: KEK Shiryōshitsu.

Kim, D.-W. 2007. *Yoshio Nishina: Father of Modern Physics in Japan*. New York: Taylor and Francis.

Koizumi, K. 1975. "The Emergence of Japan's First Physicists, 1868–1900." *Historical Studies in the Physical Sciences* 6: 3–108.

Kothari, D. S. 1985. "The Complementarity Principle and Eastern Philosophy." In *Niels Bohr: A Centenary Volume*, ed. A. P. French and P. J. Kennedy, 325–331. Cambridge, MA: Harvard University Press.

Kuki, S. 1930. *Iki no kōzō (The Structure of "Iki")*. Tokyo: Iwanami Shoten.

Low, M. 1989. "The Butterfly and the Frigate: Social Studies of Science in Japan." *Social Studies of Science* 19: 313–342.

Low, M. 2005. *Science and the Building of a New Japan*. New York: Palgrave.

Nakamura, S. 1943. *Tanakadate Aikitu sensei (Professor Tanakadate Aikitu)*. Tokyo: Chūō Kōron Sha.

National Diet of Japan. 2012. *The Official Report of the Fukushima Nuclear Accident Independent Investigation Commission: Executive Summary*. Tokyo: National Diet of Japan.

Nomura Sōgō Kenkyūjo. 1978. "Nihonjinron: Kokusai kyōchō jidai ni sonaete ("Nihonjinron: In Preparation of the Age of International Cooperation"). *NRI Refarensu (NRI Reference)*, 2. Tokyo: Nomura Sōgō Kenkyūjo.

Ogasawara, Y. 2006. *Nantonaku Nihonjin: Sekaini tsūyōsuru tsuyosa no himitsu (Somehow Japanese: The Secret of the World-Class Strength)*. Tokyo: PHP Kenkyūjo.

Oka, M. 1927. *Satoshō sonshi (The History of Satoshō Village)*. Satoshō: Oka Michio.

Rosenfeld, L. 1963. "Niels Bohr's Contribution to Epistemology." *Physics Today* 16: 47–54.

Saeki, S., and Haga, T., eds. 1987. *Gaikokujin niyoru Nihonron no meicho: Goncharofu kara Pange made (Great Books of Nihonron by Foreigners: From Goncharov to Pinguet)*. Tokyo: Chūō Kōron Sha.

Sagane Ryōkichi Kinen Bunshū Shuppankai, ed. 1981. *Sagane Ryōkichi kinen bunshū. (Sagane Ryōkichi Memorial Volume)*. Tokyo: Nikkan Kōgyō Shimbun.

Said, E. W. 1978. *Orientalism*. London: Routledge and Kegan Paul.

Shapin, S., and S. Schaffer. 2011. *Leviathan and the Air-Pump: Hobbes, Boyle, and the Experimental Life*. Paperback reissue with a new introduction. Princeton, NJ: Princeton University Press.

Tomonaga, S., et al. 1982. "Zadan Nishina sensei o shinonde ("Round-Table Talk in Remembrance of Dr. Nishina"). In *Tomonaga Sin-itiro chosakusū (Collected Works of Tomonaga Sin-itiro)*, 6:57–93. Tokyo: Misuzu Shobō.

Watsuji, T. 1935. *Fūdo: Ningengakuteki kōsatsu* (*Climate and Culture: A Philosophical Study*). Tokyo: Iwanami Shoten.

Yoshida, H., and S. Sugiyama. 1997. "Aikitu Tanakadate and the Beginning of the Physical Researches in Japan." *Historia Scientiarum* 7(2): 93–105.

Yukawa, H. 1973. *Creativity and Intuition: A Physicist Looks East and West.* Translated by John Bester. Tokyo: Kodansha International.

3

The Cultural Politics of an African AIDS Vaccine

The Vanhivax Controversy in Cameroon, 2001–2011

A new therapeutic vaccine against AIDS was presented to the world in 2001 in Cameroon. Created by Victor Anomah Ngu, a renowned professor of medicine and a former health minister, Vanhivax (for Victor Anomah Ngu HIV vaccine) rapidly became a very popular AIDS treatment in Cameroon. Ngu turned his villa in Yaoundé into a private health center, the Clinic of Hope, and hundreds of HIV-infected patients enrolled there to receive a course of treatment using the vaccine.

The Cameroonian authorities gave the "Cameroonian" vaccine a guarded endorsement, despite some obvious scientific weaknesses. At the same time, local AIDS experts and international authorities often remained silent and, when pressed to comment on Vanhivax, were always skeptical, some warning that the treatment was inefficient and dangerous. In the absence of a clear-cut official position, Cameroonian believers and doubters continued arguing, while patients at the Clinic of Hope and those seeking a cure were left to their own evaluations of the risks and promises of taking Vanhivax. This chapter analyzes this long-running (and hitherto unresolved) controversy.

The Vanhivax story is a case study of the use of culturalism in public and academic discourses on science. Vanhivax was indeed commonly described as an African or Cameroonian vaccine. According to its supporters, the vaccine demonstrated the capacity of African scientists to make major advances against all odds; it was said to be the product of a specifically African cognitive and moral approach to science and medicine. As such, it brought together an unexpected coalition of followers, including figures from the radical opposition in Cameroon, Afrocentrist activists, and philosophers of science and ethicists. The way it became entangled in the cultural politics of science, where its Africanity was defined and defended (and denounced), is the main focus of this chapter.

Studying Vanhivax is important on three levels. First, Vanhivax is one of a well-known series of AIDS "miracle cures" developed in Africa since the late 1980s, all of which have triggered similar controversies. The first African wonder drug against AIDS was MM1, a drug derived from secret plant and animal extracts invented in 1988 by Professor Ziriwabagabo Lurhuma at the University Hospital of Kinshasa, and named after President Hosni Mubarak of Egypt and President Mobutu Sese Seko of Zaire, who personally backed the initiative (Fassin 1994). It was followed in 1990 by a Kenyan biomedical AIDS treatment, Kemron, and in 1997 by Virodene in South Africa (see Fassin 2006, 92–93); in 2001 by Therastim in the Ivory Coast (Dozon 2001; Pialoux 2001); in 2002 by Metrafaids in Senegal (Bernard Taverne, work in progress); and more recently by the HIV treatment program based on "traditional medicine" developed since 2007 by Aladji Yahya Ahmed, president of the Gambia (Cassidy and Leach 2009); by the Nigerian AIDS vaccine developed by Jeremiah Abalaka, who claims to have cured more than 900 patients so far (Abalaka 2004); and by at least a dozen other lesser known neotraditional or parabiomedical AIDS treatments, currently commercialized in African countries (see Simon et al. 2007). All these episodes followed a strikingly similar pattern. As noted by anthropologist Didier Fassin (2006, 93), in all these cases the treatments were developed by scientists coming from the margins of the medical field who drew authority from both academic and traditional/political positions. In these cases also, the announcement of the discoveries circumvented usual methods of proving, evaluating, and communicating biomedical research results, based mostly on the peer-review system, by choosing direct interaction with the press and the public. Third, the wonder drugs were appropriated and exploited by nationalist politicians. Fourth, the breakthroughs were presented as a blow being struck for Africa, putting an end to a history of Western oppression, prejudice, and contempt in its scientific relationship with Africa. African culture itself was claimed to be simultaneously the source of scientific success, a badge of pride, and an explanation for the rejection of the discoveries by Western science.

Second, the Vanhivax controversy became part of renewed debate on science in Africa that peaked during 2001–2004. In Africa the 1990s was marked by aggressive restructuring of public, academic, and medical institutions, under neoliberal "structural adjustment" programs (Ferguson 2006; Nguyen 2010). This considerably weakened the universities and research institutes inherited from the postindependence years. National welfare and education systems, as well as institutions linked to former colonial powers, often collapsed (Lachenal 2011). At the same time, neoliberal reforms accelerated the rise of new types of scientific collaboration, epitomized by short-term transnational projects (Geissler and Molyneux 2011). Official initiatives and ideologies sought to reimagine the future of science in Africa and to redefine the relationships among science, the economy, and development. Optimistic visions included plans for an "African Renaissance" and the New Partnership for Africa's Development, a multilateral initiative launched in 2001 by the African Union, which combined neoliberal policies, Pan-Africanist imaginaries, and the affirmation of African pride. In times of crisis and conflict, science and technology were presented as key to the rebirth of the continent.[1]

In parallel, the definition and promotion of "African" science, and more generally of an essential African identity, became the topic of intense debate. The rise (or the persistence) of ideologies promoting "the idea of a unique African identity founded on membership of the black race" and on "re-enchanted traditions" (Mbembe 2002, 241, 267) had repercussions in the field of philosophy and history, where it translated into the scholarly and lay narratives on Afrocentrism. The controversial idea that Africa is the source of Western civilization was popularized in academic circles by Martin Bernal's book *Black Athena* (1987).[2] This debate has become tenser and more polarized in recent years and has fueled mutual misunderstanding between African and European scholars. While this particular polemic has rarely concerned the biosciences directly, its very existence has shaped the terrain where the discourse about Vanhivax took place.

Third, studying the Vanhivax controversy allows us to reflect on the use, overuse, and pitfalls of the notion of culture in the study of science, medical anthropology, and public health. Recent Africanist scholarship has contributed to the radical critique of analytical categories such as "ethnic group," "tribe," and "tradition" by demonstrating how these were historically invented and instrumented by various African and European actors, and first by colonial ethnographers and rulers.[3] Culturalism—the use of culture as an all-encompassing and tautological explanation of specific social, political, or epistemological realities—has thus been a favorite target in the field of African studies (Bayart 1995). For instance, Fassin, Jean-Pierre Dozon, and Randall Packard, among others, have pioneered the critique of public health experts and epidemiologists' "culturalist commonsense" by showing that using factors attributed to African culture to explain the AIDS epidemic on that continent were often mere reproductions of colonial clichés and hindered the proper understanding of the epidemic and the design of efficient prevention policies.[4] Drawing from these works, in this chapter I do not want to culturalize culturalism. In this case, the apparent African propensity for miracle cures for AIDS, along with their presentation as expressions of an African difference, makes it tempting to cast such culturalism as an exotic discourse—that is, using culture as a factor explaining the prevalence of culturalist pathologies in science and philosophy. Rather, this chapter aims to present the inventor of Vanhivax not as a "tropical Lysenko" considered in isolation and as the product of a cultural context but as (1) an ordinary public scientific figure (with many equivalents in Western contexts) and (2) an actor who manipulates (rather than manifests) his own Africanity pragmatically. "Culture" in this chapter is thus understood strictly as an actor's category, with performative effects, and never as an explanatory factor.

This chapter is constructed as a controversy study (Latour 2005; Pestre 2006).[5] It locates the diverse actors of the Vanhivax controversy socially and analyzes both their views and the use of science, medicine, and culture. Following this classic framework, I deal symmetrically with the orthodox and heretical poles of the controversy (those condemning and praising Ngu, respectively). As Fassin explained in his work on AIDS denialism in South Africa, maintaining this analytical posture is difficult because, in the case of contemporary AIDS controversies, one is continually pressed to take a position. Neutrality, albeit analytical, is taken as a sign of complicity with the dissidents who are threatening patients' lives and is thus considered unacceptable. Nevertheless, I follow Fassin (2006, 43, 72) in saying that this discomfort is the condition of useful anthropological and historical analysis. My project is to understand the success of Vanhivax, which requires taking seriously the political, ethical, and scientific meaning attributed to it by its supporters, promoters, and users. This task is all the more complex in my case because the positions were not as clearly drawn as they were in the case of Thabo Mbeki's denialism: since most AIDS specialists gave no opinion on Vanhivax, no scientific orthodoxy had been constructed on which to rely (which does not mean that there was no consensus, in private discussions, on the dangers of Vanhivax).

An anecdote illustrates this position. In 2006, I invited Ngu to attend a talk I gave to a local audience to communicate the final results of my doctoral dissertation on the history of medical research in Cameroon (Lachenal 2006a).[6] He had been a major actor in this history in the 1970s and 1980s, and I had interviewed him at his clinic in 2002 and again in 2003. The talk was held in the conference room of the Centre Pasteur du Cameroun, the largest biomedical research facility in Cameroon, which had been headed by French doctors since its creation. When he appeared in the room, the French expatriates looked at each other and rushed to me, obviously greatly embarrassed: "What is *he* doing here?" I explained that I had invited him as a former minister of health, although I knew very well that the creator of Vanhivax was persona non grata in this institution. During the round of questions after the talk, Ngu stood up, congratulated me—"what you have said is only the truth"—and left. Later, several Cameroonians who attended the conference told me this was the highest compliment one could have, from someone they greatly admired. In this chapter I would like to be true to both this approbation and this embarrassment.

The chapter starts with a comparison with cases of other AIDS cures developed in Africa, to highlight both the generality and the specificity of the Vanhivax controversy. I discuss in turn the four aspects identified by Fassin: the discoverer's position at the margins of medicine, the uses of biomedical evidence, the exploitation of the discovery for nationalistic ends, and the redressment narratives it inspired in Cameroon and beyond. I then recast the controversy in the shifting landscape of biomedical research in Cameroon and its associated narratives. Finally I discuss how Vanhivax was part of actual and symbolic transfers of technology and knowledge at a global level, especially with a U.S.-based biotech company, which further complicated the initial definition of Vanhivax as an African cure for AIDS.

The Outsider from Within

Given his academic and political stature, Ngu can hardly be considered on the margins of medicine in Cameroon. A closer look at his career in the last two decades does indicate, however, a growing distance from the main actors in biomedical research in Cameroon—with Ngu becoming, to quote a local researcher, "not excluded, but apart."[7] This was mainly because of his belonging to the Anglophone minority in Cameroon and because of his age:

Ngu was born in 1926 and officially retired as a professor at the University Hospital of Yaoundé in 1991.

Ngu had one of the most brilliant medicoscientific careers in Cameroon and perhaps in all postcolonial Africa. He is counted among the most prominent representatives of the first generation of doctors trained under the colonial system and taking office in an independent African nation. Born in British Cameroon, he trained initially at the University of Ibadan in British Nigeria,[8] and he received medical degrees from the University of London, where he specialized in surgery (1951–1959). He took a position at the University Hospital of Ibadan, then the flagship university of independent Africa, where he served until 1971. During the 1960s he developed his research on chemotherapy treatment for cancer, thanks notably to a Rockefeller research fellowship at Harvard University (1962–1963).[9] He made major contributions to the description, understanding, and treatment of Burkitt lymphoma (a complex, multifactorial, viral cancer affecting children in tropical Africa), which he published in major journals (including the *British Medical Journal*; e.g., Ngu 1964a, 1964b, 1967). In 1972 he along with a dozen other researchers received the Lasker Award (the "American Nobel" in medicine) for his work on chemotherapy for Burkitt lymphoma.[10] In brief, very few African scholars of his generation can match the excellence and international significance of his trajectory. In 1971 Ngu left Nigeria for Cameroon, where he was appointed professor and head of surgery at the University of Yaoundé Faculty of Medicine, which had just been created with the support of the World Health Organization (WHO), and which at that time attracted many top-level medical specialists, including French expatriates. He held this position at the University Hospital of Yaoundé until his retirement in 1991.

The second part of his career, from the mid-1970s onward, reads as an impressive progress up the administrative and political hierarchies in Cameroon. Ngu had a long history of close ties with the single-party state, dating back to his involvement in the Union Nationale Camerounaise in the 1970s (De Prince Pokam 2010). Appointed vice chancellor of the University of Yaoundé from 1974 to 1982, he became director general of scientific and technical research in Cameroon in 1982, in charge of supervising the nation's research policy and institutions. The newly appointed president of Cameroon, Paul Biya, chose him as the minister of health in 1984, a position he held until 1988 (one of the longest stay in this office under Biya). Ngu then "went back to his research" and developed Vanhivax.[11]

Despite this astonishing background, Ngu remained, in some respects, on the margins of local medicine and politics. The most obvious reason was his Anglophone origins.[12] Ngu's international career followed connections with Britain or the United States exclusively, which was exceptional in Yaoundé's academic sphere, where French institutions retained a strong influence until the 1990s (Lachenal 2009, 2011). Anglophones in Cameroon constitute a linguistic and political minority, with a history of political oppression by, and opposition to, the postcolonial state (headed by Francophones since independence). Recently the Anglophone identity has come to acquire a quasi-ethnic status, in a context of economic and social crises when ethnoregional divisions have been constructed and exploited by many political actors (Konings and Nyamnjoh 2003). The matter is all the more complex in that local policies of multiculturalism have sought to maintain a visible Anglophone presence at all levels of public administration (to disarm accusations of exclusion), which has opened career opportunities for certain high-profile Anglophone individuals, including academics. Whether this was significant or not in Ngu's trajectory, it is worth remembering that the "Anglophone question" cast a long shadow in the Vanhivax controversy—if only as a ready-made explanation for his opponents' skepticism.

Ngu was also a generational outsider. His most active period in research dated back to the 1970s, and although he never abandoned laboratory work afterward (he began to work on Vanhivax in 1989), he never regained his international connections and stature. In addition, Ngu did not become an AIDS specialist; while he was a minister of health a younger generation of medical doctors led the national response to the new epidemic, benefited from new transnational links, and took part, without Ngu, in the transformation of the biomedical sector that accompanied the AIDS crisis (Eboko 1999). The professor did not fit into the new "projectified landscape" (Whyte et al. 2013) of research, where science became more and more an activity conducted through a myriad of minimalistic, short-term but relatively well-funded transnational "projects." In the 1990s Ngu appeared more and more as a vestige of postindependence science.

The Making of Biomedical Evidence

While direct communication with the media, a touch of mystery, and criticism of the system for scientific publication as an obstacle to innovation were central in the presentation of MM1 and Virodene, Vanhivax relied on this strategy only as a secondary measure. First came scientific hypotheses and publications.

A particular feature of Vanhivax was its theoretical consistency, and the importance given to the careful presentation of the underlying immunological theory. It is worth examining this theoretical apparatus, because it had a powerful resonance in Cameroon: many people, even uneducated, were familiar with Ngu's tales of immune systems and viruses. In addition, Ngu stressed that his vaccine came as the conclusion of more than twenty years of empirical and theoretical research on immunotherapy.

This discourse was not constructed retrospectively. From the mid-1970s Ngu had worked on the immunological aspects of Burkitt lymphoma (Mc-Farlane, Barrow, Ngu, and Osunkoya 1970; McFarlane, Ngu, and Osunkoya 1970) and experimented (with limited success) with methods of cancer therapy based on serotherapy (immune serum injections) (Ngu 1967) and on the sampling, in vitro culture, and administration of patients' own leukocytes (the blood cells involved in immune defense) (Ngu et al. 1973; Ngu, Nkwanyo, and Bakare 1976). Having published his results in African journals such as the *West African Medical Journal* (a common publishing practice at that time), Ngu built a systematic approach, which he called "autobiotherapy" (Ngu et al. 1981). His idea was to boost the immune response against tumors, by using preparations of blood cells from individual patients, which were then manipulated in vitro to increase their immune potency and injected back into the same patients.

Ngu considered this approach to be indicated for tumors caused by viruses. Oncogenic viruses were one of the scientific hypes of the early 1980s, when the pathogenic role of retroviruses (e.g., human T-lymphotropic virus discovered by Robert Gallo), of the hepatitis B virus, and of the cytomegalovirus were described, retrospectively making the case of the Burkitt lymphoma, caused by Epstein-Barr virus, the first landmark in a general reinterpretation of the etiology of cancer.[13]

Drawing on these breakthroughs (in which he had played an important role through his work on Burkitt lymphoma), Ngu specified the targets of his autobiotherapy. Ngu's reflection started from the fact that all these viruses, including the newly discovered AIDS-causing HIV, were "enveloped viruses," that is, viral particles with a lipid-based envelope taken from the membranes of the cells of their human hosts. For Ngu this envelope explained their persistence in the body and their pathogenic effects: the viruses' envelope enabled them to hide their antigens and to "disguise" themselves as a part of the body's "self." Enveloped viruses thus "blackmailed" (one of Ngu's favorite expressions) the body into producing ineffective antibodies. This theoretical understanding opened a therapeutic perspective: by removing the envelope of a patient's virus in vitro and injecting the "naked viral core" back into the patient, it appeared possible to induce an immunological response to the virus and "to completely eliminate the viruses" (Ngu 1992, 20). According to Ngu, immunotherapy for cancer and a "therapeutic vaccine" against viral tumors were possible.

Ngu published this theory in 1992 in the journal *Medical Hypotheses*, a non-peer-reviewed publication specializing in such speculative propositions. The article, written in a clear, logical, textbook style, included a discussion on AIDS treatment based on the same principle, and a practical recipe for the removal of the viral envelopes in vitro: delipidation using lipid solvents such as ether or chloroform. In 1994 and 1997, Ngu published two other articles in the same journal, applying the same theory to chronic infections in general and specifically to HIV. He did not mention, however, that he had started experimenting on an HIV "therapeutic autologous vaccine" as early as 1991.

The first publications of the results of Ngu's HIV immunotherapy (which then appeared as "Vanhavix") came out in 2001 in local periodicals (Alemnji et al. 2001; Ngu, Ambe, and Boma 2001). The first was in *Le Bulletin de Liaison et de Documentation de l'OCEAC,* a bulletin published by the Organisation de Coordination et de Lutte contre les Grandes Endémies en Afrique Centrale, an interstate public health organization created in Yaoundé just after independence to coordinate vaccination campaigns in the whole Central African region, and which had since received constant support from the French aid agency Coopération. In brief, it was a minor journal, but one with a past and—at least from a local perspective—symbolic capital. The second report came in the *Journal of the Cameroon Academy of Science* or, to be more precise, on page 2 in volume 1, issue 1 of that journal, which was launched when Ngu became president of the academy. International circulation of both journals was almost nonexistent, and both were actually difficult to find even in Cameroon, but Ngu's results were "published."

The articles, and later data published on Ngu's website, were strictly based on biomedical evidence. They sought to demonstrate that Vanhivax had an effect on the immune system of the patients by following several indicators before and after treatment, such as the level of circulating immunoglobins specific to HIV, the level of CD4, and viral load.[14] The articles had obvious weaknesses: the small numbers of patients included and the absence of control subjects. Yet these flaws were readily acknowledged by the authors, and they were said to reflect the material difficulties of the Clinic of Hope and the lack of funding—in other words, the paucity of the data was one more reason to give support to the initiative. In addition, the articles took care to meet, at least in appearance, the gold standards of modern biomedical research, from p-values (indicating statistical significance) to informed consent (fulfilling ethical requirements). Not forgotten either was a detailed presentation of the protocol of vaccine preparation, and the ritual mention that the National Ethics Committee had approved the research.

Last but not least, the results were presented at a handful of international meetings: oral presentations at the International Congress of Bioethics in London in 2000 and a conference on HIV vaccines in Palm Beach, Florida (presumably the same year), and poster presentations at a prestigious Keystone symposium on HIV pathogenesis in January 2002 and at the Pasteur Institute in Paris in 2001.[15] The presentations were not always successful— ignored posters and empty rooms, according to Ngu—but, back in Cameroon, they gave Vanhivax international standing.

Compared with MM1, Therastim, and Virodene, all of which claimed scientificity but failed to go beyond a superficial use of biomedical references, Vanhivax impresses in the way it drew its legitimacy from more orthodox (albeit weak) types of proof, such as CD4 counts and a plausible (if that means anything) biological rationale. In public communications on Vanhivax with Cameroonian and African media, Ngu did use other types of arguments, such as his own clinical conviction and case histories of miraculous cures. But these remained secondary, even in the lay public's view: what people liked with Vanhivax was that it was based on sound and clear thinking—a hypothesis, a model that can be drawn, explained, and discussed; an experiment, a result. "What is great with Ngu," I was told by a football coach in Yaoundé, is that "he has a method and he sticks to it. He now applies it to cancer and hepatitis."[16] Contrary to what naive observers may expect, narratives of delipidation, naked antigens, and injection appealed to many.

A Homegrown Vaccine

The appropriation of Vanhivax as a product of "national science" by the Cameroonian government was slow and ambiguous. The first official endorsement came in November 2002, at a time when the vaccine was already widely distributed at the Clinic of Hope, was well known by the local population, and had been covered by the local media.[17] It was as a response to popular enthusiasm that the minister of public health, Urbain Olanguena Awono, organized a press conference with Ngu, minister of scientific and technical research Zacharie Perevet, and resident WHO representative He-lene Mabu Madisu to announce that the government had decided to sup-port Ngu's initiative, to enable him to develop proper clinical trials for his vaccine. Local journalists described the event as a "peace agreement" (*paix des braves*),[18] a compromise: the government recognized publicly its interest in Ngu's vaccine, while Ngu implicitly acknowledged that its efficiency had not yet been proven. The minister's discourse itself was cautious and modest, unlike the miraculous narratives concerning MM1, Virodene, and Kemron. The main message was directed to a local audience: the government wanted to prove its goodwill and to silence criticisms of its perceived opposition to homegrown science. To quote the government-owned daily, the *Cameroon Tribune*, "At last, the government of Cameroon has decided to give Prof. Vic-tor Anomah Ngu a chance."[19]

This reluctant endorsement of Ngu's vaccine coincided with the Cameroo-nian state's high-profile commitment to the international mobilization against AIDS. Significantly, the announcement of 100 million CFA francs (€150,000) to support Ngu came the week after the final ceremony launching Synergies Africaines (Eboko 2004), the First Lady's HIV/AIDS humanitarian program, in the presence of Luc Montagnier and Robert Gallo, the two codiscoverers of HIV, an event widely commented on in the international and Pan-Africanist media; Ngu appeared not to have been invited. A few days later, when most international journalists had left, the press conference on Vanhivax took place. The Cameroonian state could not have been clearer: as it prioritized consensual dimensions of the fight against AIDS, such as the prevention of mother-to-child HIV transmission, its aim was to be seen as a leading African country in the fight against AIDS, open to international partnerships and expertise. Nation-alistic dissidence, as was the case in South Africa, was not an option.[20] State support for the "homegrown" vaccine, although substantial, had to be discreet.

In this context, Ngu could only count on the private support of individual members of the government and politicians. The most prominent of these was Henri Hogbe Nlend, who was minister of technical and scientific re-search between 1997 and 2002. A respected mathematician and university professor in France, Hogbe Nlend was a leading member of the opposition party Union des Populations du Cameroun, which led the anticolonial in-surrection in the 1950s. He thus directly embodied the legacy of Cameroo-nian radical nationalism:[21] as minister of research he specialized in nationalist discourses claiming more equity in scientific collaborations (Lachenal 2005). His support for Ngu illustrates how the nationalistic use of Ngu and his work, although sometimes explicit, remained associated with political radicalism.

Quite notably, it was an Anglophone daily paper the *Yaoundé Herald* that gave the most strongly nationalistic interpretation of governmental support for Ngu—not without paradox, given its usual antigovernment stance. Following the press conference, the *Herald* explained that the agreement ended a decade of denial when Vanhivax "could not be recognized by the international scientific community and the Cameroon government because the idea . . . came from a Cameroonian." It opened a future of national glory: "While the West which discovered the AIDS virus is still grappling with a cure, Cameroon, thanks to . . . Anomah Ngu's auto-vaccine, is already light years ahead in a therapy against the scourge of the 20th century."[22]

The hesitant nationalistic appropriation of Vanhivax also reflected Ngu's own ambiguities. In interviews Ngu would frequently lament the lack of governmental support: "The government authorities tend to downplay what was made here, and they tend to put up everything that comes from France, [so that] it receives full backing,"[23] he explained to me in April 2002. At the same time, Ngu would voice a widespread discourse on Cameroon as a country whose potential was wasted by a culture of corruption, a form of ironic and populist nationalism, accusing the elite of neocolonialism and the population of passivity. Such a discourse of discouragement, which President Biya himself summed up with the famous tautology "Cameroon is Cameroon" as an explanation of his own incapacity to make significant change happen, has gained momentum in the last two decades in Cameroon. It was used by Ngu as an explanation of his difficulties:

> If I were a younger man, I won't be in Cameroon today.
>
> . . . The government should stop being corrupt. The proceeds from corrupt money from this country are enough to pay our debts three to five times over.
>
> I think we get a government that we deserve. Cameroonians are very docile and nothing upsets them. They will answer you: "*on va faire comment?*", meaning "what can you do?", implying that they are willing to lick people's boots to get by. We don't behave like people who can stand up and face situations. We prefer not to.[24]

Even after the official acknowledgment of Ngu's vaccine in 2002, the connection between Ngu and the government operated not as an alliance based on nationalism but as a deliberate misunderstanding: one the hand, Ngu could present the agreement as an official validation of his work and at the same time lament its lack of conviction; on the other hand, the government demonstrated its support for national initiatives, while it could tell its international partners that it had played its regulatory role only by acknowledging the need for further research.

An Afrocentrist Vaccine

As Fassin has reminded us, AIDS wonder drugs developed in Africa, with their lone inventors, the enthusiastic politicians promoting them, and their weak proofs, are nothing original: the AIDS epidemic saw similar stories of medical miracles and failures in Western countries too, such as the cyclosporin scandal in France, where three Parisian doctors announced publicly, together with the minister of health, Georgina Dufoix, the discovery of an AIDS treatment—without any evidence other than their clinical instinct and their commitment to advancing French science (Dodier 2003).

There was, however, an element specific to the Afro-biomedical AIDS "cures": their interpretation as historic victories. These African discoveries, it was said in African media reports, proved Africa's scientific capacity and even superiority; they were products of African science, by African scientists, for the health of African populations; they enabled Africa to strike a blow against a past of scientific imperialism, exploitation, and the marginalization of African scientists. Such discourses were dramatically amplified by the rise of the Internet and the "network society" (Castells 1996) in the 1990s: Thabo Mbeki's affirmation that HIV was not the cause of AIDS and later Ngu's Vanhivax were first and foremost phenomena of the cybersphere (Fassin 2006). On the web, the stories circulated among very diverse audiences, took unexpected meanings, and brought together a global coalition of followers, from Virginia to Venezuela.

The global coverage of Vanhivax followed the rhythm of the press conferences organized in Cameroon (beginning with the 2002 agreement with the government), which resulted in press articles in French and English, mostly from Cameroonian and Pan-African journals. The articles then began their second life, copied and pasted, transferred through mailing lists, discussed on websites, blogs, and forums. Although impossible to track, this circulation mainly concerned two audiences. The largest was made up of users of the websites, forums, and mailing lists specializing in AIDS and for people living with HIV, where there was frequent discussion of whether the Vanhivax "cure" was reliable or not. The second audience concerned more specifically the "African diaspora" taken in the widest sense: websites linked to

Afrocentrist movements from France; U.S., African, and French-Caribbean (Antillais) community websites; reggae culture websites; conspiracy theory websites; and most important, news sites and discussion groups related to Cameroon and Nigeria.[25] It was in this universe (where the well-designed Vangulabs.com website made Vanhivax visible) that Ngu became an African hero of science. Although controversial, since it was not unusual to read adverse reports, Vanhivax became a matter of African pride and anti-imperialist romance.

It was not surprising, therefore, to find among his supporters the Cameroonian singer Longue Longue, also known as Le Libérateur, who had made a name for himself with (rather excellent) Makossa hits denouncing white power in Africa; he was the guest star of a Vanhivax fundraising gala in Yaoundé in 2004.[26] More recently, Vanhivax found an unexpected echo in Venezuela, where the story fitted well with the Third Worldist mood in the country.

Ngu's Pan-Africanist and Afrocentrist recognition took on an institutional dimension when he was invited, according to Ngu-related sources, to Washington, DC, in October 2003 to receive the Leon H. Sullivan Lifetime Achievement Award. Leon Sullivan was a Baptist minister and social entrepreneur whose aim was to empower African Americans through self-help and small businesses; he was notably, according to his official biography, the "first Afro-American" to sit on the board of a major corporation in the United States (General Motors). The Leon Sullivan Foundation was created after his death in 2001 to develop humanitarian projects in Africa and to build partnerships between African elites and the "global Diaspora."[27] The fact that it (presumably) honored Ngu as a great "African scientist" illustrates the resonance that the Vanhivax story found in certain African American Christian (if not explicitly Afrocentrist) circles, which was facilitated by the fact that Ngu constantly referred to his Christian faith as the main inspiration for his work. Ngu was also honored in Africa, where the West African College of Surgeons named its annual lecture after him. In 2005, it was said that he even met Thabo Mbeki to propose his treatment for use in South Africa.[28]

Lastly, the "Africanity" of Vanhivax was constructed in sophisticated ways by Godfrey Tangwa, a philosopher and ethicist based at the Department of Epistemology and Philosophy of the University of Yaoundé I. Based at the same university, and sharing with Ngu an Anglophone background, Tangwa had been aware of Ngu's therapeutic experimentation in the early 1990s.[29] As he developed his research on the theme of bioethics in developing countries, he saw in Ngu a good case to study and defend. According to Tangwa, Ngu embodied an alternative to the ultratechnical approach dominating Western biomedicine. His research was based on a strict hypotheticodeductive method—a fact that could only please an epistemologist—and simple technology. Ngu epitomized the potential of an African approach to the AIDS epidemic, based on what Tangwa calls "traditional African values," and it illustrated the deterring effects of Western prejudices toward indigenous African research.

Ngu and Tangwa collaborated and coauthored a presentation at the Fifth World Congress of Bioethics in London in 2000, which was the very first communication on Vanhivax outside Cameroon. They presented Ngu's method and its merits regarding ethical issues, since the "autoimmunotherapy" strategy (a vaccine or, more properly, an immunological therapy, prepared from the patient's own blood) avoided the usual pitfalls of vaccine research (e.g., exposing healthy patients to potentially harmful inactivated viruses). A more developed version of Tangwa's philosophical case for Vanhivax appeared in a special issue of the *Journal of Medicine and Philosophy* (Tangwa 2002). His claim that "traditional African values" of empathy and *ubuntu* "should be enlisted to fight the [AIDS] pandemic]" (Kopelman and van Niekerk 2002, 141) apparently convinced the editors, but they were embarrassed enough by the fact that half of his paper was a detailed promotion of Vanhivax that they included a comment in the issue's introduction that his paper was worth consideration "whether or not one agrees with his claims that particular studies have been overlooked" (Kopelman and van Niekerk 2002, 141). With Tangwa's intervention, the affinity of Ngu's work with the culturalist literature on "African knowledge systems" materialized.

The Africanity of Ngu's vaccine was thus constructed through the process of its global dissemination; its positioning in "African culture" defined by specific ethical values and specific ways of thinking was actively produced through a process of translation, publication, and appropriation (and rejection) at an international level. This framing of Vanhivax, which was absent from the early presentations of the discovery, eventually enabled Ngu to present his work as a specifically African approach to biomedicine. According to him, his approach to the vaccine was in stark contrast to the reductionist "Western-European approach to the virus." Ngu claimed his African approach was more holistic and more intellectual—more scientific: "Let's put it this way: Western scientists think that [their] methods are the only ones that should be used. Unless you do this, you cannot get valid results, you cannot have valid ideas. ... I would say [that] one should not exclude all our methods. And I would say that they have the advantage of a lot of material and equipment, but they have the disadvantage that the equipment they have has made them lose perspective. If I can express it in a crude way, 'equipment has taken over their thinking.'"[30]

Through such discussion of culture and with the help of professional epistemologists, Vanhivax became part of the promotion of an African bioscience, presented both as an alternative to and a potential victim of Western scientific hegemony. Although anchored in Cameroonian idioms of nationalism and minority politics, the "Cameroonian vaccine" was thus representative of a more general debate on science and culture in Africa.

Vanhivax as a Vestige: The Clinic of Hope and the Shifting Landscape of Biomedicine in Cameroon

The success of Vanhivax cannot be reduced to the appeal of Ngu as an African hero. It can be estimated that more than a thousand HIV-infected patients enrolled in the vaccine program at the Clinic of Hope (although many of them did not complete the course of treatment): they signed up for therapy not to boost national pride but at a time (until 2003) when the policies of the fight against AIDS in Cameroon prioritized prevention and epidemiological surveillance instead of access to the treatments available in the North since the late 1990s.

More fundamentally, Vanhivax also embodied another way of practicing biomedicine, which met the needs and expectations of many patients. This other way was precisely an idealized version of the type of medicine inherited from colonial and postcolonial development, which disintegrated in the 1990s in Cameroon following economic crises and structural adjustment. Vanhivax is thus part of another historical and memorial narrative, which is distinct from stories of African redress and renaissance: a story of lost institutions, lost values, and lost expectations, which the Clinic of Hope seemed to reactivate (while at the same time, paradoxically, it epitomized the ongoing privatization of the health system). It was thus as a vestige that the Clinic of Hope attracted many, and reciprocally as a countermodel of an outdated style of science that it was derided by the actors of the new biomedicine. The Vanhivax debate opposed different models of science: diverging concepts of how medical research should be conducted, of what makes a proof valid, of what makes a clinical trial ethically acceptable, and of what makes scientific collaboration equitable.

In the last two decades, the development of AIDS research has led to the transformation of the scientific landscape in Cameroon, which has gradually made Ngu (who retired in 1991) look like a figure from another epoch of science. Although global in its causes and manifestations, this transformation has been especially visible and radical in Cameroon. Considered as the "cradle of HIV," Cameroon indeed became a global hotspot for retrovirology during the 1990s, with dozens of international teams setting up research projects in Yaoundé, often limited to the extraction of blood samples to the North with the help of local scientific personalities (Lachenal 2006b).

In this context, the Clinic of Hope stood as an anomaly. It embodied therapeutic research disconnected from international priorities, based on a genuine theory of viral infection and immune reaction, legitimized by the seniority, altruism, and commitment to the nation of a solitary innovator (Ngu said he sacrificed his pension for his research). The mainstream AIDS research projects were, on the other hand, limited to diagnosis and sample collection, self-justified by their insertion into international networks and by their evaluation by modern ethical and scientific procedures. Ngu's work embodied a philosopher-friendly science made of "ideas" and basic laboratory operations, oriented toward therapy and relief and guided by a humanitarian (and explicitly Christian) ethic and aesthetic of self-sacrifice and urgency: "At my age, I do not expect anything from life. I would be happy to leave the hope of an AIDS cure to posterity. In the meantime, I will not let people die that I am able to save."[31] On the other hand, virology projects consisted in complex, high-tech practices integrated into global flows of samples and DNA sequences, aimed at obtaining patents, high-profile scoops, and press releases about emerging viruses, and were morally legitimized by formal, secular, and international ethical guidelines and review boards. The contrast between these "ideal-typical" approaches to medical research manifested at several levels. The old Mercedes and the austere allure of Ngu himself contrasted with the big four-wheel-drive cars of the AIDS "big men" working for international projects; the Clinic of Hope, with its well-cared-for green plants and TV screen in the waiting room contrasted with the decaying premises of most Yaoundé's hospitals (and a fortiori of provincial health institutions); the Clinic of Hope ambulance was seen everywhere in Yaoundé, driving patients or transporting samples. A material detail completes the picture: the Vanhivax method consisted precisely in *injecting blood back into the patients*, while the virological research carried out in Cameroon was in practice based on sampling, freezing, and exporting blood—which led activists from Act-Up to denounce the "vampirization" of patients and others to denounce it as "DHL research"[32] (a form of research which would be limited to posting samples of blood to Western partners through DHL or Fedex).

Price mattered too: a total course of therapy at the Clinic of Hope, including the initial sampling and testing, the elaboration of the vaccine from patient's samples (delipidation and cell culture), and several injections of vaccine preparations, was priced at 10,000–30,000 CFA francs (€15–45) per month, for six months or more—a considerable sum when the minimum wage was approximately 50,000 CFA francs per month, but one that compared very favorably to the cost of imported antiretroviral treatments (ART) until 2002–2003.[33] In brief, to borrow a local journalist's own categories, Ngu's research corresponded to "a true revolution in the treatment of AIDS patients," which he contrasted with "the abstract fight against AIDS."[34] Drawing on Nicolas Dodier's work on the social and political configurations of biomedical research in the twentieth century, it may be said that the work of Ngu embodied the "clinical tradition": a declining mode of doing and regulating medical science, "based on the overpowering legitimacy of the ethical and cognitive judgement of clinicians concerning their own patients, both in the experimental field and as regards ordinary health care" (Dodier 2005, 205), which Dodier contrasted with the rise of "transnational therapeutic modernity" in the 1980s, where the ethical and epistemic evaluation of experimental medicine is institutionalized and codified through ethics committees and peer-reviewed journals.

In Cameroon, where the transition to transnational modernity in the 1990s coincided with the collapse of medical and scientific institutions, Ngu was an object of nostalgia. Admiration for the man and his work reveals a longing for medicine that cures and for science that discovers. Significantly, Ngu drew considerable prestige in Cameroon from the fact that he was a former student of Alexander Fleming, the inventor of penicillin. Ngu indeed often pointed out that Fleming was his mentor back in the 1950s at St. Mary's Hospital in London and that he was, just like him, a lone and inspired innovator who had to overcome the skepticism of his peers. On the street, people would add that Ngu had even been a "personal doctor of the Queen of England"—a fact that I have been unable to check and that might be an interpretation of Ngu's affiliation with the Royal College of Surgeons. That Ngu did not, contrary to some of his local colleagues, publish discoveries of new viruses in *Nature* or *Science* was clearly irrelevant: he was the (living) embodiment of a past that was better, in many respects, than the present.

Our Kind of Charlatan: The Relationship between Ngu and "the Rest"

Although symbolically opposed, the two universes were not impermeable: people, techniques, and samples circulated between Ngu's side and mainstream science. Ngu's group was deeply connected, if only through informal acquaintance, with the rest of Yaoundé's biomedical scene. In addition, the binary opposition between the Clinic of Hope and the rest of the AIDS researchers became blurred after 2003, when the first large-scale programs to distribute ART were implemented in Cameroon with the support of government subsidies. In most cases, the institutions involved in virological research became leaders in ART. In other words, the era of ART weakened Ngu's monopoly on AIDS therapy and contributed to a significant reduction in the intensity of the debate surrounding Ngu following its peak in 2001–2003. But therapeutic vaccination at the Clinic of Hope continued to attract patients, many of them following ART treatments elsewhere. Ngu generally encouraged patients to take ART, especially when they had a high viral load; he also offered high-quality (and relatively costly) follow-up and treatment for opportunistic infections.[35]

Mutual criticism was integral to the relationship between the Clinic of Hope and the rest of the AIDS sector in Cameroon. Ngu liked to point out how foreigners wanted to use Cameroonian scientists "as a service providing raw material." "They are only interested in raw material, and I refuse to become a provider of such material. Even if they give a lot of money, they make your life impossible," he explained to me in an interview, and also criticized the French scientists' neocolonial attitude in Cameroon.[36]

The criticisms addressed to Ngu were rarely frontal attacks. A few French experts intervened directly on websites and forums where Vanhivax was discussed to denounce it and warn patients in Cameroon and elsewhere in Africa of its dangers. In April 2004 a group of Cameroonian doctors associated with the Douala Antiretroviral Initiative, which had been the first pilot program enabling access to ART in Cameroon in 2000, in a press conference openly warned against the "false claims" made about Vanhivax. Significantly, the accusation was based not on statistics or trials but on personal conviction and clinical narratives, a doctor (requesting anonymity) explaining that he "came across AIDS patients Anomah Ngu was administering treatment to, and who still ended up dying." But such an accusation was exceptional. The general response, when local and international experts were asked for their opinion in the press, was silence. Nobel Prize winner Luc Montagnier explained in 2006 that "the knowledge I have of [Ngu's] work does not allow me to take a position. Therefore I will not comment on that today."[37] Similarly his historical rival, Robert Gallo, said in a 2002 interview during his stay in Cameroon:

JOURNALIST: Professor Victor Anomah Ngu's auto-retroviral vaccine is a near breakthrough. Any comment?

ROBERT GALLO: It is unfortunate I have not yet studied the approach you are talking about. I happened to hear about it for the first time earlier this year in the United States of America. I met Professor Ngu just yesterday and we didn't have the chance to talk about it. It is indeed difficult to appreciate the effort for the moment.[38]

In the private interviews I conducted, most researchers agreed that "something had to be done against Vanhivax," that the government had been too ambiguous, but that such action was the responsibility of "international and local authorities."

Such silent consensus on the inefficiency, if not on the hazards, of Vanhivax has no straightforward explanation.[39] Ngu's ties with several high-level political officials and his popular status made it difficult for individual researchers to denounce his work publicly. Another factor is that several major Cameroonian researchers had, in the 1990s, developed and commercialized their own cures for AIDS—chemical treatments or plant-based preparations—and that any attack on Vanhivax could become embarrassing for them and their international partners (who included the WHO and major French institutions). In addition, figures such as Montagnier and Gallo may have found it difficult to give lessons in scientific and medical ethics, given their own history of prophecies that failed to bear fruit (Montagnier repeatedly announced an imminent AIDS vaccine in the 1980s and 1990s) and repeated allegations of scientific misconduct (in the case of Gallo).[40]

Still, it may be suggested that the lack of criticism of Vanhivax did, in fact, play a role in the local AIDS sector—or, to put it differently, that Ngu was *useful* in the debate on medical research and transnational scientific collaboration in Cameroon. Ngu served as a counterexample, which helped to present the move toward transnational neoliberal modernity (and the moral and institutional norms and forms associated with it) as inherently good. Ngu, the "charlatan" who disliked foreigners, was a perfect enemy.

From Cameroon to California: The Aging Innovator and the
Biotech Company

"Like Saint John the Baptist in this season of Advent we too want to say that
there is HOPE—for HIV." On December 16, 2003, Ngu opened his press conference at the Hilton Hotel in Yaoundé with a solemn prophecy. The news
was that "four special patients" had been cured of HIV: since 1989, Vanhivax
had led to four cases of seroreversion (HIV-positive patients who had become
seronegative). But what caught the attention was the conference's second
message: Vanhivax had been hijacked, Ngu explained. The aim of the press
conference was to "inform our Cameroon public about our European and
American 'friends and collaborators' who visited and with whom we signed
agreements of confidentiality and collaboration." Ngu further explained that
an American biotech company, Lipid Sciences, Inc., and the French phar-
maceutical company Aventis (a global leader in the vaccine industry) "have
BORROWED WITHOUT OUR PERMISSION our discovery and will shortly
claim it for themselves." According to Ngu, both firms had pretended to be
interested in collaborating with the Clinic of Hope, to obtain Ngu's "secrets
at all costs," before "melting into thin air." Names, dates, and publications as
evidence of the fraud were provided. Ngu concluded that, fortunately, his
group was "a few steps ahead" but that further financial support was needed
to "preserve that lead." He proposed, as a conclusion, to "copy from the French
as well" and to organize a telethon for Vanhivax.[41]

The fundraising gala for Vanhivax took place a few weeks later at the Hilton
Hotel, but without much success.[42] However, the revelation that the Cameroonian invention had been misappropriated by the French and Americans
reverberated in the press and on the web. The news fitted all too well into the
narrative: the very fact that the discovery could be copied and stolen confirmed a posteriori its efficiency and its "truth"—piracy was a proof of concept.
Moreover, the cast was perfect: Cameroonians as great scientists, victims of
Western greed; Western scientists as unimaginative, fraudulent villains. The
reports, for that reason, circulated widely. It summed up the "Vanhivax story,"
clarified the meaning of the whole narrative, and epitomized the worldview
and the cultural politics underpinning it. It gave credit to Vanhivax and a key
point to Ngu's supporters, one that would be used for years in public debate.[43]
It confirmed, at the same time, the incompatibility of Ngu's style and ethos
with the practices of transnational scientific partnerships.

Much of Ngu's account was true.[44] In 2002, one of his assistants, Dr.
Gaston Boma, had presented a poster at the Keystone symposium on HIV

pathogenesis. Among the many world-renowned specialists in HIV and vaccine research present at the prestigious Colorado ski resort, Aftab Ansari, a professor of pathology at Emory University, found the presentation interesting.[45] Ansari knew that a newly founded biotech company based in California, named Lipid Sciences, Inc., was developing strategies for therapies for viral infections and cardiovascular diseases based on technologies targeting lipids, such as the delipidation of serum. In other words, Lipid Sciences was working independently on an approach to HIV therapy very similar to Ngu's (published) protocol of autologous vaccine preparation. That was sufficient reason for Ansari to get in touch with both Lipid Sciences and Ngu and to set up a visit to Cameroon. In June 2002, Ansari and Lipid Sciences CEO Philip Radlick met Ngu in Yaoundé and signed a collaboration agreement. On June 12, a joint press conference took place in Yaoundé, which was reported on the Cameroon Radio Television network and in the *Cameroon Tribune*. Ansari praised the qualities of Vanhivax, announced forthcoming trials of the vaccine, and left for the United States.

The collaboration never materialized. According to Ngu, the Americans "reneged on their agreement" once they had gathered the information they needed on his breakthrough and gave him only one dishonest excuse—that their board did not "allow them to collaborate with institutions outside of the USA."[46] According to Ansari, many reasons led Lipid Sciences to abandon the idea of a trial in Cameroon: the differences in the technology they used, which posed a problem given the investment the American firm had already made in developing its own (patented) delipidation protocol; the costs of the potential trial; Ngu's own limited expertise in HIV research; the weak data he had on the patients undergoing treatment; and the control Ngu demanded over the research as the principal investigator.[47] Although they were convinced by the method itself, the Americans did not see in Ngu a promising partner.

A little more than a year later, on December 3, 2003, Lipid Sciences issued a press release on the results of their first study testing the delipidation technique: it showed that the delipidation of retroviruses made them more immunogenic than live viruses when administered to mice. The full study was presented in February 2004 at the Conference on Retroviruses and Opportunistic Infections (the most important yearly conference on basic HIV science); although based on a rudimentary model, the "approach, utilizing delipidated viruses, has potential application in designing more effective vaccines against enveloped viruses such as HIV" (Ansari et al. 2003). The scientist who led the study was Ansari, who after playing the go-between with Ngu, now sat on the scientific advisory board of Lipid Sciences.[48] According to Ngu, this was sufficient evidence that his work was being plagiarized and that the time had come to seek the help of intellectual property lawyers; it was, albeit indirectly, final proof that his method worked. Lipid Sciences' success story was beginning: between mid-November 2003 and mid-February 2004, the stock price of Lipid Sciences rose by more than 300 percent on the NASDAQ stock exchange.[49]

Epilogue and Conclusion

At the time of writing the paper on which this chapter is based, in 2011, I tried to check a few quotes on the websites of the Clinic of Hope and Lipid Sciences. Back in 2004 both were impressively designed websites—one sold its vision to patients and the other to investors[50]—which I used extensively (and archived). In 2011 both sites were closed. Lipid Sciences had wound up its activities in 2008, after having spent more than $10 million in cash every year since 2001 (more than the annual budget for Cameroon's Ministry of Research).[51] The success story, like so many Silicon Valley startups, was brief. Although the company applied for two patents on the delipidation method and its use in HIV vaccines (Lipid Sciences 2003, 2008), and although the first study testing the vaccine approach on nonhuman primates was successful, Lipid Sciences could not find investors to allow it to further its research. The Clinic of Hope, to the best of my knowledge, is still active, although Ngu died in 2011. In Yaoundé, it is still said that patients are cured of AIDS at the Clinic of Hope, although there is no sign that any progress has been made to publish evidence of it. The latest news on the patent application, which Ngu had deposited in the United States in 2005, was that it was still under examination.

The "hope" spoken of in Cameroon and California was never far from sheer hubris. In his press conferences, Ngu liked to say that the Vanhivax approach was close to being tested as a cure of hepatitis B, hepatitis C, and cancer. In 2004 Lipid Sciences boasted that the delipidation approach would bring a cure to influenza and SARS (severe acute respiratory syndrome), which was then considered a top global health priority. With their "autologous delipidated therapeutic vaccines" the aging innovator and the biotech company may be recounting the same familiar story. From Yaoundé to Palo Alto, they tell of the promises, the pretenses, and the failures of global biomedicine and the way it generates considerable profits from current dispossessions and afflictions (Sunder Rajan 2006).

Vanhivax, I argue, took part in this fundamentally global, neoliberal, and contemporary configuration of biomedicine—a highly problematic one in my view. At the same time, in Cameroon it inspired local narratives of nationalism, redress, and nostalgia. This study shows that there is no contradiction between the two sides of the story: the culturalist, Afro-nationalist framing of Vanhivax is not the manifestation of any African culture or of any local atavism. Rather, it must be understood as the signature of the practical as well as imagined connections of Cameroon with the *global*, as one more piece of evidence of Africa's long history of extraversion (Bayart 2000).

Notes

My research on the Vanhivax controversy was part of a wider study of the history and politics of AIDS research in Cameroon. It is based on fieldwork in Yaoundé, Cameroon, on five occasions during 2002–2009. Sources include oral interviews and informal conversations with AIDS researchers and policy makers in Yaoundé, as well as a systematic study of scientific and media publications on AIDS in Cameroon. It received the support of the French Institute de Recherche pour le Dévelopement and the Agence Nationale de la Recherche sur le SIDA et les Hépatites Virales. All translations are my own. I thank Richard Kennedy for his help with language editing.

1. For a critical perspective on the "African Renaissance," see Ferguson 2006, 113–115.

2. For examples of academic contributions to the debate on Afrocentrism, see Howe 1998 and Chrétien, Fauvelle-Aymar, and Perrot 2000.

3. See Hobsbawm and Ranger 1983, Amselle and Mbokolo 1985, Mudimbe 1988, and Chrétien and Prunier 2003.

4. See Dozon and Fassin 1989, 2001, Packard and Epstein 1991, Fassin 2000, and Dozon and Fassin 2001.

5. On methodological conversations with pragmatic sociology and microhistory, see Revel 1996 and De Blic and Lemieux 2005.

6. Published results include Lachenal 2005, 2006b, 2009, and 2011.

7. Cameroonian scientist, interview by the author, 2002.

8. British Cameroon, although a separate administrative entity as a UN trusteeship territory, was de facto administrated as a part of the neighboring colony of Nigeria.

9. Ngu's curriculum vitae, available at the time at the Vangulab website, www.vangulabs.com, and at www.vangu.8m.net/index.html, accessed June 1, 2007 (discontinued). A less precise curriculum vitae is on the new version of the website at vanlabs.org, accessed August 15, 2013.

10. Lasker Award website, www.laskerfoundation.org/awards/1972_c_description.htm#ngu, accessed April 11, 2011.

11. V. A. Ngu, interview by the author, 2002.

12. Cameroon is a bilingual country since the unification in 1972 of the federation formed by the former British Cameroon and French Cameroon, themselves inherited from the partition of the German colony of Kamerun after World War I. On the political history of Cameroon, see Bayart 1979, Joseph 1977, and Mbembe 1996.

13. On the investigations of viral cancers in the late 1970s, see Grmek 1995.

14. Viral load results were published on Ngu's website and in the patent application deposited in 2005 at the U.S. Patent Office (application 10573724).

15. Keystone symposia are prestigious conferences held mostly at the Keystone ski resort in Colorado. See www.keystonesymposia.org/index.cfm?e=Web.Mission Statement, accessed August 15, 2013.

16. Author's fieldwork notes, Yaoundé, 2003.

17. For example, "Recherches: La piste camerounaise," *Mutations* (Yaoundé), July 22, 2002.

18. *Cameroon Tribune* (Yaoundé), November 26, 2002.

19. *Cameroon Tribune* (Yaoundé), November 26, 2002.

20. For a compared approach of African states' responses to AIDS, see Eboko 2005.

21. On the development of the Union des Populations du Cameroun since independence, see Deltombe, Domergue, and Tatsitsa 2011.

22. Gerald Ndikum, "Anomah Ngu Beats Western Medics in AIDS Cure, Government Recognizes Breakthrough at Last!," *Yaoundé Herald,* November 27, 2002.

23. Ngu interview, 2002.

24. "Prof. Anomah Ngu: If I Were a Younger Man, I Won't Be in Cameroon Today," *L'Effort Camerounais* (Yaoundé), May 1, 2006.

25. The complete list is hard to gather. A Google search of "Vanhivax" and "Anomah Ngu" returns more than seven thousand pages.

26. Marie-Noëlle Guichi, "Recherche sur le Vanhivax: Le téléthon donne 10 millions à Anomah Ngu," *Le Messager* (Douala), January 21, 2004; "Le Vanhivax d'Anomah Ngu sans soutien," *Mutations* (Yaoundé), January 21, 2004.

27. Leon Sullivan Foundation website, http://www.thesullivanfoundation.org, accessed April 11, 2011.

28. Valentin Zinga, "Cameroun: SIDA: l'espoir en Clinique," Radio France International, December 25, 2002, http://www.rfi.fr/actufr/articles/048/article_25889.asp, accessed April 11, 2011.

29. Godfrey Tangwa, interview by the author, 2003.

30. V. A. Ngu, interview by the author, 2003.

31. Ngu interview, 2003: "A mon âge, je n'attends plus rien de la vie. Je serai heureux de laisser l'espoir de la guérison du SIDA à la postérité. En attendant, je ne vais pas laisser mourir des gens que je peux sauver."

32. On the importance of vampires stories as ways of understanding and critiquing biomedicine and colonialism, see White 2000 and Geissler and Pool 2006.

33. "Un vaccin suscite beaucoup d'espoir chez des sidéens au Cameroun," PANA Press, February 30, 2003.

34. Quoted in Roland Tsapi, "Vanhivax, quel avenir . . . ," *Le Messager* (Douala), January 5, 2004.

35. A study of the prevalence of *Pneumocystis jirovecii* pneumonia, an opportunistic infection associated with AIDS, indicates comparatively better follow-up and prevention of such infections at the Clinic of Hope than at other Yaoundé health centers (Nkinin et al., 2009).

36. Ngu interview, 2002.

37. "Prof. Luc Montagnier: Le Sida ne vient pas du Cameroun," *Mutations*, July 3, 2006.

38. *Cameroon Tribune* (Yaoundé), November 18, 2002.

39. For a similar discussion in the case of the president of Gambia's miracle cure, see Cassidy and Leach 2009.

40. The best account of the Gallo/Montagnier controversy by far is Seytre 1993.

41. Ngu's discourse was reproduced on his website: www.vangu.8m.net/photo.html, accessed May 21, 2004. Also see "SIDA / Le Pr. Anomah Ngu persiste: Le chercheur a réaffirmé les vertus curatives de son vaccin lors d'une conférence de presse hier," *Cameroon Tribune* (Yaoundé), December 17, 2003.

42. Ten million CFA francs was collected (€15,000). Guichi, "Recherche sur le Vanhivax"; "Le vanhivax d'Anomah Ngu."

43. See, e.g., "Lutte contre le SIDA: Le VANHIVAX menacé de piraterie!," Cameroon -info.net, December 16, 2003, http://www.cameroon-info.net/reactions/@,13838,7,1utte -contre-le-sida-le-vanhivax-menace-de-piraterie.html, accessed April 12, 2011.

44. Though I have not been able to confirm or contradict any involvement of Aventis in this story.

45. Aftab Ansari, personal communication to the author, 2003.

46. Venant Mboua, "Vih/Sida Anomah Ngu réinvente l'espoir," *Le Messager*, December 4, 2003.

47. Ansari, personal communication.

48. "Lipid Sciences Forms Viral Scientific Advisory Board," *Business Wire*, April 12, 2002.

49. Data from NASDAQ: Lipid Sciences (LPD) stock rose from $2.25 on November 17, 2003, to $7.30 on February 16, 2004.

50. Lipid Sciences home page, www.lipidsciences.com, accessed June 22, 2006. After I wrote this chapter, a new website was launched for the Clinic of Hope: vanlabs.org/. I thank Richard Kenney for finding this.

51. See the Lipid Sciences quarterly reports at the U.S. Securities and Exchange Commission, http://www.sec.gov/cgi-bin/browse-edgar?action=getcompany&CIK =0000071478&owner=exclude&count=40, accessed April 12, 2011.

References

Abalaka, J. O. A. 2004. "Attempts to Cure and Prevent HIV/AIDS in Central Nigeria between 1997 and 2002: Opening a Way to a Vaccine-Based Solution to the Problem?" *Vaccine* 22: 3819–3828.

Alemnji, G. A., E. N. Lumngwena, T. Asonganyi, and V. A. Ngu. 2001. "Humoral Immune Response in HIV/AIDS Patients on Immunotherapy." *Le Bulletin de Liaison et de Documentation de l'OCEAC* 34: 8–13.

Amselle, J. L., and E. Mbokolo, eds. 1985. *Au coeur de l'ethnie: Ethnie, tribalisme et État en Afrique*. Paris: La Découverte.

Ansari, A., F. Villinger, J. E. K. Hildreth, A. Mayne, and J. B. Maltais. 2003. "Use of Delipidated SIV to Augment SIV Specific Cd4+ and Cd8+ T Cell Memory Responses in Mice—A Model for a New Therapeutic Vaccination Strategy against HIV Infection."

Poster presented to the 11th Conference on Retroviruses and Opportunistic Infections, San Francisco, February 8–11.

Bayart, J.-F. 1979. *L'Etat au Cameroun*. Paris: Presses de la Fondation Nationale des Sciences Politiques.

Bayart, J.-F. 1995. *L'illusion identitaire*. Paris: Fayard.

Bayart, J.-F. 2000. "Africa in the World: A History of Extraversion." *African Affairs* 99: 217–267.

Bernal, M. 1987. *The Fabrication of Ancient Greece, 1785–1985*. Vol. 1 of *Black Athena: The Afroasiatic Roots of Classical Civilization*. New Brunswick, NJ: Rutgers University Press.

Cassidy, R., and M. Leach. 2009. "Science, Politics and the Presidential AIDS Cure." *African Affairs* 108: 559–580.

Castells, M. 1996. *The Rise of the Network Society*. Oxford: Blackwell.

Chrétien, J.-P., F.-X. Fauvelle-Aymar, and C.-H. Perrot, eds. 2000. *Afrocentrismes: L'histoire des Africains entre Égypte et Amérique*. Paris: Karthala.

Chrétien, J.-P., and G. Prunier, eds. 2003. *Les ethnies ont une histoire*. Paris: Karthala.

De Blic, D., and C. Lemieux. 2005. "Le scandale comme épreuve: Éléments de sociologie pragmatique." *Politix* 18: 9–38.

Deltombe, T., M. Domergue, and J. Tatsitsa. 2011. *Kamerun! Une guerre cachée aux origines de la Françafrique, 1948–1971*. Paris: La Découverte.

De Prince Pokam, H. 2010. "La participation des universitaires au processus de construction/reconstruction de l'espace public au Cameroun." *L'anthropologue africain* 17, nos. 1–2: 81–116.

Dodier, N. 2003. *Leçons politiques de l'épidémie de SIDA*. Paris: Editions de l'Ecole des Hautes Études en Sciences Sociales.

Dodier, N. 2005. "Transnational Medicine in Public Arenas: AIDS Treatments in the South." *Culture, Medicine and Psychiatry* 19: 285–307.

Dozon, J. P. 2001. "Que penser de la publicité faite à l'invention d'un remède anti-SIDA en Côte d'Ivoire?" *Transcriptases Sud* 7. Accessed July 22, 2016. http://www.pistes.fr/transcriptasesud/7_159.htm.

Dozon, J.-P., and D. Fassin. 1989. "Raison épidémiologique et raisons d'Etat. Les enjeux socio-politiques du SIDA en Afrique." *Sciences sociales et santé* 7: 21–36.

Dozon, J.-P., and D. Fassin, eds. 2001. *Critique de la santé publique: Une approche anthropologique*. Paris: Balland.

Eboko, F. 1999. "Logiques et contradictions internationales dans le champ du SIDA au Cameroun." *Autrepart* 12: 123–140.

Eboko, F. 2004. "Chantal Biya: Fille du peuple et égérie internationale." *Politique africaine* no. 95: 91–106.

Eboko, F. 2005. "Patterns of Mobilization: Political Culture in the Fight against AIDS." In *The African State and the AIDS Crisis*, ed. A. S. Patterson, 37–58. Burlington, VT: Ashgate.

Fassin, D. 1994. "Le domaine privé de la santé publique: Pouvoir, politique et SIDA au Congo." *Annales: Histoire, Sciences Sociales* 49: 745–774.

Fassin, D. 2000. "Entre culturalisme et universalisme." In *Les enjeux politiques de la santé: Etudes sénégalaises, équatoriennes et françaises*, 129–150. Paris: Karthala.

Fassin, D. 2006. *Quand les corps se souviennent: Expériences et politiques du SIDA en Afrique du Sud*. Paris: La Découverte.

Ferguson, J. 2006. *Global Shadows: Africa in the Neoliberal World Order*. Durham, NC: Duke University Press.

Geissler, P. W., and S. Molyneux, eds. 2011. *Ethos, Ethnography and Experiment: The Anthropology and History of Medical Research in Africa*. Oxford: Berghahn.

Geissler, P. W., and R. Pool. 2006. "Editorial: Popular Concerns about Medical Research Projects in Sub-Saharan Africa—A Critical Voice in Debates about Medical Research Ethics." *Tropical Medicine and International Health* 11: 975–982.

Grmek, M. D. 1995. *Histoire du SIDA*. 2nd ed. Paris: Payot.

Hobsbawm, E. J., and T. O. Ranger. 1983. *The Invention of Tradition*. Cambridge: Cambridge University Press.

Howe, S. 1998. *Afrocentrism: Mythical Past and Imagined Home*. London: Verso.

Joseph, R. A. 1977. *Radical Nationalism in Cameroun: Social Origins of the U.P.C. Rebellion*. Oxford: Clarendon Press.

Konings, P., and F. B. Nyamnjoh. 2003. *Negotiating an Anglophone Identity: A Study of the Politics of Recognition and Representation in Cameroon*. Leiden: Brill.

Kopelman, L. M., and A. A. van Niekerk. 2002. "AIDS and Africa: Introduction." *Journal of Medicine and Philosophy* 27: 139–142.

Lachenal, G. 2005. "Les réseaux post-coloniaux de l'iniquité: Pratiques et mises en scène de la recherche médicale au Cameroun." *Outremers* 93: 123–149.

Lachenal, G. 2006a. "Biomédecine et décolonisation au Cameroun, 1944–1994: Technologies, figures et institutions médicales à l'épreuve." PhD diss., Université Paris 7 Denis-Diderot.

Lachenal, G. 2006b. "Scramble for Cameroon: Virus atypiques et convoitises scientifiques au Cameroun, 1985–2002." In *L'épidémie de SIDA en Afrique subsaharienne: Regards historiens*, ed. C. Becker and P. Denis, 309–330. Louvain-la-Neuve: Academia Bruylant.

Lachenal, G. 2009. "Franco-African Familiarities: A History of the Pasteur Institute of Cameroon." In *Hospitals beyond the West*, ed. M. Harrison, 411–444. New Delhi: Orient Longman.

Lachenal, G. 2011. "The Intimate Rules of the French Coopération." In *Ethos, Ethnography and Experiment*, ed. W. Geissler, 373–401. Oxford: Berghahn.

Latour, B. 2005. *Reassembling the Social: An Introduction to Actor-Network-Theory*. Oxford: Oxford University Press.

Lipid Sciences, Inc. 2003. "Modified Immunodeficiency Virus Particles." U.S. patent application 7407663.

Lipid Sciences, Inc. 2008. "Modified Viral Particles with Immunogenic Properties and Reduced Lipid." U.S. patent application 12118335.

Mbembe, A. 1996. *La naissance du maquis dans le Sud Cameroun, 1920–1960*. Paris: Karthala.

Mbembe, A. 2002. "African Modes of Self-Writing." *Public Culture* 14: 239–275.

McFarlane, H., R. O. Barrow, V. A. Ngu, and B. O. Osunkoya. 1970. "Excretion of Immunoglobulins in Burkitt's Lymphoma." *British Journal of Cancer* 24: 258–259.

McFarlane, H., V. A. Ngu, and B. O. Osunkoya. 1970. "Immunoglobulin Deficiencies in Burkitt's Lymphoma." *African Journal of Medical Sciences* 1: 401–407.

Mudimbe, V. Y. 1988. *The Invention of Africa: Gnosis, Philosophy and the Order of Knowledge*. Bloomington: Indiana University Press.

Ngu, V. A. 1964a. "The African Lymphoma (Burkitt Tumour): Survivals Exceeding Two Years." *British Journal of Cancer* 19: 101–107.

Ngu, V. A. 1964b. "The Chemotherapy of Cancer: A Preliminary Report of Its Scope in West Africa." *West African Medical Journal* 13: 171–182.

Ngu, V. A. 1967. "Host Defences to Burkitt Tumour." *British Medical Journal* 1: 345–347.

Ngu, V. A. 1992. "Human Cancers and Viruses: A Hypothesis for Immune Destruction of Tumours Caused by Certain Enveloped Viruses Using Modified Viral Antigens." *Medical Hypotheses* 39: 17–21.

Ngu, V. A. 1994. "Chronic Infections from the Perspective of Evolution: A Hypothesis." *Medical Hypotheses* 42: 81–88.

Ngu, V. A. 1997. "The Viral Envelope in the Evolution of the Human Immunodeficiency Virus (HIV): A Hypothetical Approach to Inducing an Effective Immune Response to the Virus." *Medical Hypotheses* 48: 517–521.

Ngu, V. A., F. Ambe, and G. Boma. 2001. "Effective Vaccines against and Immunotherapy of HIV: A Preliminary Report." *Journal of the Cameroon Academy of Sciences* 1: 2–8.

Ngu, V. A., J. Nkwanyo, and J. K. Bakare. 1976. "The Immunobiotherapy of Cancer: A Preliminary Report of a Study of Leucocytes Cultured with B. C. G. or Tuberculine on Malignant Disease." *Nigerian Medical Journal* 6: 9–17.

Ngu, V. A., B. O. Osunkoya, O. O. Ajayi, and M. S. Khwaja. 1973. "The Immunotherapy of Advanced Human Cancer with Neuraminidase." *Nigerian Medical Journal* 3: 15–19.

Ngu, V. A., V. P. Titanji, D. Muna, J. Nyoth, C. Lekeuagni, and C. Yaoundé. 1981. "Clinical Experience with Autobiotherapy of Malignant Tumor Disease: A Preliminary Report." *Journal of the National Medical Association* 73: 927–937.

Nguyen, V.-K. 2010. *The Republic of Therapy: Triage and Sovereignty in West Africa's Time of AIDS*. Durham, NC: Duke University Press.

Nkinin, S. W., K. R. Daly, P. D. Walzer, E. S. Ndzi, T. Asonganyi, N. Respaldiza, F. J. Medrano, and E. S. Kaneshiro. 2009. "Evidence for High Prevalence of *Pneumocystis jirovecii* Exposure among Cameroonians." *Acta Tropica* 112: 219–224.

Packard, R. M., and P. Epstein, 1991. "Epidemiologists, Social Scientists, and the Structure of Medical Research on AIDS in Africa." *Social Science and Medicine* 33: 771–794.

Pestre, D. 2006. *Introduction aux Sciences Studies*. Paris: La Découverte.

Pialoux, G. 2001. "Affaire Therastim: L'espoir et les doutes." *Transcriptases Sud* 7. Accessed July 22, 2016. http://www.pistes.fr/transcriptasesud/7_150.htm.

Seytre, B. 1993. *SIDA, les secrets d'une polémique: Recherche, intérêts financiers et médias*. Paris: Puf.

Simon, E., M. Egrot, A. Traore, B. Taverne, and A. Desclaux. 2007. "L'expérimentation des thérapies néo-traditionnelles en Afrique: Questions d'éthique." Communication 113, 4ème Conférence francophone HIV/SIDA, Paris, March 29–31.

Sunder Rajan, K. 2006. *Biocapital: The Constitution of Postgenomic Life*. Durham, NC: Duke University Press.

Tangwa, G. 2002. "The HIV/AIDS Pandemic, African Traditional Values and the Search for a Vaccine in Africa." *Journal of Medicine and Philosophy* 27: 217–230.

White, L. 2000. *Speaking with Vampires: Rumor and History in Colonial Africa*. Berkeley: University of California Press.

Whyte, S. R., M. A. Whyte, L. Meinert and J. Twebaze 2013. "Therapeutic Clientship Belonging in Uganda's Projectified Landscape of AIDS Care." *When People Come First*, edited by J. O. Biehl and A. Petryna, 140–165. Princeton NJ: Princeton University Press.

4

Worrying about Essentialism
From Feminist Theory to Epistemological Cultures

A couple of years ago, Amy Chua, a second-generation Chinese American and Yale Law School professor, published a hilarious and outrageous memoir of parenting "the Chinese Way." The book, *Battle Hymn of the Tiger Mother*, caused a huge uproar, perhaps especially among Asian American mothers. Here is an excerpt: "The Chinese mother believes that (1) school work always comes first; (2) an A minus is a bad grade; (3) your children must be two years ahead of their classmates in math; (4) you must never compliment your children in public; (5) if your child ever disagrees with a teacher or coach, you must always take the side of the teacher or coach; (6) the only activities your children should be permitted to do are those in which they can eventually win a medal; and (7) that medal must be gold" (Chua 2011).

Chua put these practices to work in the rearing of her own children. She recounts calling one daughter "garbage" when she failed to pass muster and threatening to dispatch her other daughter's dollhouse to the Salvation Army if she didn't master a certain song on the piano. As Gish Jen (2011a), a distinguished novelist who is another second-generation Chinese American, wrote in a review of the book, "The question of whether Chua's methods amount to child abuse is a hot button"—as, of course, is the question of what, if anything, Chua's practice of the "Chinese way" of parenting tells us about China. In the *Boston Globe*, Jen (2011b, n.p.) notes the obvious: "Not all Asian mamas are Asian Mamas, of course, and I hardly need to point out that you don't have to be Asian, or even a mama, to be an Asian Music Mama. Brahms had one, after all, as did Beethoven and Mozart. Still, I suspect that aspects of Chua's approach may have resonance for some readers of Asian heritage. As much as I wanted to condemn Chua's calling her children 'garbage,' for example, I found that it felt like condemning, well, my Asian mama." In the end, Jen gives Chua kudos, not for succeeding in raising her daughters to become virtuoso musicians (which she did), or even for getting past their inevitable resentment (in fact, they helped her write the book), but for "tromping where few dare tread"—for daring to raise the issue of cultural differences in maternal practices in a climate in which multiculturalism has degenerated into a maelstrom of difference so capricious and so labile as to be effectively indistinguishable from universalism.

Does it make sense—is it even possible—to speak of a "Chinese way" of parenting? If by this term is meant a way of parenting practiced by all parents of Chinese extraction, today, yesterday, and tomorrow, the answer of course is no. At the same time, however, to refrain from acknowledging, describing, and studying different styles of parenting—prevailing in different communities at different times—is tantamount to denying the existence of such differences in style. As Jen remarks, Chua's style of parenting owes as much to her family's experience as outsiders desperate for economic and social success—first in the Philippines and later in the United States—as it does to her Chinese extraction. Furthermore, styles of parenting in China today (at least in urban areas) are undergoing change, apparently becoming more permissive. But even so, she concludes, "It would be unfair to say that Amy Chua's Chinese-ness is not Chinese: Ethnic enclaves have been known to preserve a home culture in a far more pristine state than does the country of origin. But perhaps we should keep Chua's personal context in mind before extrapolating too much. Her 'Chinese way' tells us at least as much about migration and identity and yes, America, as it does about China today" (Chen 2011a, 7). The challenge that faces Chua, and the territory she dares to tread, is in fact the challenge and territory of this book.

We have been invited not to study differences in the styles and practices among parental cultures but to study variations among different cultures of scientific practice. Yet despite the difference in focus, we inevitably confront the same challenge that faced Chua: acknowledging the importance of (and of giving due credit to) cultural differences in scientific practice without succumbing to the pitfalls of cultural essentialism, particularly hazardous when linked with national, racial, or sexual identity. This is what we here call "culturalism," and the challenge referred to in the title of our book, *Cultures without Culturalism*. The obvious question is, why is it so difficult to avoid these pitfalls?

Some might say that the difficulty inheres in the very word "culture." Once said to be the unifying concept of anthropology, the meaning of the term has become so contentious as to be altogether avoided by many contemporary anthropologists. For starters, the concept of culture is plagued by an entrenched polysemy—referring in some contexts to that which distinguishes civilized from uncivilized humans; in others, to that which distinguishes humans as a category from other biological organisms; in still other contexts, to a particular way of life shared by a (any) group of individuals (see, e.g., Francisco 2007); and finally, as the property of an identifiable and enduring ethnic, gender, or national collectivity. Anthropologists have been especially sensitive to the chronic slippage in uses of the term (a slippage to which the very multiplicity of meanings makes us prone).

For us here, however, the problem at issue inheres more specifically in the notion of "cultural essentialism," a term that came into use in the late 1980s following the rise of identity politics in the United States. "Cultural essentialism" expresses a concern with (and challenge to) the very assumption that ethnic, gender, or national identity can be characterized by fixed properties of any sort. Here the main issues of contention have come to focus on the questions, first, of whether or not it is possible to speak of culture as if it were a discrete (and stable) thing, and second, of whether or not it is possible to meaningfully attach such a notion of culture to a discrete, identifiable, and stable social/ethnic group.

Lessons from Feminism

One of the first arenas in which the critique of cultural essentialism took root was that of feminist theory, especially as developed in the United States in the middle to late 1980s. Indeed, I would argue that modern feminism has derived its own identity (or identities) from its various confrontations with charges of essentialism. Perhaps its history may provide us with some usable lessons.

Initially, the principal issue of debate revolved around the meaning of gender and its relation to the biological category of sex, and the charge of essentialism that arose in the late 1970s in response to early efforts to disrupt the universality of "man" (and consequent marginalization of women) by focusing on generalized differences between male and female. For the most part these efforts treated "women" as a universal category, ignoring the importance of differences among women in age, class, historical era, race, or ethnicity. In other words, many of these discussions seemed to repeat the universalizing move on the level of women rather than of humans, with the consequent effect of marginalizing non-Western, nonwhite, and non-middle-class women. Here essentialism implied biological essentialism, as if the mere possession of two X chromosomes sufficed to define the meaning of "woman." By the 1980s, however, American feminist theory had largely come together around Simone de Beauvoir's credo, "one is not born but becomes a woman," adopting as a shared analytic starting point the assumption of a clear and stable distinction between sex (which many if not most authors accepted as a stable and transcultural biological category) and gender (taken as a cultural product). In this framework, gender difference, even when still privileging differences between masculine and feminine, clearly also allowed for such other axes of difference as Western versus non-Western or race, class, and ethnicity.

It did not take long for recognition to surface that once again feminism was at risk of replicating the errors of universalizing, although now on the level of cultural essentialism rather than of gender or biological essentialism. Chandra Mohanty (1991, 70), for example, complained of the widespread tendency to assume that " 'women' have a coherent group identity within the different cultures discussed, prior to their entry into social relations. Thus, Omvedt can talk about 'Indian women' while referring to a particular group of women in the State of Maharashtra, Cutrufelli about 'women of Africa,' and Minces about 'Arab women,' as if these groups of women have some sort of obvious cultural coherence." In a similar vein, and stressing the parallels between biological (or gender) and cultural essentialism, Uma Narayan (1998, 878) complained a few years later that

> Seemingly universal essentialist generalizations about "all women" are replaced by culture-specific essentialist generalizations that depend on totalizing categories such as "Western culture," "Non-western cultures," "Western women," "Third World women," and so forth.
>
> Although often motivated by the injunction to take differences among women seriously, such moves fracture the universalist category "Woman" only slightly, because culture-specific essentialist gen-

eralizations differ from universalistic essentialist generalizations only in degree or scope, and not in kind. The resulting portraits of "Western women," "Third World women," "African women," "Indian women," "Muslim women," or the like, as well as the pictures of the "cultures" that are attributed to these various groups of women, often remain fundamentally essentialist. They depict as homogenous groups of heterogeneous people whose values, interests, ways of life, and moral and political commitments are internally plural and divergent.

Indeed, by the last decade of the twentieth century, the very notion of "woman" had come under such intense dispute that any generalization about women—either as an all-inclusive group or as delineated subgroup—came to seem suspect. What, after all, is a woman? What is a Chinese woman? A middle-class African American woman? I vividly recall an early reminder of just how intellectually and politically paralyzing such challenges can be when, in 1979, at what may well have been the first international conference on feminist theory to be convened, a young woman yelled out, "I am a Yugoslav woman with seven children—who is speaking for me?"

Harvard anthropologist Michael Herzfeld (2001, xiii) has noted that "in English the use of the singular with a definite article ... is a clear play for conceptual closure," and he suggests that "pluralization is a fine way to destabilize the authority of received categories." But the efficacy of pluralization—a strategy that has become extremely popular in academic discourse over the last couple of decades—is clearly limited. Shifts from "feminism" to "feminisms" or from "feminist theory" to "feminist theories" were clearly helpful, but clearly not so a shift from talk about "woman" to talk about "women." Furthermore, sometimes a shift back from plural to singular (or, perhaps more accurately, gerundial) form can also be helpful (I am thinking of bell hooks's [1981] felicitous suggestion that we talk of "feminist movement" rather than of either "the feminist movement" or "feminist movements"). In any case, the subsequent history of modern feminism has made it abundantly clear that no simple grammatical strategy (certainly neither pluralizing nor gerundizing) can suffice for healing the rifts that had been opened.

For modern feminism, the consequence of such antiessentialist critiques was a conspicuous fracturing (if not a reeling) of the very project to build feminist theory. Programs in women's studies soon morphed into programs in gender studies (or of women and gender), with subconcentrations in LGBT (lesbian, gay, bisexual, transsexual) studies, queer studies, postcolonial or cultural studies, and their connection to the political stimulus that had originally spawned feminist theorizing grew ever more difficult to discern. At the same time, even this proliferation of categories, notwithstanding the manifest interest in acknowledging and respecting heterogeneity that had driven them, failed to provide immunity from criticism. It rapidly became evident that the charge of slighting difference could be invoked at any level of generalization: by their very nature, all generalizations could be said to tend toward a kind of universalism, just as all political agendas could be said to depend on "normative assessments of the salience and weight of particular kinds of 'differences'" (Narayan 1998, 104). The question thus arises, what kind of intellectual or political agenda can survive without generalization, without, that is, the erasure (or slighting) of some differences?

One answer to (at least the political part of) this question was offered by Gayatri Chakravorty Spivak with her introduction of the eminently commonsensical notion of "strategic essentialism": the provisional denial or neglect of internal differences and conflicts that nationalities or ethnic or minority groups can advantageously use to achieve certain goals (see, e.g., Spivak 1988). And indeed, it is precisely by such provisional denial that political movements maintain their coherence. But as Spivak herself came to realize, the history of political movements also reveals this strategy as one that never escapes contestation, misuse, and even abuse. And perhaps that is the point: just as boundaries of nationalities, ethnicities, and genders can be useful despite their persistent defiance of clear articulation, so too, the strategic use of such boundaries for particular goals (be they political or intellectual) continues to be an effective resource despite the persistent invitation of such boundaries to contestation, misuse, and abuse.

Anthony Appiah (1992, 58) reminds us that what the critics of universalism "truly object to—and who would not?—is Eurocentric hegemony posing as universalism." So too, it is useful to recall that what critics of essentialism object to is the insidiously normalizing thrust of generalizations about group attributes. As Cressida Heyes (2002, n.p.) puts it, "Generalizations made about particular social groups in the context of identity politics may come to have a disciplinary function within the group, not just describing but also dictating the self-understanding that its members should have." Equally insidious is the slide from difference to inequality, a slide that inevitably reinforces the very hegemony that identity politics was intended to combat. The solution to the problem of normativity, according to Heyes (as well as to numerous others), is not the abandonment of identity politics but ongoing alertness in its practice: "The idea of a dominant identity from which the oppressed may need to dissociate themselves remains, but the alternative becomes a more fluid and diverse grouping, less intent on guarantees of internal homogeneity and more concerned with identifying 'family resemblances' than literal identity." Perhaps a similar alertness and a similar fluidity and diversity in our groupings can help protect against the slide into hegemonic hierarchies.

Cultures without Culturalism in Scientific Practice

I have barely scratched the surface of the enormous literature these issues have generated over the last couple of decades, but my hope is that even this very brief gloss will suffice for asking what, if anything, we can learn from this history that may be of use in our own current efforts. Many of the concerns raised by critics of cultural essentialism are obviously relevant for our analyses of scientific practices, just as are many of the concerns that underlie the emergence of identity politics in the first place. For the former, the primary worry is erasure of heterogeneity (and the normative implications of that erasure); for the latter, it is "hegemony posing as universalism." And all of us will worry about the use of notions of a Chinese, Indian, or Arabic science as different from Western science to support the hegemony of the latter. But therein lies another concern that looms especially large when thinking about scientific practices: the problem of epistemological relativism, close kin to essentialism. Lorraine Code (1998) has written insightfully about the use of claims of "universal sameness to marginalize the experiences of minority populations, and about the perception by many feminists of relativism as an antidote to such tendencies that permits those who are privileged 'to claim to have access to the one true story.'" An obvious counterpart of cultural universalism is to be found in the epistemological universalism of traditional discourses of science, and here the sense of threat posed by any hint of relativism (either in feminist critiques of science or in science studies more generally) has been loudly declaimed—most notably in the responses of the "science warriors" of the 1990s.

One of the first and most successful efforts to disrupt the universality of scientific stories came from Ian Hacking's ([1992] 2002) discussion of styles of scientific reasoning, and it is interesting to view his efforts in the light of concerns both about essentialism and about relativism. Hacking thanks Alistair Crombie for inspiring his own investigation of different styles of reasoning, and he in fact begins with Crombie's own taxonomy and his list of "styles of scientific thinking in the European tradition." An early version of this list (the version on which Hacking drew) reads as follows:

(a) the simple postulation established in the mathematical sciences,

(b) the experimental exploration and measurement of more complex observable relations,

(c) the hypothetical construction of analogical models,

(d) the ordering of variety by comparison and taxonomy,

(e) the statistical analysis of regularities of populations and the calculus of probabilities, and

(f) the historical derivation of genetic development (Crombie 1981, 284).

Later, Crombie was explicit in his acceptance of some degree of cultural relativism in the methods of science: "When we read a text we must ask to what questions the author was giving an answer.... We must approach our subject with an explicit cultural relativism. As ourselves products of a particular time and culture, we may then expose ourselves to the surprise that effective thinking could be based on assumptions and have aims and motivations so various and so different from our own" (Crombie 1994, 1:5). Even so, his taxonomy must now appear to many as somewhat categorical, ahistorical, and even arbitrary. For example, it includes postulational thinking but not other kinds of mathematical activity, assuming moreover that all kinds of postulational thinking are the same. Furthermore, the title of his magnum opus, *Styles of Scientific Thinking in the European Tradition*, makes his focus on Western science equally explicit, reflecting his understanding of science as uniquely based on the "principles of causality and proof" as these were "established by the ancient Greeks" (22).

But Hacking is not troubled; indeed, he repeatedly takes Crombie's list as canonical. At one point he invokes Clifford Geertz's metaphor of "continents of meaning" in defending (and elucidating) Crombie's taxonomy: "When a legendary Thales 'discovered the continent of mathematics', we began postulational reasoning, the deduction of or speculation about the consequences of precise assumptions. That way of thinking has grown, reconceived itself, abandoned old aspirations and achieved new heights. It is the accumulation not so much of knowledge ... as of ways of finding out" (1992a, 132). In the sense that they are historical achievements, styles of reasoning—ways of "finding out"—are born, discovered, perhaps even forged, but always in particular social and historical contexts. Hacking argues, however, that if successful they ultimately outlive their origins. They arise at particular points of time, but if useful they stick around, travel, and even take over. Despite the fact that they arise in particular communities, their reach extends ever outward, until, as Hacking puts it, "they run the world" (132).

Hacking's investigations of styles of reasoning—most notably his work on statistical reasoning and the "laboratory style" (a melding of Crombie's second and third categories, "experimental" and "hypothetical") have been extremely influential, and for good reason. They introduced entirely new kinds of questions into the history and philosophy of science and contributed hugely to the growth of historical epistemology as a field of inquiry. For us, his contribution was critical, but like any investigation, it has its limits. And part of those limits can be traced to his occasional employment of the assumption (shared with Crombie) that a style of reasoning (whether statistical, postulational, or a "continent of mathematics") is effectively homogeneous, that in its maturity stabilizes into "a rather timeless canon of objectivity" ([1992b] 2002, 188). (In his most recent writings, he plays with the idea that styles of reasoning are rooted in innate modules of the human mind that are universally available, even if only provisionally deployed; see, e.g., Hacking 2009.) Here Hacking can indeed be said to verge on a form of strategic essentialism—an essentialism that is undoubtedly useful for some purposes (e.g., in a foray against the unity of science) while at the same time remaining problematic for others. But one of Hacking's great strengths is his characteristic irreverence for fixed strategies of any kind. His project is complex, and his goals are multiple; as such, they invite a variable deployment of (as well as distancing from) a number of assumptions that he himself refers to as myths.

Similar issues arise in relation to my own efforts. I turn therefore to my initial aims in introducing the concept of epistemological culture (Keller 2002) and to the pitfalls to which this concept may have fallen prone. My inquiry grew out of a desire to understand what scientists meant by a "satisfying explanation." From my own experience as a scientist—first in theoretical physics and later in mathematical biology—I knew that the answer to the question was not simple, that what provided explanatory satisfaction to one scientist would often leave another entirely unsatisfied. I well understood that most philosophers regarded the criteria of scientific explanation as fixed and therefore as something that could be clearly articulated, but my own experience contradicted this assumption. On numerous occasions—most conspicuously when speaking about my work in mathematical biology to audiences of biologists—I had found my expectations for a satisfying explanation to be in manifest conflict with those of my interlocutors. And again, differences in what is taken as a satisfying explanation—what leads a reader or listener to say something like, "Ah, now I understand"—loomed large when

I tried to make sense of Barbara McClintock's sense of alienation from her colleagues within biology (see, e.g., Keller 1983). Differences were clearly real, but the problems I struggled with were how to describe them, how to map them, and how to account for them. These questions are obviously intertwined, the resolution one attempts of one question clearly influencing the ways in which it then makes sense to address the others.

Furthermore, it soon became evident that one could proceed in numerous directions, and in fact, I did attempt a variety of quite different descriptive taxonomies. For example, initially I tried to sort logical, psychological, and instrumental satisfaction from one another, but eventually I found a disciplinary taxonomy easier to work with—as it were, more satisfying. (Fortunately, and notwithstanding the popular reading of my McClintock book as a manifesto for a feminine science, the idea of trying to sort explanatory preferences by gender did not even occur to me.) Narrowing the focus of my inquiry even further to explanations of biological development (i.e., to accounts of how a fertilized egg cell developed into an adult organism, the core question of embryology) helped decide the issue, for this particular puzzle had exposed (or so it seemed to me) a fault line that could be nicely mapped onto a difference between the epistemological values of physicists and those of experimental biologists. Hence, the idea that physicists and biologists inhabited different epistemological cultures, characterized by different attitudes toward the value of mathematical models, different understandings of generality (and correspondingly different expectations of the utility of generalization), different attitudes toward special or exceptional cases, and so forth. And in trying to develop this line of thought—in trying to identify (and characterize) the dissimilarities in the assumptions, values, and expectations that these two kinds of disciplinary practice inculcated, and in trying to understand how such differences might have arisen from distinctions between their respective objects of study—I walked straight into at least one of the pitfalls of cultural essentialism, seeming to efface the heterogeneities so manifest in each disciplinary practice.

But does that mean that my effort was thereby misguided? I would say no. Certainly, I did not at the time see any other way I might have proceeded, and I am not at all sure I see an alternative today. Taking the word "culture" in its most pragmatic sense, to refer to "the way we do things around here" (Francisco 2007, 1), it is obvious that the epistemological culture of physicists—even of theoretical physicists, or even of theoretical physicists thinking about biological development—is not homogeneous, any more than is the epistemological culture of experimental developmental biologists. My aim in introducing this notion was not to suppress or slight internal heterogeneity but to bring into focus epistemological heterogeneity at a higher level, that is, at the level of natural science writ large. But does not the very act of making distinctions at one level inevitably invite suppression of difference at another? And does not the converse also hold, that generalization by its very nature invites the slighting of variation? Indeed, the basic idea of this book is in its very wording a protest against the hegemonic implications of assuming universality in styles and cultures of scientific practice. We sought to give legitimacy to the multitude of styles and modi vivendi in which science is, and has been, practiced in different contexts, periods, and regions of the world. But at the same time, the title of the book was intended to serve as a warning against simplistic stereotypes.

Spivak suggested the notion of strategic essentialism as an often effective strategy for nationalities or ethnic or minority groups in their political struggles against the hegemony of universalism. Our struggle here is not against political hegemony but against the intellectual hegemony of universalism, and it may well be that some form of strategic essentialism is an equally necessary strategy for us. The difficulty, just as it is in the political domain, is to know when such a strategy is more useful than harmful. And that, of course, will depend (and necessarily so) on the nature of our questions, our goals, and the particular context in which we seek to make an intervention.

Culturalism, Relativism, and the Problem of Truth: Cultures without Culturalism in Scientific Practice

Our starting point in organizing this project was far removed from any notion of the epistemological unity of scientific practice. It was different from Hacking's, as it was different from that of my earlier inquiry in *Making Sense of Life* (Keller 2002). Rather than a reaction against assumptions of unity, our initial point of departure assumed the heterogeneity of scientific practices as given; our aim was to give form to this heterogeneity. But like our subject, our aim, too, has evolved. Largely in response to the manifest heterogeneity in the goals of the participants in this project, our questions, and our responses, have fractured and diversified, feeding on new metaphors as they respond to new challenges. Given how easy it is to forget the temporal variability of culture (or style or form), the original metaphor of "culture" may just have been too static to do justice to the dynamic nature of our project (just as my reference to the aim of giving form to heterogeneity surely is as well). Hans-Jörg Rheinberger (1995, 318) argues that "an epistemology that tries to assess scientific thinking in its dynamicity must be as plastic, as mobile, as fluid, and as risky as scientific thinking itself." It must credit both the regionalization and the historical mobility of knowledge, and he reminds us that, long before Hacking and Crombie, Gaston Bachelard had also worked these fields: "Bachelard often also talks of cantons, regions, or domains of knowledge within the city of science, such as the 'relativistic canton' in the 'city of mechanics.' . . . These cantons are islands of scientific culture; they create their own cultural codes and forms of emergence, which only an intimate knowledge of the respective region allows one to judge" (318). Also, Rheinberger reminds us that, for Bachelard, the historicity of scientific knowledge is precisely what "marks it as a peculiar culture of truth. . . . Knowledge *is* an evolution of the spirit, it is nothing that is accomplished once and forever, it is grounded in its very own discardability, and not in the timeless unity of a thinking ego" (319).

Returning to the struggles in the 1980s and 1990s over the possibility of an identity politics—even over the very meaning of "identity"—lessons about the slighting of internal differences were well learned. We came to recognize that identity is always and already multiple, and that all of our referential categories (or cultures) are forever shifting. For evidence we need only look at how dramatically Western categories of gender have themselves shifted over the last decades. But it is possible that we may have learned our lessons too well, losing sight of the force that even fuzzy, provisional, and porous boundaries can have in political and intellectual life, and of the extent to which both political and intellectual activity in fact depends on drawing boundaries and marking categories, however provisional they must be. It may need constant reminding that neither identities nor cultures are natural kinds. Even so, recognizing their inherent heterogeneity and variability ought not prevent us from simultaneously recognizing the enduring stability and force of at least some categories—women, for example.

Much the same can be said about my own efforts to articulate a set of distinctions that separated the epistemological cultures of mathematical physics and experimental biology. Here too, the boundaries I was trying to work with were both porous and labile. The epistemological values of particular scientific disciplines change all the time (as, indeed, do their very boundaries). Nevertheless, for the purpose of trying to understand the failure of repeated efforts to build a mathematical biology, the categories were useful (and widely recognized as such by those who had attempted to inhabit that porous and labile boundary). Today, little over a decade later, the interface between physics and biology has become so populated and so well funded that it is easy to forget that any sort of boundary ever existed. Nevertheless, its presence was a significant force in the history of twentieth-century biological thought; indeed, even though diminished, that force continues to be exerted for many practitioners to this day.

Finally, I would like to end with the question of what, if anything, makes epistemological cultures a rather special subset of human cultures, a question Bachelard also calls to our attention. For Bachelard, cantons of scientific knowledge are indeed peculiar, and what makes them so is their relation to truth. There are, however, those in science studies who would reject the assumption that scientific cultures can be substantively distinguished from other kinds of human culture. Others might quibble with the very notion of truth, preferring to speak in terms of robustness, durability, or portability. But either way, the issue must be faced, and it is an issue inextricably intertwined with the spectre of cultural relativism. Indeed, it was the frequently drawn link between cultural relativism and identity politics that fueled many of the attacks on feminist critiques of science in the science wars of the 1990s and, at the same time, what (at least in the eyes of the science warriors) linked these critiques with the relativism attributed to (and at least sometimes, apparently espoused by) social constructionists.

Cultural relativism is said to be a mainstay of twentieth-century anthropology, but it is a concept with more than one meaning. By one interpretation, most contemporary historians of science would be in full accord: cultures can be understood only in their own terms. But there is also a stronger meaning of the notion of relativism that invites discord—in anthropology as in science studies: "All cultures are equally valid interpretations of reality" (Herzfeld 2001, 184). Not surprisingly, the difficulties of this view arise with particular force when applied to cultures of scientific practices, and especially to their relation to the concrete materiality of the world (or worlds) in which we live.

To be sure, human cultures—including epistemological cultures—may be separated by large spans of time and space and thus may tempt us to view them as effectively local. But today, material developments have exposed our world as one we are all obliged to share, forcing us to recognize the limitations of viewing knowledge as strictly local. Perhaps the most dramatic example of such limitations is provided by the accumulating evidence of the impact of local industrial practices on climes far removed in both space and time.

Anthropologist Kay Milton (1997, 487) was one of the earliest to invoke the warming of our planet as a case ad rem, arguing that, if only for practical purposes, the environmental problems of the contemporary world demand the unequivocal rejection of "extreme cultural relativism":

> It may be possible to sustain the belief that all cultures are equally true for the purposes of academic argument, but for the purposes of living in the world—of achieving goals and solving practical problems—a choice has to be made; one needs knowledge that works. . . . This point can be illustrated with reference to a contemporary environmental issue. Scientists say that the world is getting warmer and that human activities, specifically those which release certain gases into the atmosphere, are causing this to happen. But, as we know, the scientific view is just one among many cultural perspectives on the world. Other cultural perspectives may deny that global warming is happening, or, if they accept it, may attribute it to the actions of spirits, or a divine creator. . . . Each interpretation suggests a different solution: to reduce carbon emissions, to appease the spirits or to revive the ancient traditions. An approach which treats all cultural perspectives as equally true has no basis for choosing between them, and therefore cannot select a solution. In the face of global warming, . . . an extreme cultural relativist is paralysed by logic.

More recently, some early proponents of radical constructivism have expressed similar concerns. Harry Collins (2009, 30), for example, writes that "the prospect of a society that entirely rejects the values of science and expertise is too awful to contemplate. What is needed is a third wave of science studies to counter the skepticism that threatens to swamp us all." In a similar vein, Paul Edwards (2010, 437) concludes his book on climate science with the worry that constructivist arguments had clearly gone too far: "It was as if people thought we could stop bothering about the iron ore, the trees, and the gypsum and just make skyscrapers directly from blueprints, mortgages, and contracts. . . . Science became little more than ideology or groupthink, within which any belief at all might come to count as 'knowledge.' . . . As a result, all too often STS scholars characterized all sides in a scientific controversy as equally plausible, and saw knowledge simply as the outcome of struggles for dominance among social groups." Such a position "entails a relativism that soon becomes, if not entirely incoherent, at least useless for practical analysis" (438). Edwards seconds Collins's call for "a third wave of science studies," one that "would recognize and respect the value of scientific evidence, the tacit knowledge gained from disciplinary experience, and the wisdom of expert communities," but, he emphasizes, such a development would also "require of scientists that they 'teach fallibility, not absolute truth'—recognizing the provisional character of all knowledge" (438).

So much seems obvious. Yet it not only needs saying but also bears repeating—especially for a project such as ours. However variable, however diverse, in whatever guise, style, or culture, what scientific investigations have in common—indeed, what defines them as scientific—is the aim of providing reliable knowledge about the material world in which one lives. To the extent to which those worlds are stable, continuous, and connected, and to the extent to which the questions pursued are commensurable, judgment about the effectiveness, adequacy, or utility of different answers cannot be ignored. Cultures may be separated by time and space, yet in ways that become ever more evident, their worlds belong to a single planet—indeed, to a planet that grows smaller by the day. I conclude, therefore, with a plea that we include in our considerations the challenges of connecting the different forms of scientific practice we encounter, the different kinds of questions posed and the different kinds of answers these practices provide to those questions that can be shared across cultures.

References

Appiah, K. A. 1992. *In My Father's House: Africa in the Philosophy of Culture*. New York: Oxford University Press.

Chua, A. 2011. *Battle Hymn of the Tiger Mother*. New York: Penguin Books.

Code, L. 1998. "How to Think Globally: Stretching the Limits of Imagination." *Hypatia* 13(2): 73–85.

Collins, H. 2009. "We Cannot Live by Scepticism Alone." *Nature* 458: 30.

Crombie, A. C. 1981. "Philosophical Perspectives and Shifting interpretations of Galileo." In *Theory Change, Ancient Axiomatics and Galileo's Methodology*, ed. J. Hintikka, C. D. Gruender, and E. Agazzi, 271–286. Dordrecht: Reidel.

Crombie, A. C. 1994. *Styles of Scientific Thinking in the European Tradition: The History of Argument and Explanation Especially in the Mathematical and Biomedical Sciences and Arts*, Vol. 1. London: Gerald Duckworth.

Edwards, P. N. 2010. *A Vast Machine: Computer Models, Climate Data, and the Politics of Global Warming*. Cambridge, MA: MIT Press.

Francisco, S. 2007. "The Way We Do Things around Here: Specification versus Craft Culture in the History of Building." *American Behavioral Scientist* 50(7): 970–988.

Hacking, I. 1992a. "Statistical Language, Statistical Truth and Statistical Reason: The Self-Authentication of a Style of Reasoning." In *The Social Dimensions of Science Studies in Science and the Humanities from the Reilly Center for Science, Technology, and Values*, ed. E. McMullin, 130–157. Notre Dame, IN: University of Notre Dame Press.

Hacking, I. [1992b] 2002. "'Style' for Historians and Philosophers." In *Historical Ontology*, 178–199. Cambridge, MA: Harvard University Press.

Hacking, I. 2009. "Humans, Aliens, and Autism." *Daedalus* (Summer): 44–59.

Herzfeld, M. 2001. *Anthropology: Theoretical Practice in Culture and Society*. Malden, MA: Blackwell.

Heyes, C. J. 2002. "Identity Politics." In *Stanford Encyclopedia of Philosophy*. Accessed February 5, 2011. http://www.illc.uva.nl/~seop/entries/identity-politics/.

hooks, b. 1981. *Feminist Theory: From Margin to Center*. Boston: South End Press.

Jen, G. 2011a. "Mother Superior: How Chinese Is the 'Chinese Mom'"? *New Republic*, February 17, 5–7.

Jen, G. 2011b. "Tao of Tough Love." *Boston Globe*, January 16, 2011. Accessed February 1, 2011. www.boston.com/ae/books/articles/2011/01/16/tao_of_tough_love/.

Keller, E. F. 2002. *Making Sense of Life: Explaining Biological Development with Models, Metaphors, and Machines*. Cambridge, MA: Harvard University Press.

Milton, K. 1997. "Ecologies: Anthropology, Culture and the Environment." *International Social Science Journal* 49(154): 477–495.

Mohanty, C. T. 1991. "Under Western Eyes: Feminist Scholarship and Colonial Discourses." In *Third World Women and the Politics of Feminism*, ed. C. T. Mohanty, A. Russo, and L. Torres, 51–80. Bloomington: Indiana University Press.

Narayan, U. 1998. "Essence of Culture and a Sense of History: A Feminist Critique of Cultural Essentialism." *Hypatia* 13(2): 87–106.

Rheinberger, H. J. 1995. "Gaston Bachelard and the Notion of 'Phénoménotechnique.'" *Perspectives on Science* 13(3): 313–328.

Spivak, G. C. 1988. "Subaltern Studies: Deconstructing Historiography." In *Selected Subaltern Studies*, ed. R. Guha and G. C. Spivak, 3–32. London: Oxford University Press.

II. DISTINGUISHING THE MANY DIMENSIONS OF ENCULTURED PRACTICE

5

Hybrid Devices

Embodiments of Culture in Biomedical Engineering

Research laboratories have long been sites for ethnographic research into so-cial, cultural, and material practices of scientific research. Much less atten-tion has been given to them as sites for investigating the development and use of sophisticated cognitive practices. Most existing studies reflect the long-standing divide between "cognitive" and "cultural" accounts of science practice. For several years I have been leading an interdisciplinary research group that has been conducting ethnographic and cognitive-historical studies of lab-oratories in the engineering sciences, in part to understand how cognition and culture are intimately entwined in research practices. Thus far we have investigated five research labs in bioengineering sciences, but the account developed here focuses on two that are situated in biomedical engineering (BME): tissue engineering and neural engineering.

Engineering scientists in the BME field are a breed of researcher—often interdisciplinary—whose aim is to make fundamental contributions to "basic science," as well as creating novel artifacts and technologies for medical ap-plications. However, even their basic science research, such as on vascular biology, is approached largely from the perspective of engineering assumptions, principles, and values. Most often engineering scientists investigate real-world phenomena through building and running models—physical or computational. For engineers, "to engineer" means conceiving, designing, and building artifacts in iterative processes. In BME "engineering" extends to biological ("wet") artifacts through which to carry out research. The reason for focusing on these two labs is that each laboratory engineers physical simulation models—locally called "devices." Devices are in vitro models that serve as sites of experimentation on selected aspects of in vivo phenomena of interest. In effect, the researchers build parallel worlds in which devices mimic specific aspects of phenomena they cannot investigate directly due to issues of either control or ethics. In the literature on models, "simulation" is customarily reserved for computational modeling. However, respondents in our investigation variously use "simulate" and "mimic" in explaining how their physical models perform dynamically, that is, are "run" under experimental conditions. We thus use "physical simulation model" to refer to such bioengineered devices. Simulation by means of devices is an epistemic activity that comprises open exploration, testing and generation of hypotheses, and inference. It is central to an engineering science "style of reasoning" in the BME area (Hacking 1992). Physical simulation modeling forms the basis for explanation, prediction, and understanding in these communities.

Early observations in the labs directed our attention to the simulation devices as loci of integration of cognition and culture. They simultaneously constitute the material culture of the community, give rise to social practices, and perform as cognitive artifacts in their problem-solving practices. It is not possible to fathom how these communities produce knowledge by focusing exclusively on one or the other aspect. I have argued that the framework of distributed cognition provides a means of analyzing the problem solving in these labs in an integrated fashion (Nersessian 2006). However, as developed in this chapter, the analysis we aim at cannot be done simply by applying the current cognitive science framework of distributed cognition (Hollan, Hutchins, and Kirsch, 2000; Hutchins 1995; Kirsh and Maglio 1994). Furthermore, examination of the practices of these labs in building the device components of the distributed system opens the possibility for a rapprochement between two quite different notions recently advanced in science studies as important for investigating the particularities of different scientific subcultures: epistemic cultures and epistemological cultures.

"Epistemic culture" was introduced by Karin Knorr Cetina (1999) in part to capture the significant differences in the "machineries" of the manufacture of knowledge among different scientific subcultures. As Knorr Cetina details through case studies of experimental physics and molecular biology practices, these machineries comprise sociocultural structures as well as the technologies of research. Analysis of epistemic cultures does not attend to the differences in epistemological assumptions underlying the practices of knowledge-making subcultures. However, as Evelyn Fox Keller (2002, 4) has pointed out, there are differences in "the norms and mores of a particular group of scientists that underlie the particular meanings they give to words like theory, knowledge, explanation, and understanding, and even to the concept of practice itself" that are equally significant for individuating subcultures and understanding their practices. Her notion of "epistemological culture" is meant to capture the role of these differences in knowledge construction. Accounting for the research in the labs we investigated requires attending to the devices both as machineries of making knowledge, located in environments of "construction of the machineries of knowledge construction" (Knorr Cetina 1999, 3), and as artifacts that embed, and through which one can begin to discern, the epistemological assumptions, norms, and values of the subculture of BME and the further subdivisions into tissue and neural engineering. From a distributed cognition perspective, a device has a dual nature: it serves as a site of simulation not only of biological processes (machinery) but also of the researchers' epistemic norms and values.

In this chapter, I examine this dual nature of physical simulation models as participants in distributed cognitive systems. To preview my conclusion, I hope to show that focusing on how researchers build these cognitive-cultural artifacts enables understanding how culture is built into cognitive processes in science and vice versa.

The Framework of Distributed Cognition

The study presented here is framed, broadly, in terms of distributed cognition (DC): cognitive phenomena such as problem solving are "distributed" (Hutchins 1995) or "stretched" (Lave 1988) across a complex system comprising humans and artifacts. Within this framing, to understand how problem solving is achieved requires examining the generation, manipulation, and propagation of representations across media (human and artifactual) within the system and how these do cognitive work. Distributed cognition is best construed as an interpretive framework in the sense proposed by Ryan Tweney (1989, 344): "A framework is an attempt to reconstruct a model of the world that meets criteria other than testability as such. An adequate framework is one that is consistent with the details of the real-world process, is interestingly related to our theories of the world, and reduces the complexity of the real-world process in a way that permits anchoring the framework to the data." The DC framework is especially suited to thinking about integration of cognition and culture in investigations of science practices.[1]

As I have argued elsewhere, the cognitive-cultural divide that has long dominated science studies is an artifact of the interpretive frameworks we have brought to bear on science and not authentic to practice (Nersessian 2005). Producing scientific knowledge requires sophisticated cognition that only rich social, cultural, and material environments can enable. The basic premise of DC is that cognition and culture are coconstitutive. As Edwin Hutchins (1995, xvi) has argued, "Humans create their cognitive powers by creating the environments in which they exercise these powers." Cognitive powers consist in not just the natural (biological) cognitive capacities of humans but also those that are created through the coordinated efforts of people and the artifacts they create to attain various goals. Culture is understood as a process, as Hutchins continues, "that takes place both inside and outside the minds of people." Viewed in this way, engineering scientists do not just *use* the devices they build; they *think and reason through* them—mind and device are *coupled*. They not only create knowledge within the lab; they create the lab, both materially and as a way of doing science (Nersessian 2012). Each generation (~five years in a lab) provides methods, artifacts, and ways of thinking that serve as cognitive-cultural ratchets (Tomasello 1999) that scaffold the next generation in their problem-solving activities.

The primary unit of analysis of DC is the sociotechnical system, which comprises people, sociocultural structures, and the technologies (artifacts) that perform cognitive functions. In the major studies that have contributed to developing the framework thus far, although the technologies have a history, they remain stable during the problem-solving processes. Most important, the DC framework has been constructed, thus far, on studies of well-structured task environments such as the cockpit of a plane or the bridge of a naval ship. Cognitive powers arise from existing technologies. The primary role accorded to a person in the system is to provide coordination among representations (Hutchins 1995, 316). Often established roles and behaviors align with existing representations. One important critique raised within the framework is that of Rogers Hall, Reed Stevens, and Anthony Torralba (2002), who argue for a more active, ongoing construal of cognitive processes. They note that "'distributed'... is an adjective, past tense, and a modifier of already existing cognition," whereas "'distributing' is a verb, operating in an on-going present, and shifts our attention to studies of how cognition ... is produced historically out of human activity" (180). But even these studies, primarily of science practices, do not examine the processes of how people distribute cognition through building the technologies that participate in the cognitive system (Chandrasekharan and Nersessian 2015).

By contrast, problem solving of the sort my research group has been studying is ill structured and messy. The distributed cognitive systems customarily studied can be cast as dynamic but largely synchronic; that is, the situation changes over time, but the technological components are stable. In the case of the research laboratories, however, although there are loci of stability, problem-solving processes often involve iterations of designing, building, and redesigning the technologies through which they distribute cognition. These selective, hybrid representations of biological phenomena are manipulated to generate outcomes that propagate representational states within the problem-solving systems (comprising humans and artifacts) of the lab. Further, as we have found in these cutting-edge communities, all researchers are learners, even the lab director. So, researchers and technologies have intersecting developmental trajectories. We thus cast these distributed cognitive systems as dynamic and diachronic (Nersessian et al. 2003).

Indeed, the main aspect of the BME research lab that is novel for DC analysis, and thus can serve to extend the framework, is the signature practice of building cognitive powers through creating the technological components of the distributed systems (Chandrasekharan and Nersessian 2015). That is, in these labs researchers do not just use cognitive-cultural artifacts; they create them. The main way the researchers build the distributed cognitive system that is the lab, as well as the distributed cognitive subsystems of research projects, is through creating hybrid devices. The situation for the DC studies noted earlier is akin to building the plane while it is flying—or, in our studies, building the lab while it is generating knowledge (Nersessian 2012). Studying researcher practices in relation to engineering devices affords a window into how "humans create their cognitive powers" through creating their material culture. By unfolding what goes into building, we can discern how creating the machineries of knowledge making embeds the epistemological "norms and mores" of these communities in the distributed cognitive-cultural systems of the research laboratory.

The Investigation

We have conducted a five-year investigation of two university research laboratories in the area of BME: we began in the first year with a tissue-engineering laboratory and added a neural engineering laboratory in the second year. Intensive data collection was conducted in each laboratory for two years, with follow-up of the participants, their research, and questions pertaining to our research for an additional two years.

We selected these sites in accord with multiple aims for our National Science Foundation–sponsored research. Salient features of the sites to be selected were that they were (a) rich locales for investigating the interplay of cognition and culture in creative research practices, (b) significant sites of learning, and (c) interdisciplinary communities where concepts, methods, discourses, artifacts, and so forth, from different disciplines interact.

Cognitive-Historical Ethnography

Following researchers in distributed and situated cognition who have adapted ethnographic methods from anthropology to study cognitive processes in naturally situated practices (Goodwin 1995; Hutchins 1995; Lave 1988), we conducted "cognitive ethnographies" of the research laboratories. Several members of our research group became participant observers of the day-to-day practices in each lab. The ethnographic part of the study (observations and interviews) sought to uncover the activities, tools, and meaning making that support research as these are situated in the ongoing practices of the community. We took field notes on our observations, audio taped interviews, and video and audio taped research meetings (full transcriptions are completed for 148 interviews [largely unguided] and 40 research meetings). Early observations directed our attention to the custom-built simulation models as hubs for interlocking the cognitive and cultural dimensions of research. Because of this, the research meetings (monthly or less for lab A and weekly for lab D), though useful, assumed a lesser importance than they have in other research on cognitive practices in laboratories (see esp. Dunbar 1995). We needed to elicit from researchers their understanding and perceived relation to simulation artifacts, and to see how they functioned within the life of the labs, which was better addressed through interviews and extensive field observations. As a group we estimate we did over 800 hours of field observations. Our focus was on the daily work within the lab, not on their interactions with the outside world, except as noted by researchers in their interviews.

Significantly, these laboratories are evolving systems, where the custom-built technologies are designed and redesigned in the context of an experiment or from one research project to another. In these circumstances histories of the devices within the evolving research program of the lab become a vital resource for present-day research. In conducting research, history is appropriated hands-on as meaningfully related to working with devices (Kurz-Milcke, Nersessian, and Newstetter 2004). To capture this and other historical dimensions of research in these laboratories, we also used interpretive methods akin to those used in cognitive-historical analysis (Nersessian 1992). In our investigations, cognitive-historical analysis examines historical records to advance an understanding of the integrated nature of cognition and culture. Data collection for this part of our study included publications, grant proposals, dissertation proposals, PowerPoint presentations, laboratory notebooks, e-mails, technological artifacts, and interviews on lab history.

Field Site Descriptions

The labs we investigated reside in a BME community that places high value on what it calls "interdisciplinary integration." In both labs, this manifests in their building hybrid bioengineered artifacts where living cells and cellular systems interface with nonliving materials in model-based simulations. The interdisciplinary nature of the research requires lab members who start out primarily as engineers to develop equal facility with, for instance, cell culturing, as with engineering design, and to develop a hybrid BME identity though their practice. Further, both these lab subcommunities see themselves as cutting-edge, frontier researchers. The lab ethos is infused with an open-ended sense of possibility, as well as a tinge of anxiety about how little is known in their area and whether PhD research projects will work out. The researchers place high value on innovation in methods, materials, and applications. They value high-risk research that requires what the tissue engineering lab director called taking the "big gamble." Failure is omnipresent, as are support structures for dealing with it as a common feature of the research. The social structure in each lab is largely nonhierarchical—a feature we attribute to the interdisciplinary nature of the research, where none consider themselves *the* expert and knowledge is distributed across the community. Opportunities to innovate are provided to everyone. In the short time we were conducting our investigation, we saw a few instances where "big gambles" led to high payoffs, which sustains this attitude, despite the fact that most of the researchers engaged in the high-risk research are doing it for their dissertation projects. All the students during our investigation graduated and went on to excellent positions in academia and industry.

LAB A. Lab A is a tissue engineering laboratory that dates to 1987, when the director moved to a new university to take advantage of the opportunity to "take the research in vitro," mainly because problems of control with animal studies were limiting research possibilities. During our investigation, the main members included the director, one laboratory manager, one postdoctoral researcher, seven PhD graduate students (three graduated while we were there; the other four, after we concluded formal data collection), two MS graduate students, and four long-term undergraduates (two semesters or more). Of the graduate students, two were male and five were female; the postdoc was female. Additional undergraduates from around the country participated in summer internships and international graduate students and postdocs visited for short periods. The laboratory director's background was in aeronautical engineering, but he was by then a senior, highly renowned pioneer in the field of BME. All of the researchers came from engineering backgrounds, mainly mechanical or chemical engineering. Some had spent time in industry prior to joining the lab. The lab manager had an MS in biochemistry. Researchers frequently consulted with a histologist located in the building and traveled to other institutions mainly to collect animal cells and to run gene microarray analyses, and in one case to conduct an animal (baboon) experiment using a "wet" device designed in lab A.

Lab A's overarching research problems were to understand mechanical dimensions of cell biology, such as the effect of forces on gene expression in endothelial cells, and to engineer living substitute blood vessels for implantation in the human cardiovascular system. The dual objectives of this lab explicate further the notion of an engineering scientist as having both engineering and scientific research goals. Examples of intermediate problems that contributed to the daily work during our investigation included designing and building living tissue blood vessel wall models—locally called "constructs"—that mimic properties of natural blood vessels, creating endothelial cells (highly immune-sensitive) from adult stem cells and progenitor cells through applying mechanical forces, designing and building environments for mechanically conditioning constructs for experiments, and designing means for testing their mechanical strength. The lab had the look of a biology lab with flasks, pipettes, a sterile workbench with a hood, incubators, and hazardous waste containers. Researchers were often seen working with cells. The graduate student researchers had individual projects but were often assisted by undergraduates. During the period we spent there all projects were directed toward aspects of designing and building the "construct" device for moving it both toward a more "physiologic" model as a site of basic vascular research and toward a viable vascular implant (an alternate name for their construct model, "tissue-engineered vascular graft," signals this dual goal).

LAB D. As an institution, the neural engineering laboratory was in existence for a few months and still very much in the process of forming when we entered in 2002. During our study the main members included a director, one laboratory manager, one postdoctoral researcher (who left for a tenure-track job after around six months), four PhD graduate students in residence (one left after two years, three graduated after we concluded formal collection), one PhD student at another institution who periodically visited and was available via video link, one MS student, six long-term undergraduates, and one BS volunteer for nearly two years, not pursing a further degree, who helped out with breeding rats. Of the graduate students, two were female and three were male; the postdoc was male. When we began, the laboratory director was a new tenure-track assistant professor, fresh from a postdoc in a biophysics laboratory that develops techniques and technologies for studying in vitro network cultures of neurons. He already had achieved some recognition as a pioneer. His background was in chemistry and biochemistry, with his engineering experience largely self-taught, though highly sophisticated. He described himself, in an early interview, as someone who had always loved "to tinker and build things," starting as a child with his father, and in the postdoc lab had even built a two-photon microscope for real-time optical imaging of in vitro dishes of neurons. The backgrounds of the initial researchers in lab D were more diverse than lab A and included mechanical engineering, electrical engineering, BME, physics, microbiology, and a joint degree in life sciences and chemistry, though all were now bioengineering students. The student with the microbiology background worked on optical imaging in the lab but left after two years to pursue a "more cognitive science"–oriented PhD.

Lab D's overarching research problems were to understand the mechanisms through which networks of neurons learn and, potentially, to use this knowledge to develop aids for neurological deficits and, more visionary, "to make people smarter," as the lab director described it. Problems of control in in vivo studies were a major motivation for in vitro research for the lab D director, as it was for the lab A director. At the time he began research, the major approach to neuron learning was through single-neuron studies on living animals, where one might gauge responses to stimuli but not control supervised learning. Examples of intermediate problems that contributed to the daily work included developing ways to culture, stimulate, control, record, and image neuron arrays; designing and constructing feedback environments (robotic and simulated) through which the "dish" of cultured neurons could learn; and using electrophysiology and optical imaging to study plasticity. Unlike lab A, lab D's lab space looked like a computer lab. Aside from a large covered microscope and an incubator (hiding the dish from sight) that could be mistaken for a mini fridge, the most striking features of the lab were the copious wires crisscrossing the space carrying electrical signals produced by the neurons to the researchers' computers. As in lab A, the researchers had dual scientific and engineering agendas, and all graduate students had individual projects. However, since there was virtually no knowledge of "dish" behavior and how to control and use it to study learning when the lab started, there was significantly more interaction among the researchers as they got the infrastructure around the dish up and running, and in their research projects over time, than we witnessed in lab A.

Our research findings are rich, and the account and interpretations presented in this chapter are largely restricted to those pertaining to a central practice of crossing the in vivo–in vitro divide and the associated activities of designing, building, and experimenting with devices. Each of these labs constitutes what Hans-Jörg Rheinberger (1997; see chapter 11 in this volume) calls an "experimental system." On one level, these systems belong to differing experimental cultures—one of tissue engineering, the other of neural engineering. There are interesting differences in labs and in these cultures that are pertinent for some of the research questions of our National Science Foundation project. On another level, however, and for the purposes of this chapter, there are also interesting commonalities—ways of doing research that robustly transfer across both sites. Here I unfold some of these engineering norms and mores embedded in conducting in vitro research through building bioengineered physical simulation models. Two interrelated dimensions of this practice are taken up in turn: interdisciplinary engineering design culture and the concomitant primacy of engineering assumptions, norms, and values.

Interdisciplinary Engineering Design Culture

The fundamental orientation of these labs is toward designing simulation artifacts for use in various kinds of experiments. The interdisciplinary engineering design orientation of BME relates directly to the epistemic saliency of the devices. A device is designed "to predict what is going to happen in a system [in vivo]. Like people use mathematical models . . . to predict what is going to happen in a mechanical system? Well, this [the construct model system she was designing] is an experimental model that predicts—or at least you hope it predicts—what will happen in real life."[2] One dimension of interdisciplinarity is the hybrid nature of the devices as what the lab members call "model systems." As one respondent explained: "When everything comes together I would call it a 'model system.' . . . I think you would be very safe to use that [notion] as the integrated nature, the biological aspect coming together with an engineering aspect so it's a multifaceted modeling system." A model system can be a single device (as in "the dish," discussed below) or devices conjoined in a simulation ("flow loop" and "construct," discussed below). These hybrid model systems are designed for addressing fundamental biological questions, as well as potential medical applications. Some of the research in these labs is hypothesis driven (e.g., to test the hypothesis that shear stresses contribute to vascular disease), but as is often the case in engineering design, much is problem-driven "tinkering" or just "playing around." Indeed, the neural engineering researchers often used the expression "playing with the dish" to characterize their attempts to stimulate, record, and train the living neural networks in the absence of any understanding of how the "dish" behaves in the lab's initial period of research.

The overarching design/redesign agenda that characterizes the ethos of the research extends to a design orientation in how the lab members understand the biological systems and processes they investigate, as described by the director of the tissue engineering lab (indeed, as reflected by its very name):

DIRECTOR: What was really motivating me by 1970–1971 was the fact that those characteristics of blood flow [mechanical forces] actually were influencing the biology of the wall of the blood vessel. And even more than that, the way the vessel was designed.

INTERVIEWER: So this was influencing the characteristics of the biology.

DIRECTOR: Yes, right, and influencing biological processes that were leading to disease. The way the blood vessel is designed is, it has an inner lining called the endothelium. It's a monolayer; we call it a monolayer of cells because it's one cell thick. But it's the cell layer that is in direct contact with flowing blood. So it made sense to me that if there was this influence of flow [a force] on the underlying biology of the vessel wall, that somehow that cell type had to be involved, the endothelium.

The idea that the pressures and forces of blood flowing through the lumen (the space inside a blood vessel) could be causing changes in the endothelial cells that were responsible for disease processes, such as atherosclerosis, stemmed directly from his design orientation and from his early research on the effects of forces on the cardiovascular system during reentry from space flight. He maintained it took nearly twenty years to convince biologists that the biomechanics of the vascular system was worth looking into. At the outset of this research program, he proposed a novel hybrid concept, arterial shear, that is, the frictional force of blood as it flows through the lumen of an artery, and had spent forty years articulating it. In the 1970s vascular biologists were focused on biochemical processes, and he could find no collaborators among them. So he ended up in a lab of a veterinary physiologist, and together they created animal models (surgically modified cows) to investigate the nature and effects of arterial shear. Problems of control were significant in these in vivo modeling practices and led him to set up the program of in vitro research in his own lab, which required designing the simulation devices.

The initial configuration of research in lab A centered around the "channel flow device," locally called the "flow loop," which is designed to simulate normal and pathological forces of blood flow through the lumen, that is, a model of hemodynamics. The flow loop design is based on the fluid mechanics of a long channel with a rectangular cross section and accompanying flow-inducing components (using a liquid approximating the viscosity of blood) that can approximate the shear stresses during blood flow in an artery. When cells mounted on slides are "flowed" under different conditions, changes in cell morphology can be related directly to the controlled shear stresses. This device began as a large cumbersome artifact on a stand. Contamination was a constant problem because it could not be assembled under the sterile workbench hood. Within a few years it was revised into a compact design that fits under the hood and in the incubator. The flow loop, when we entered the lab, could be the locus of experiments on cells and cellular systems or just one step in a multimodel experimental process. Also, as discussed below, when we entered it was being used with a "wet" device, the "endothelial cell–seeded vascular blood vessel wall model"—or "construct"—that only recently had been designed by the lab as a "more physiological model" of the blood vessel than "cells in plastic" on slides.

During our investigation, in both labs cellular systems were seen as having design possibilities that feed into numerous design options. Wet devices are simulation models for which design involves manipulating living cell cultures. Research revolves around a central wet model in each lab: the construct in lab A and the dish in lab D. The construct device is grown first on a specially designed structure ("mandrill") comprising tiny silicon tubes that allow cells to attach and grow on them. Its design is based on what is currently understood of the biological environment of endothelial cells and vascular biology, the kinds of material available, and the bioengineering techniques developed thus far. Their ongoing research advances all these aspects. The construct forms a family of models in that different variations of it as a representation of the blood vessel wall are built depending on the objectives of a given experiment. For instance, it can be seeded with smooth muscle cells and endothelial cells, or just the latter, and the components of the collagen scaffolding can vary. Figure 5.1 shows a dish of constructs on specially designed mandrills and the flow loop device setup. The rectangular figure in the foreground of figure 5.2 is the flow chamber in which either cells on slides or constructs are placed and are subjected to various fluid flow conditions.

The "dish" of lab D comprises neurons and glia (support cells) harvested from an embryonic rat cortex, dissociated, and plated on a specially designed petri-dish-style arrangement of electrodes (called the multielectrode array, or MEA) that poke up into the neurons to allow stimulation and recording of the developing neural networks. They are plated in "one single layer of neurons. We try to get them down to a monolayer. . . . It's a simpler system to study then." A more pragmatic reason for a monolayer is that it is easier to "feed" than a brain slice. A sugary cocktail of biological chemicals feeds the cells, and the lid of the dish is a thin film of Teflon that allows oxygen and carbon dioxide through, so the cells can breathe, and keeps contaminants out. While constructs are created anew for each experiment, the dishes are intended to live a long time (the longest was just over two years). Figure 5.3 shows the MEA on which the neurons are plated and figure 5.4 the completed dish, which is then placed in an incubation chamber connected to various pieces of equipment for stimulating and recording.

Devices in general are sites of long-term investment in a line of research. Although we witnessed instances of de novo design, existing designs are

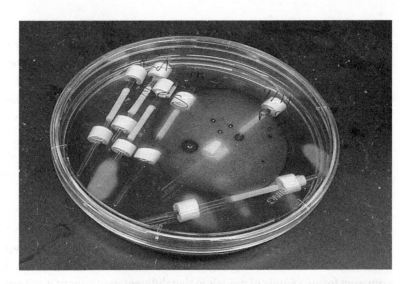

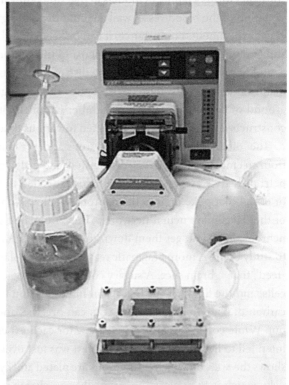

FIGURE 5.1 AND FIGURE 5.2. Tubular construct devices on mandrills in culture (5.1) and the flow loop device setup (5.2).

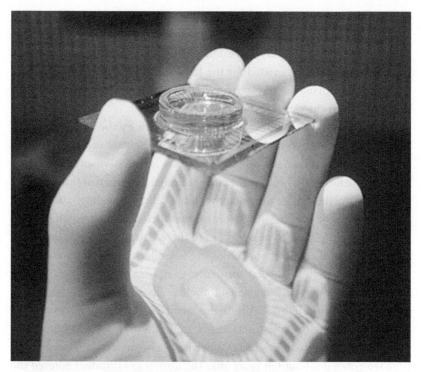

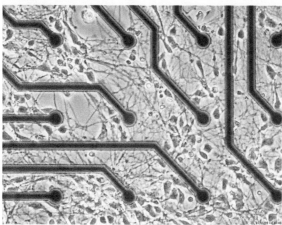

FIGURE 5.3 AND FIGURE 5.4. The multielectrode array on which neurons are plated (5.3) and the neuron dish device (5.4).

often revised to fit the needs of the evolving research program. It is important to the design agenda for the researchers to know and pass along the history of the devices. The current design is understood as conditioned on the problem situation as it existed for the laboratory at a prior time. It is thus important in moving the research forward that researchers know what, how, and why design changes have been made to a device.

Experimental possibilities are constrained both by the cellular systems and by the devices themselves. For example, in the case of the construct–flow loop model system the tubular constructs are sliced open before insertion into the flow loop because the current design requires that the construct lay flat for the fluid to be in contact with the cells. Interestingly, one of the early members of the lab, with whom we conducted a lab history interview, noted that his "job was to move us out of these slab studies [cells on slides] into tubular studies. And more so trying to understand how we can apply physical forces to these tubular constructs to stimulate the cells inside there, namely, smooth muscle, but also endothelial cells, to recognize—just to appreciate their surroundings." Although a small design change was needed to accommodate the flat constructs, redesigning the flow loop to accommodate tubular constructs would have been a major undertaking, and at the time we entered (about ten years after the former grad student entered the lab), they were still cutting the constructs open to flow. The justification they gave us for the legitimacy of that practice in that the flat construct cell wall "is not an approximation" from "the cell's perspective. . . . The cell [in the in vivo artery] sees basically a flat surface. The curvature is maybe over one centimeter, whereas the cell is a micrometer. . . . So to the cell, it has no idea that there's actually a curve to it." Only near the end of our investigation did one researcher attack the problem of how to redesign the flow loop to flow the tubular construct, because she needed it in that shape for an animal experiment.

Finally, for these labs, the notion of design extends beyond artifacts to the researcher/learners themselves, most of whom (graduate and undergraduate) are participants in a new interdisciplinary educational program designed explicitly to move beyond the customary model of collaboration between a biologist or medical doctor and an engineer to creating an interdisciplinary individual: "integrative" biological-and-medical engineers who can conduct their own research across these areas, as well as collaborate. Training in the research lab aims at integrating biology and engineering in the context of medical problems, requiring that learners master intricacies of cell culturing and Western blots, MATLAB and other modeling resources, and engineering design, as well as achieve a measure of conceptual integration of related biological and engineering knowledge. The objective is to create a new kind of expertise, which by design would mitigate problems of collaboration among disciplinary individuals, enable these engineering scientists to conduct basic biological research themselves, and potentially shorten the span between laboratory research and medical application. The push for such educational programs and for the emergence of BME as an "interdiscipline" has come from engineering pioneers in field who express considerable frustration at not being able to form collaborations with biologists, in particular, and from an implicit (though sometimes explicitly expressed in our interviews) belief that people who go into biology do not have the capability to think quantitatively or adopt an engineering mind-set—a belief we surmise leads to the situation where nearly all the graduate researchers in BME have engineering undergraduate backgrounds.

The Primacy of Engineering Assumptions, Norms, and Values

The design of the simulation models is dominated by norms and values associated with engineering, such as control, abstraction, approximation, quantification, simplification, and a good measure of pragmatism—compromises with respect to what it is feasible to do vis-à-vis engineering. Our own research group has been working for many years in an engineering institution, and we recognize these norms and values as widely in use across the various engineering fields and as inculcated in engineering courses. Many are shared with what Keller (2002) portrays as the values and norms of mathematical physics, which she contrasts with those of experimental biology. Within the context of our investigation these epistemic norms and values did not seem to be undergoing hybridization—at least, we have no evidence to support such a claim. Perhaps this kind of epistemic hybridization takes place more slowly and the BME community is still young. By examining what considerations go into the design of the devices, we come to grasp how an engineering epistemological culture is built into the machineries that participate in knowledge building through distributed cognitive processes.

An overarching goal of the labs is to determine what is the appropriate—or best feasible—abstraction of the in vivo phenomena to sufficiently address their major research questions. In the design of a device, there is tension between the constraints on its design and functionality that derive from biology and from engineering. Given the hybrid nature of the simulation models, the first challenge for the researchers is to develop selective, integrative understanding of both biological and engineering concepts, methods, and materials. Here "selective" means that researchers need to integrate in thinking and experimenting those dimensions of biology and engineering relevant to their research problems and goals. For example, lab A researchers need to develop an integrated understanding of the endothelial cell biology in terms of the stresses and pressures of fluid dynamics of blood flow in an artery. In lab D the notion of plasticity, a biological property, needs to be understood in terms of quantitative measures based on recorded spikes of electrical activity of the neuronal network. Further, in designing and conducting experiments with devices, researchers need to consider what constraints these devices possess that derive from their design and construction and what limitations these constraints impose on the simulation and subsequent inferences and interpretations. Both labs, for example, isolate cells for "simplicity" and to achieve experimental control.

Since the endothelial cells are the ones immediately in contact with the blood flow, lab A began with studying their behavior in response to flow independent of the rest of the blood vessel components and environment. In lab D, the embryonic rat neuron connections are broken apart (so all learning will be de novo) and plated on the MEAs as a monolayer, leaving out other parts of the brain usually thought to be implicated in learning. As the director explained, "It's a simplified model. I say that the model is not—it's artificial. It's not how it is in the brain, but I still think the model would answer some basic questions because the way the neurons interact should be the same whether it's inside or outside the brain. . . . I think that the same rules would apply."

This tendency to isolate and control is, however, counterbalanced by the systems thinking also common to engineering. After a period of investigation with simple models, researchers are willing to take the risk of building back in some of the biological complexity that might matter. Although examining the responses of endothelial cells on slides to shear stresses for several years provided significant insights about cell morphology, lab A's director decided to take the "big gamble" of trying to engineer a more physiological blood vessel system in which to embed endothelial cells so they can communicate with their "natural neighbor," smooth muscle cells. Lab D researchers realized that they were taking risks with the design of the MEA dish, but felt they would have established something important by achieving recognizable learning (which they did during our investigation—see Chandrasekharan and Nersessian 2009). As the director noted, "I'm much more interested in what happens when you take a bunch of cells and have them interact with each other, what sort of emergent properties pop up. So, it probably isn't necessary to include [in the model design] all the details you would find in a single neuron model, but maybe, so that's part of our job to find out the details of the biology that are important to these sorts of network properties and network phenomena and which are sort of incidental." After building a basic understanding of a simplified model—primarily through developing a control structure for supervised learning—they could go on to design a more complex model system.

One form that building complexity took in lab D was designing systems for "embodied learning" in the dish. More realistic modeling of animal learning requires a feedback loop that feeds sensory input to the dish and outputs behavior. "In the traditional way to do in vitro physiology . . . the closest thing to behavior is little waves on the oscilloscope screen. It has nothing to do with behavior." Researchers realized they needed not only to conduct open-loop physiology experiments where they stimulated the dish with electrical impulses but also to create closed-loop simulations in which they provide the dish with a sensory feedback loop. To explore embodiment, they created two different kinds of model system: "animats"—simulated animals whose activity takes place in a computational world, using the dish as its "brain"; and "hybrots"—hybrid robotic devices connected to the dish that perform in the real world.

In conducting experiments, the interlocking constraints of both the device qua in vitro model and device qua engineered system need to be taken into account. The MEA dish, for example, provides only a monolayer network of neurons instead of the rich three-dimensional connections found in vivo. Here, the researchers have reduced the complexity of the dish to a minimum along one dimension. The choice to use only monolayer cultures satisfies not only the constraints mentioned above but also an engineering constraint: the recording technology is limited to the 8-by-8 grid of electrodes embedded in the bottom of a dish; any neurons that are not in the bottom layer would not be accessible to recording. So, the monolayer provides a reasonable reduction of information for data analysis, while (it is hoped) still maintaining the salient qualities of interneuron communication.

An extended example of constraint interaction is provided by a series of interviews with lab A members about the flow loop as a simulation model. Their reasoning reveals further how engineering epistemic norms and values predominate in building devices to simulate biological processes. For a simulation to enable prediction about the in vivo phenomena, the flow loop and construct model system needs to behave as though blood were flowing through an artery to a satisfactory degree of approximation. Taking first the flow loop, as a model it represents the shear stresses during blood flow in an artery to a "first-order approximation of a blood vessel environment . . . as blood flows [through] the lumen," and enables a "way to impose a very well-defined shear stress across a very large population of cells such that their aggregate response will be due to [it], and we can base our conclusions on the general response of the entire population." The correspondence between the mathematical and the physical models is established through the geometry of the flow channel of the device, which has been designed with appropriate dimensions within a physiologically meaningful range. Using this device, changes in cell morphology can be related directly to the controlled shear stresses. But in conducting flow loop simulations, the researchers are doing "something very abstract because there are many in vivo environments and many in vivo conditions within that environment. Things change constantly in our bodies over our lifetimes, including physiological flow rates. . . . So I don't think we are trying to mimic the exact conditions found in vivo." These changes were a significant problem in the earlier animal studies, and a motivation for the move in vitro. Researchers make this approximation in design because "as engineers, we try to eliminate as many extraneous variables as possible. So we focus on the effects of one or perhaps two, so that our conclusions can be drawn from the change of only one variable." However, some circumstances could lead to a need for a revision to simulate higher-order effects, for instance, in vivo "blood sloshes around in the blood vessel" (i.e., is turbulent rather than laminar flow), and this would need to be simulated in the event that in vivo "there's a whole different pattern of genes that are upregulated in pulsatile shear."

In conducting experiments with model systems, devices are arranged in interlocking configurations, and the constraints of the various devices interact. The original flow loop was designed to flow cells on slides, but "cell culture is not a physiological model," and although much still is learned by just using cells, "putting cells in plastic and exposing them to flow is not a very good simulation of what is actually happening in the body. Endothelial cells . . . have a natural neighbor called smooth muscle cells. . . . These cell types communicate with one another." So, the lab director formed the aim "to use this concept of tissue engineering to develop better models to study cells in culture. . . . So we had the idea: let's try to tissue engineer a better model system [construct] using cell cultures." Unlike cell cultures on slides, constructs are three-dimensional surfaces in which cells are embedded; the construct "behaves like a native artery because that's one step closer to being functional." The thickness of the construct required a small modification of the design of the flow chamber to accommodate them.

However, as discussed earlier, the constraints of the flow loop chamber (designed originally for cells on slides) require that the tubular constructs be cut open and flowed flat (unlike an in vivo artery). The justification provided for this simulation is that since the endothelial cells that line it are so small, flowing the liquid (representing blood) over the flat construct is a manipulation that should provide accurate information. So, until there was the possibility of doing an implantation experiment with an animal, there was thought to be no need to flow constructs in tubular form, which would require major changes to the flow loop. When we entered the lab, a new researcher described herself as the one designated to "take the construct in vivo." Her experiment would implant the construct in an external shunt connecting the femoral artery and vein of a baboon, and she initially proposed to redesign the flow chamber to accommodate the tubular construct. As it turned out four years later, the researcher had the insight that it should be possible to design an "external shunt" for the flow loop—by analogy with the baboon shunt—and attach the construct to it for flowing. Although this required some redesign, it proved to be much simpler than redesigning the chamber and flow components.

A major issue for these researchers is whether they have made the right abstractions pursuant to the goal of engineering simulation models that selectively, in the words of the researchers, "parallel," "mimic," or "simulate" features of the phenomena thought relevant for their investigational purposes. If a first-order approximation will do, there is no need to capture the full complexity of the in vivo system. A good way to couch the desired representational relation between device and in vivo phenomena is that researchers seek to exemplify aspects of the in vivo phenomena thought salient to their investigation. As introduced by Nelson Goodman (1968), and extended to scientific models by Catherine Elgin (1996, 2009), a representation exemplifies certain features if it "both is and refers to" those features; that is, "exemplification is possession plus reference" (Goodman 1968, 53). One of Goodman's examples is that of a tailor's fabric swatch, which is a "sample of color weave, texture, and pattern; but not of size, shape, or absolute weight or value" (53). Similar to the way refining a piece of ore to consist only of iron creates an artifact that provides epistemic access to just those properties, devices afford selective epistemic access to properties of biological phenomena. In the case of physical simulation devices, the representational relations are built into the design that aims to capture the features thought relevant to epistemic goals. The dish does not just refer to neural processing; it is doing such processing. The flow loop, in performing, not only refers to blood flow through the lumen of a vessel but also flows a liquid with selected properties of blood (e.g., viscosity) in normal and pathological ranges over the construct device that has been designed to have selected properties of the vessel. The BME researchers we have been studying engineer physical simulation models to the degree of specificity they believe sufficient to parallel the in vivo phenomena, or to the degree that the lack of knowledge of the phenomena or the nature of the design materials constrains them.

Devices are successful as exemplifications if, indeed, they possess the relevant features of phenomena under investigation, and much of the research is directed toward determining this. The major related questions are whether the device/model system exemplifies what is important about the in vivo phenomena, and whether the properties found through experiments with the model system are the ones that would be found in vivo. Answering these questions provides, of course, the major challenge of this kind of in vitro research. Some boosts in confidence come from correlating various findings with related findings from other areas of biological research. Some come from achieving significant goals such as devising a control structure through which the dish reliably demonstrates learning. But perhaps, as with engineering more broadly, the final verdict will be rendered in how successful they are with in vivo medical applications, such as creating a viable replacement artery or controlling neurological disorders such as Parkinson's disease through techniques used to quiet unwanted "bursting" (spontaneous dish-wide electrical activity) in the dish.

Drawing on the notion of exemplification serves as a reminder of the deeply sociocultural nature of representation. In line with Goodman's observation that a paint chip is usually taken to exemplify a color, but in certain circumstances might be taken to exemplify a geometrical shape, what a simulation model exemplifies depends on the goal, purposes, and context. A paint chip exemplifies color within a particular set of social norms surrounding the practice of selecting paint for one's walls or house. In their simulation modeling practices, the researchers of lab A and lab D are drawing on a repertoire of practices and the conventions of the communities surrounding these.

Discussion

A central research practice of the BME sciences is crossing the in vivo–in vitro divide through designing, building, and experimenting by means of physical simulation models. Each lab we have studied can be characterized by a signature device or model system. For lab A, this is the construct–flow loop model system, which plays a role in nearly all experimentation; for lab D, it is the MEA dish model system, which can be experimented with on its own or as part of a larger, embodied system. Further, as signature devices, these model systems provide both an internal and external identity for the research lab. Lab members often identify their place in the research community through them ("we do flow loop studies"), and others use them to identify their pioneering contributions to the field: lab A as the flow studies and tissue-engineered model pioneer; lab D as the MEA dish pioneer. Within the labs, these signature objects are the hubs around which most practices and activities center. As with transportation hubs, where many services interlock at central stations, devices interlock many of the cognitive-social-cultural dimensions of practice. They are central artifacts around which social practices form, such as mentoring of newcomers; they are sites of learning; they perform as cultural ratchets (Tomasello 1999), which enable researchers to build upon the results of previous generations and thus move the problem solving forward. The devices perform as cognitive-cultural artifacts in the distributed model-based reasoning processes. Through briefly unraveling some of the epistemic norms, values, and assumptions that go into building these devices, rather than just considering how they are used in experimentation, it is possible to discern how an epistemological culture is embodied in the technologies that participate in creating knowledge.

The nature of the research questions in BME dictates the hybrid nature of the model systems and that researchers attain a degree of hybridization as bioengineers. Indeed, the interdisciplinary culture in which these labs reside self-consciously refers to itself as "hybrid" and "integrative," with respect to both the design of its research artifacts and its learner/researchers. But as we have seen, in the epistemic commitments of the researchers, engineering assumptions, norms, and values predominate. The biological phenomena are to be explained and understood through extending engineering concepts (shear forces on endothelial cells, electrical noise in neuron cultures) and applying engineering methods to screen, isolate, and control the messiness and complexity. Experimental setups need to provide outcomes that can be turned into mathematical form. Although cell cultures that need to be created, cared for, and sustained are at the core of the research, they are thought about as opportunities for design into devices. As evidenced in our interviews, the design of the simulation models is dominated by norms and values associated with engineering, such as abstraction, approximation, control, quantification, constraint satisfaction, and simplicity.

Physical simulation models are the means through which researchers extend their natural cognitive capacities into distributed cognitive systems that perform problem solving and inference. These artifacts thus simultaneously constitute the material culture and perform cognitive functions within the experimental research programs of both the BME labs we investigated. One researcher aptly characterized what they do in experimenting by means of designing and building physical simulation model as "putting a thought into the benchtop and seeing whether it works or not." The primary epistemic activity the devices afford is model-based inference (Nersessian 1992, 2008, 2009). In the language of distributed cognition, inferences are derived from constructing, manipulating, and propagating representations, in this case both mental (researcher) and artifact models. The distributed bioengineered model system is a conceptual and an in vitro physical system representing the in vivo system under investigation. It is an abstraction—selective and schematic in nature—that represents dimensions of biological phenomena of interest to the researchers. Models are interpretations of target phenomena (e.g., forces within the human vascular system, learning in the brain) constructed to satisfy constraints drawn from the domain of the target problem (e.g., the biology and physics of the vascular system, the information processing carried out by the brain) and, often, one or more source domains (e.g., the flow loop's cardiovascular and fluid engineering domains, the dish's neuroscience, electrophysiology, and engineering domains). Performing a simulation can lead to new constraints, or to recognizing previously unnoticed constraints. Inferences made through simulations can provide new data and hypotheses that play a role in evaluating and revising models to comply with constraints or provide potential new insights into in vivo phenomena.

Different kinds of abstractive processes underlie model construction. These abstractions provide different means of selectively focusing on features relevant to the problem solving while suppressing biological information that could inhibit the process. Suppression and selective highlighting of features provide ways of representing the problem in a cognitively and physically tractable manner. Approximation, for example, provides a means of discounting the relevance of features of the in vivo system, such as the fact that blood flow is turbulent. A standard approximation in engineering is the first-order approximation used when applying a mathematical representation, as with the force of the flow modeled by the flow loop device. Basically it makes the assumption that any higher-order effects are likely to be irrelevant or to be so complex as to make the analysis intractable. A laminar flow—one without currents or eddies—provides a first-order approximation sufficient for solving problems pertaining to many fluid dynamical phenomena. The choice of lab A was to go as far as the research would take them using a steady laminar rather than pulsatile flow and the approximation that this flow is two-dimensional over the surface of constructs cut open and laid flat.

Some abstractions are made because the devices themselves possess engineering constraints that often require simplification not deriving specifically from the biological system they are modeling, as we saw with design and construction of the dish model system. The dish provides only a monolayer network of neurons instead of the rich three-dimensional connections found in vivo. Here, the researchers have reduced the complexity of the dish to a minimum along one physical dimension. The choice to use only monolayer networks overcomes two hurdles. First, the recording technology is limited to a monolayer 8-by-8 grid embedded in the bottom of the dish. Any neurons that were not in the bottom layer of the dish would not be accessible for recording. Second, from the perspective of data analysis, the monolayer provides a reasonable reduction of information, while still maintaining (it is hoped) the salient qualities of interneuron communication. This limiting case abstraction makes the interpretation of the electrical response tractable.

In sum, the BME researchers we have been studying design and create hybrid in vitro model systems paralleling the in vivo phenomena they wish to study to the degree of specificity that they believe sufficient or to the degree that lack of knowledge of the phenomena or the nature of the design materials constrains them. One main feature of the model systems is that they are designed to abstract away potentially irrelevant features, thereby focusing attention on those features salient to the problem-solving context and enabling controlled experimentation. The expectation is that a properly designed model system warrants the researchers in pursuing where the experimental outcomes of the in vitro world might lead, in lieu of being able to carry these out in the in vivo world. These machineries of knowledge making embed the researchers' current understanding of the biological phenomena and their assumptions about knowledge and knowledge making. Understood in this way, a device provides a simulation not just of the biological phenomena but also of the epistemic assumptions, values, and norms of the researchers.

Conclusion

Conducting experiments with physical simulation model is an epistemic activity that enables inference through selectively creating objects, situations, events, and processes that parallel phenomena of interest. From a distributed cognition perspective, experimentation and inference are conditioned upon a fabric of interlocking models—across space, time, people, and artifacts, and connecting mental and physical representations and processes. What I have tried to establish in this chapter is that the processes of conceiving, designing, and building physical simulation models ("engineering devices" in the process sense) embed epistemic assumptions, norms, and values deeply in the machineries through which they strive to make knowledge ("engineering devices" in the artifact sense). Thus, the machineries of the manufacture of knowledge (devices of the epistemic culture) have the values, norms, assumptions, explanatory possibilities, and so forth of an epistemological culture embedded in their design and execution. Further, I have argued that shifting attention from using to building "cognitive" artifacts in DC helps us understand how culture is built into cognitive processes and vice versa.

Notes

I gratefully acknowledge the support of U.S. National Science Foundation grants RECO106733, DRLO411825, and DRLO909971 in conducting this research. This analysis derives from research conducted with Wendy Newstetter, Elke Kurz-Milcke, Lisa Osbeck, Ellie Harmon, and Christopher Patton. I thank the members of the research labs for allowing us into their work environment, letting us observe them, and granting numerous interviews. I also thank participants in the workshops for their comments on previous drafts, especially Karine Chemla, Hans-Jörg Rheinberger, and Emmylou Haffner. Finally, I thank the Fondation Les Treilles for its workshop support and for the ideal environment in which to carry out sustained intellectual discourse.

    1. Knorr Cetina (1999) states several times that high-energy experimentation at CERN can be construed as a "kind of distributed cognition" but does not develop the idea and does not cast molecular biology experimentation in a similar manner. On my account it should also be. As Giere (2002) has pointed out, she seems to be conflating DC with "collective cognition," à la Émile Durkheim.

    2. Quotes are taken from interviews with lab members, unless stated otherwise.

References

Chandrasekharan, S., and N. J. Nersessian. 2009. "Hybrid Analogies in Conceptual Innovation in Science." *Cognitive Systems Research* 10: 178–188.

Chandrasekharan, S., and N. J. Nersessian. 2015. "Building Cognition: The Construction of External Representations for Discovery." *Cognitive Science* 39(8): 1727–1763.

Dunbar, K. 1995. "How Scientists Really Reason: Scientific Reasoning in Real-World Laboratories." In *The Nature of Insight*, ed. R. J. Sternberg and J. E. Davidson, 365–395. Cambridge, MA: MIT Press.

Elgin, C. Z. 1996. *Considered Judgment*. Princeton, NJ: Princeton University Press.

Elgin, C. Z. 2009. "Exemplification, Idealization, and Understanding." In *Fictions in Science: Essays on Idealization and Modeling*, ed. M. Suárez, 77–90. London: Routledge.

Giere, R. N. 2002. "Models as Parts of Distributed Cognitive Systems." In *Model-Based Reasoning: Science, Technology, Values*, ed. L. Magnani and N. J. Nersessian, 227–242. New York: Kluwer/Plenum.

Goodman, N. 1968. *Languages of Art*. Indianapolis: Hackett.

Goodwin, C. 1995. "Seeing in Depth." *Social Studies of Science* 25: 237–274.

Hacking, I. 1992. "'Style' for Historians and Philosophers." *Studies in the History and Philosophy of Science* 23: 1–20.

Hall, R., R. Stevens, and T. Torralba. 2002. "Disrupting Representational Infrastructure in Conversation across Disciplines." *Mind, Culture, and Activity* 9: 179–210.

Hollan, J., E. Hutchins, and D. Kirsch. 2000. "Distributed Cognition: Toward a New Foundation for Human-Computer Interaction Research." *ACM Transactions on Computer-Human Interaction* 7(2): 174–196.

Hutchins, E. 1995. *Cognition in the Wild*. Cambridge, MA: MIT Press.

Keller, E. F. 2002. *Making Sense of Life: Explaining Biological Development with Models, Metaphors, and Machines*. Cambridge, MA: Harvard University Press.

Kirsh, D., and P. Maglio. 1994. "On Distinguishing Epistemic from Pragmatic Action." *Cognitive Science* 18: 513–549.

Knorr Cetina, K. 1999. *Epistemic Cultures: How the Sciences Make Knowledge*. Cambridge, MA: Harvard University Press.

Kurz-Milcke, E., N. J. Nersessian, and W. Newstetter. 2004. "What Has History to Do with Cognition? Interactive Methods for Studying Research Laboratories." *Journal of Cognition and Culture* 4: 663–700.

Lave, J. 1988. *Cognition in Practice: Mind, Mathematics, and Culture in Everyday Life*. New York: Cambridge University Press.

Nersessian, N. J. 1992. "How Do Scientists Think? Capturing the Dynamics of Conceptual Change in Science." *Minnesota Studies in the Philosophy of Science* 15: 3–44.

Nersessian, N. J. 2005. "Interpreting Scientific and Engineering Practices: Integrating the Cognitive, Social, and Cultural Dimensions." In *Scientific and Technological Thinking*, ed. M. Gorman, R. D. Tweney, D. Gooding, and A. Kincannon, 17–56. Hillsdale, NJ: Erlbaum.

Nersessian, N. J. 2006. "The Cognitive: Cultural Systems of the Research Laboratory." *Organization Studies* 27: 125–145.

Nersessian, N. J. 2008. *Creating Scientific Concepts.* Cambridge, MA: MIT Press.

Nersessian, N. J. 2009. "How Do Engineering Scientists Think? Model-Based Simulation in Biomedical Engineering Research Laboratories." *Topics in Cognitive Science* 1: 730–757.

Nersessian, N.J. 2012. "Engineering Concepts: The Interplay between Concept Formation and Modeling Practices in Bioengineering Sciences." *Mind, Culture, and Activity* 19: 222–239.

Nersessian, N. J., E. Kurz-Milcke, W. Newstetter, and J. Davies. 2003. "Research Laboratories as Evolving Distributed Cognitive Systems." In *Proceedings of the Cognitive Science Society 25*, ed. D. Alterman and D. Kirsch, 857–862. Hillsdale, NJ: Erlbaum.

Rheinberger, H. J. 1997. *Towards a History of Epistemic Things: Synthesizing Proteins in the Test Tube.* Stanford, CA: Stanford University Press.

Tomasello, M. 1999. *The Cultural Origins of Human Cognition.* Cambridge, MA: Harvard University Press.

Tweney, R. D. 1989. "A Framework for the Cognitive Psychology of Science." In *Psychology of Science: Contributions to Metascience*, ed. B. Gholson, W. R. Shadish Jr., R. A. Neimeyer, and A. C. Houts, 342–366. New York: Cambridge University Press.

6

Glass Ceilings and Sticky Floors
Drawing New Ontologies

How do social scientists come to recognize and establish a phenomenon, or group of related phenomena, to become a bona fide object of study for their fields? In such activity, how do facts get separated out and coalesce together in domains that are dominated by community-level values? And how do different kinds of knowledge, held by different kinds of communities with different values, play a role in defining those phenomena?

The notion that there is a recognizable divide between contexts of discovery and those of justification has long seemed a straw man, blown away in practice by studies such as Steven Shapin and Simon Schaffer's (1985) account of uncovering the nature of a vacuum or Hasok Chang's (2004) more recent account of discovering the characteristics of boiling water. Such accounts have found favor within all three of the history, philosophy, and sociology of science communities. In such cases, as in the case discussed here of the glass ceiling, discovery and justification are knit together. Discovering why women were not making it into top jobs depended upon social scientists in the late twentieth century working to establish a whole set of middle-level facts in which processes of justification were involved all along the way. But the distinction is not quite without merit, for it alerts us to see how this normal science activity involved not just establishing lots of middle-level facts but revealing, labeling, and conceptualizing these materials as evidence of a genuine phenomenon or phenomena. It was only after these many middle-level findings were successfully woven together in an account that related the main characteristics of "the glass ceiling" and outlined the (admittedly long list of) causes involved that social scientists become confident in talking about the glass ceiling as a phenomenon, singular, and without putting the tentative quote marks around it. These difficulties, both in grasping the nature of the phenomenon and in weaving together an account of it, stem in large part from the complex and open nature of the territory under investigation.

Discovery is a two-sided process involving both recognition that there is something to be investigated and the construction of an initial account of that something, which is where and why the elements and context of discovery and justification come together. (Of course, the phenomenon might turn out to be an artifact, or the initial account might turn out to be wrong and need a lot more investment of normal science to establish it beyond doubt, but this is so in any discovery process.) Recognition is associated here with labeling: the introduction of a new terminology to describe the specific institutional processes experienced by women that resulted in unequal workplace outcomes. Since the social scientists used these labels to frame the phenomenon for themselves, I use that labeling to tell an account of that process, and I begin the story with events in the United States.

The attention to labels and names in this case also provokes some broader reflections about the ways in which the circulation of facts within a community depends on the values held by that community and how the circulation of values within a community depends on the facts held by that community. Facts and values are constitutive of a culture, as depicted by the illustrations presented in this chapter. Of course, cultures exist at different levels. In this case, transformation of values in the broader society prompted investigation of previously taken-for-granted facts about women's employment. But within the larger society, there were a number of different social science knowledge communities, each holding its own facts and each with its own values, where these labels and names provided an important means by which the glass ceiling phenomenon came to be described and recognized. I return to this matter at the end of the chapter.

Recognizing the Glass Ceiling and Labeling Its Elements

The glass ceiling was originally portrayed as a perception or experience of individual women in the world of business. The first use of the term "glass ceiling," according to the *Oxford English Dictionary* and most of the potted histories on and off the web, occurred in a remark in the U.S. magazine *Adweek*, March 15, 1984: "Women have reached a certain point—I call it the glass ceiling. They're in the top of middle management and they're stopping and getting stuck." But this was not a throw-away remark, for its author, Gay Bryant, by this stage ex-editor of the magazine *Working Woman*, had used the term in the same year in a book-length analysis of the position of working women in America (Bryant et al. 1984, 19). This book not only drew on a wide variety of evidence, academic and public, but also provided advice and information, including an extremely useful appendix telling their readers about the laws on discrimination and how to file complaints under them. Bryant is not sure whether she heard the term elsewhere or made it up—it does not seem that the term was widespread in the mid-1980s, although it is difficult to know because the materials that were saved then, and are searchable now, exclude much of the ephemeral and popular literature of the period.[1]

A more conscious and widely noticed use of the term appeared in 1986 in a thirty-two-page special report in the *Wall Street Journal* on "The Corporate Woman" in the United States, and this is where the social science story really begins. The cover story/lead article, titled "The Glass Ceiling," drew on interviews with male and female managers and on a variety of research projects undertaken in universities, by nonprofit institutions, and by executive recruitment agencies.[2] The lead article's writers, Carol Hymowitz and Timothy D. Schellhardt, described the glass ceiling as "an invisible barrier that blocks them [women] from the top jobs."[3] Their article detailed how few women there were in top management and blamed this glass ceiling on two things. One was the beliefs of male managers, whose views about the performance of women managers often were not borne out by research (also quoted) on comparative male-female performance, which found either little difference in the performance of male and female managers or, that in some respects, women were more committed as managers. The second was male managers' personal preferences not to work with women. But the writers were wary about pronouncing how the glass ceiling was instituted and maintained, preferring to quote the program director at Catalyst, the main nonprofit organization working in this field:[4] "Up to a certain point, brains and competence work. But then fitting in becomes very important. It's at that point that barriers against women set in."[5] And, as the article's writers suggested, "not only do senior women managers frequently find themselves pressing up against a glass ceiling, they also find themselves on display under glass,"[6] as we see in their accompanying illustration (figure 6.1).

A book-length research report into the corporate glass ceiling undertaken by Morrison et al. (1997) with the Center for Creative Leadership (another nonprofit), defined the glass ceiling as "a transparent barrier, that kept women from rising above a certain level in corporations. . . . The glass ceiling applies to women as a group who are kept from advancing higher *because they are women*" (13). Their research was based on a substantial interview-based survey of women who were high up in, but not at the top of, the management tree in very large American corporations during 1984–1985 (and the survey design was similar to a previous survey undertaken of male executives so that the results could be compared). Female interviewees were asked what was required in individual women to make it through the glass ceiling into general management (even if they then "hit the wall," which stopped them becoming CEO or chief financial officer) and what "derailed" them. The assumption seemed to be (held by both investigators and subjects) that there were some automatic, even natural, paths of promotion inside corporations, and the individual women's voiced experiences were reported at considerable length to illustrate how women were stopped on such paths.

Between the first and second edition of Morrison et al.'s book in 1992, as remarked in their new foreword, the "'glass ceiling' has become a household term" (xi). This household recognition was doubtless part of the public sphere in which the term had gained currency, but it also reflected a flurry of social science research activity of various kinds, sponsored and undertaken by a variety of agents, in different fields, for different reasons, and reported in a variety of formats.

The most significant of these public developments in the U.S. context was the Glass Ceiling Initiative, created in 1989 by Elizabeth Dole (and then taken over by Lynne Martin) at the U.S. Department of Labor, apparently in response to these various earlier research projects. This initiative followed the department's own report, *Workforce 2000* (Johnston and Packer 1987), which had suggested that the rapid growth of the labor force of white males was slowing down and so future economic growth would depend on the way in which ethnic minorities and women might increase their share of the U.S. workforce.[7]

It is important to this story that the Department of Labor was also responsible for ensuring that government contractors were compliant with

FIGURE 6.1. Douglas Smith's illustration for a special supplement of the *Wall Street Journal* titled "The Glass Ceiling" (March 24, 1986, 1D). (© Douglas Smith. Republished by permission.)

the equal employment opportunity laws via its Office of Federal Contract Compliance Programs.[8] Its mandate was to ensure not only that contractors "do not discriminate" but also that they take "affirmative action to actively recruit" from all sectors of the labor force and "to provide training and advancement opportunities for all employees" (U.S. Department of Labor 1991, 4). The department undertook a set of "compliance reviews" under the Glass Ceiling Initiative, including a pilot study looking at nine individual companies taken from the Fortune 500 list and representing different sectors of the economy.

The aim of the Department of Labor reviews was activist: to "identify systematic barriers to the career advancement of minorities and women" and then to work with the companies to eliminate them (U.S. Department of Labor 1991, 3–4). The reviews themselves were strongly shaped by the requirements of compliance. Whereas previous research had found the outcomes—namely, that there were very few women in top management—and inferred that there must be barriers, these compliance reviews set out to look for, and at, the barriers directly. The review teams thus spent a lot of time figuring out how each corporation was organized and how it worked. They recognized that these corporate cultures all differed, yet they found evidence to identify a number of common institutional barriers: in promotion systems, in mentoring, in training, and in lack of high-level accountability within the company for equal employment opportunity issues. Identifying these barriers not only gave them the grounds to lay out recipes for compliance from those companies[9] but also created benchmarks for further compliance investigations that would reveal more about these barriers.

The department's 1991 report documenting its work is critical to this story of recognizing the phenomenon of the glass ceiling in two related respects. It developed the definition: the compliance reviews established the glass ceiling as "those artificial barriers based on attitudinal or organizational bias that prevent qualified individuals from advancing upward in their organization into management level positions" (1). "Artificial" can best be understood by contrasting with what might be termed "natural" barriers based on qualifications or career breaks that affected years of experience. At the same time, including "organizational bias" moved the frame beyond a sole or primary dependence on individuals and their work relations (the "attitudinal" elements) and toward the organizational or systemic barriers within which individuals acted. This shift is particularly noticeable compared with Morrison et al.'s 1987 survey of the experiences reported by women, who tended to locate the glass ceiling problem as lying in their own hands and/or in their relation to other individuals in their organization. Over 50,000 copies of the 1991 report were distributed within the first year of publication (U.S. Department of Labor 1992, 2), which, like the earlier 1986 *Wall Street Journal* special section, now appears as a seminal moment in the documentation and acceptance of the glass ceiling phenomenon. As one reporter noted, "Now even a conservative administration has certified that the glass ceiling is as real as steel."[10]

A direct outcome of the Glass Ceiling Initiative was the 1991 Glass Ceiling Act, introduced by Bob Dole (as part of the civil rights legislation of that year), which established the Glass Ceiling Commission as a high-profile platform in the American political sphere.[11] The commission produced two reports in 1995. The substantive one reporting its research, evidence, and analysis outlined a list of "glass ceiling barriers" (U.S. Department of Labor, 1995a, 7–8, 25–36) and grouped these barriers into three different sorts: (1) those based in society, (2) those under the control of business, and (3) those that were the responsibility of government. The second group was focused on what the report called "internal structural barriers," which stopped those who had got into the organization (i.e., who were in the pipeline) from progressing up in those organizations. These reports were the result of commissioned research studies, focus groups, oral witness hearings, and so forth, that were all part of the commission's work. At the same time, a further fifty-three compliance reviews were undertaken in 1993–1994 by the Department of Labor.[12]

The early 1990s saw an explosion of research and a plethora of new terms associated with such glass ceilings. A 1992 *Wall Street Journal* article reported on the "glass walls" found in corporations (by the nonprofit research group Catalyst), which identified the horizontal barriers that prevent women moving laterally in corporations.[13] A labor economist, Myra Strober from Stanford University, interviewed for this 1992 article pointed out that this was associated with a well-known phenomenon known as occupational segregation. But once again, it was the switch of attention that mattered here, for Catalyst's research focus was not the occupational segregation per se but the nature of the barriers—between occupations but within the same organization—that mattered for vertical moves. Catalyst's research found that corporate women needed to gain the right kind of lateral-move experience—usually into the profit centers of an organization (in manufacturing and selling) rather than service centers such as human resources (where they had tended to be hired and remain)—before they could gain vertical promotions. Bridges sideways were needed in order for women to reach vertical ladders, but in many companies there were no such bridges: the glass walls functioned to hold up the glass ceilings. In retrospect once again, such glass walls had been noticed—indeed, they were signaled both in Morrison et al. (1987) and in the 1986 *Wall Street Journal* special report—but not fully recognized, labeled, or systematically researched in those earlier analyses.

Another significant term coined in the same year was Catherine Berheide's (1992) account of "sticky floors" in a report for the Center for Women in Government at the State University of New York. This research was done in response to all the focus on glass ceilings, for as she said in an interview the following year, "most women should be so lucky to have the glass ceiling as their problem. Many [women are] mired in . . . the sticky floor" (quoted in Laabs 1993, 35). Based on a statistical study of women and ethnic minorities in state and local government, Berheide showed how most individuals in these employment fields were stuck in jobs with low pay with very limited possibilities for upward or sideways movements. There were no career ladders within organizations for those stuck to the floor.

I focus on these three terms—glass ceilings, glass walls, and sticky floors—as the most significant, but there were others. For example, "glass escalators" fast-tracked male candidates through the glass ceiling in predominantly female occupations, while the "double trap doors" of gender stereotyping caught unwary women who behaved in too masculine or too feminine a way.[14] And so it went on. They were all captured in Mary E. Guy's (1994) account of "organizational architecture" that drew together the wide range of social science and public literature that had built upon the separate elements.[15] What these terms and labels had in common was that they were the result of research that stopped looking at the individual men and women of organizations as individuals, and looked at the barriers—the elements of the organizations' social and cultural architectures—that stopped women's (and ethnic minorities') employment mobility. This is not to say that these social scientists thought individual attitudes did not matter; rather, they argued that those attitudes were set within organizational structures and cultures that allowed individuals more or less freedom to affect—for good or bad—the career paths of themselves and of others.

Normal Science in the Midst of Discovery

Of course, I do not intend to suggest from my short history of the terminology and of the research that suddenly glass ceilings were a new phenomenon in the 1980s. Most of the experiences clustered under the umbrella terminology of glass ceilings were commented upon earlier, and even analyzed, but—significantly—they did not use these labels, and only rarely were they framed and understood in quite these ways.

In retrospect, it has become clear that the earlier study most salient for glass ceiling researchers (such as Morrison et al. [1987])—precisely because it did make such a framing—was Rosabeth Moss Kantor's *Men and Women of the Corporation* (1977), which won the C. Wright Mills Award, an eminent social science book prize, in the same year that it was published. This "ethnographic study of a corporation" as she described it in the opening page of her original preface, grew out of her discipline of sociology and her feminist experience, as well as her work that crossed over the academic, consultancy, and public service divides, a profile that was matched in the different sets of researchers working on the glass ceiling materials. Kantor's study focused on the relation of the individual to the organization, and she discussed the various ways that men and women in various jobs experienced ceilings, got stuck, found themselves in dead-end positions that had no ladders, and so forth. More important, she stressed that the main programs of the time (the 1970s) that attempted to resolve the problems of unequal workforce representation were bound to fail because they focused on changing individuals (making individual women more confident, individual men less prejudiced). In her view, these projects were based on a faulty analysis: those who assumed that "the factors producing inequalities at work are somehow carried inside the individual . . . are using the wrong model and drawing the wrong conclusions" (261). The "wrong model" assumed that such inequalities were due to either nature or nurture: either women were different from men by nature (they were not as striving or competitive), or women were nurtured differently from men (so that they preferred not to be competitive or in positions of leadership): nature or nurture could equally explain why women did not make it to the top, but acting on either analysis with curative interventions would not solve the inequalities.[16] For Kantor, such problems could be resolved only at the organizational level, for it was organizations that made individuals fit into their jobs and created behavior patterns in them.

Clearly these barriers were not new, but there were two important, connected, contingent contexts that prompted their recognition as barriers in America at this time. One lies in the legislative context in American society. Following the earlier equal opportunity rights in employment legislation of the 1960s, there was an expectation that change would follow—that women and ethnic minorities would make it into the workforce, creating a "pipeline" that would over time feed them into positions across the occupations and up the hierarchies in proportions commensurate with their numbers in society. That is, the cohort effect would work itself out in the new generation of workers at all levels, to resolve the problem of unequal outcomes and to equalize not just employment opportunities but employment experiences. That ambition, that hope, appeared thwarted: yes, women and ethnic minorities had made it into the workforce in much greater numbers, and even into middle management. The development of professional personnel managers using this new legislation had made a difference to the work experience of women and minorities.[17] But the *Workforce 2000* report commissioned by the Department of Labor (Johnston and Packer 1987) showed what was already widely known by experience—that those two groups were still mostly stuck in the sticky floor jobs.

The second contingency lay in the flowering of the feminist movement in America during the 1970s and 1980s, with its catchwords that the "personal is political" and its associated reliance on the value of personal experience.[18] Personal experience when voiced did not just raise consciousness within women as groups but was understood to have the potential to transform society's values and so the position of women in society. While the Department of Labor was concerned that the country would run out of the right kind of worker, namely, white males, it was the feminist movement that made the position of women in the labor market seem nonnatural, a problem to be addressed rather than a normal state of affairs. Thus problematized, women's employment experience became an issue for society at large and so belonging on the political agenda.

Even when the phenomenon of the glass ceiling (as opposed to hierarchical or occupational segregation) appears to have become widely accepted as a social scientific phenomenon of many different communities and so recognized as part of the genuine experience of women (and ethnic minorities) in the labor market, rather than, say, a "figment" of women's imagination, this was not the end of the matter but the beginning. The problem with these glass ceilings and walls was that these barriers could not be *seen* from below or from above. The men above could no more see them as institutional barriers than could the women below. And for the research community studying the issues, even while the facts of limits to occupational mobility reported in these investigations were taken to confirm that such barriers must exist, the constitution of these barriers remained difficult to define and characterize when moving beyond the site in which they had been located, namely, corporate America. This required a widening and deepening of research to identify the barriers and to see how they operated in other parts of the economy and even in other countries.

Research in this widening and deepening process was rarely straightforward. It was often unclear if researchers were giving different and inconsistent interpretations of evidence, or defining complementary characteristics for the same elements, or really had different accounts because they were looking at slightly different things. As one small example, two different academic studies, each using the term "glass cliff," were picked up by the press and internet on successive days in 2010.[19] This term had come up in earlier labeling but lacked a stable meaning. One study apparently described the glass cliff as the fact that women in jobs traditionally held by men are watched more closely and when they make a mistake are blamed more severely. The other apparently suggested that women were more likely to be chosen in a crisis if several men had already failed to clear it up, but would then be given less credit for solving it. Were these two sides of the same element, or in fact different characteristics of the broader pattern that made up the glass ceiling?

Yet the very lack of stabilized understandings may be why the normal muddle of normal science stimulates—as here for the glass ceiling—a variety of accounts of the phenomenon that taken together developed scientific understanding of that phenomenon. A more extended example shows how this occurs, and it too starts from a labeling moment. While economists had been involved in the early 1990s work following up glass ceiling barriers to promotion in a variety of fields, and not always coming to the same conclusion, a new front opened up in the late 1990s and early 2000s that we can trace by following the path of one particular influential paper. This paper, by the Australian labor economist Alison Booth working in Britain in conjunction with two collaborators, Marco Francesconi and Jeff Frank, was an analysis of British household survey data relating to the early 1990s. This representative data set covered all occupations and sectors and, though small in numbers surveyed, was rich in detail about the individuals. The title of the original working paper of 1998 was "Glass Ceilings or Sticky Floors," and the abstract expressed the authors' surprise that women were more likely to be promoted than men, though that difference disappeared when more information about the individuals was taken into account. But they also found that while women in this survey did not experience a promotion glass ceiling, they did not get such a large pay rise on promotion as men did; that is, women got through the glass ceiling but were stuck to that ceiling as a floor. They labeled the latter experience the "sticky floor," apparently unaware that the term was already in circulation, in America and Australia, to denote a broader problem with somewhat different characteristics—namely, being stuck at the bottom of the wage and promotion pile of the whole distribution.

It is possible to trace the way in which this second notion of the sticky floor developed as these findings were taken up by PhD students, research assistants, and those who worked in close contact with the original collaborators. This insider's map can be used to trace the process by which the facts found in the paper traveled around the academic community, taking this version of the sticky floor label along with the model that it utilized, to other economists working on women's employment in those years.[20] When the paper was finally published (Booth, Francesconi, and Frank 2003), the authors were no longer expressing surprise: by this time the mathematical model they used to explain their statistical facts had become accepted by the disciplinary community, and so the facts themselves no longer seemed peculiar.[21] Yet when the further trajectory of this work is followed, it is evident that the issue of pay became more seriously studied than promotion (probably because there are much better data sets on pay).[22] By investigating the male-female pay differentials first at the top of the distribution of all workers and then at the bottom of the distribution, in Europe and then elsewhere, economists added a set of characteristics to the glass ceiling phenomenon that had previously been concerned primarily with promotions. In the process, Booth, Francesconi, and Frank's (1998) sticky floor notion in labor economics was revised to become coherent with—if not fully coextensive with—the original label, which referred to the pay differential and lack of promotion prospects at the bottom of the whole distribution rather than at or near its top. At the same time, however, economists' glass ceiling characterization became refocused on the pay gap at the top of the scale, rather than on job titles or promotions or levels of responsibility.

It is evident that labeling the different barriers was only a prelude to exploring causes, unraveling relationships between the elements, and searching for better descriptions, as social scientists mapped the glass ceiling's aspects and characteristics. This involved investigators from psychology, economics, sociology, management, and those in related "applied" or professional fields of public administration, personnel management, industrial relations, and so forth. These diverse specialists circled around and over the various elements that made up the phenomenon, collecting evidence and testing hypotheses, to pin some elements down more firmly, disagreeing about others, and perhaps unpicking others in accordance with their own fields' standards of evidence and conceptual and theoretical predilections. The fact that these researches continue is evidence that the phenomenon has been accepted as a genuine one for social scientists to investigate, to theorize about, to gather evidence on, and to develop specialized accounts of in different fields.

Labeling and Naming

We can see many of these same elements of recognition of the phenomenon, the initial accounts constructed to account for it, and the further research that followed if we look at the Australian experience, which paralleled many of the elements of the U.S. story but also proves revealing in somewhat different ways. In Australia, too, these glass ceiling researches were prompted by feminist activism to focus on assessing the outcome of earlier legislative moves and with a similar expectation that the cohort effect would do away with "the problem" of women not making it up the promotion ladders—but this expectation had proven wrong just as it had in the United States. So, in common with the American story, there was parallel set of private-public sector activities both to investigate the phenomenon and to take action to alter the situation, although in Australia the net of sites for investigation focused not on corporate life but on service industries such as finance, education, and particularly government services. The Australian literature took up and used the American terminologies of glass ceilings, glass walls, and sticky floors and referred to a wide selection of the American literature on these topics that has been discussed above.

Australian social scientists also brought something different to the investigations: an attitude that was more openly argumentative in the academic sphere and openly aggressive in the public domain (see figure 6.2). For example, Limerick and Lingard's (1995) volume of essays on gender in education management, appeared at first sight—from its list of contents and its foreword—to be the work of feminist activists: "What has been our collective experience as women in the reform movement? Three decades and one of the most significant feminist movements in recorded history, a restructured public sector with affirmative action policies and anti-discrimination watchdogs. But has there been a fundamental change?" (iii). In fact, the volume's contributors were those that are usually associated with belonging to "the establishment."

One such was Leonie V. Still, an authoritative writer on corporate women in Australia and the first female dean of a business school in that country and, at that time, deputy vice chancellor of Edith Cowan University. She was Australia's equivalent of Rosabeth Moss Kantor and in 1997 delivered an

FIGURE 6.2. Cartoon by Peter Nicholson from the newspaper *The Australian*,
July 30, 2004, www.nicholsoncartoons.com.au (© Peter Nicholson, Rubbery Figures
P/L. Republished by permission.)

important research report for Australia's Human Rights and Equal Opportunity Commission titled *Glass Ceilings and Sticky Floors: Barriers to the Careers of Women in the Australian Finance Industry*. This was one of the largest industries employing women, and the research used several different social science methods to analyze the workforce, understand the barriers, assess the affirmative action reports, and uncover the range of attitudes of those involved. Still (1995) had already drawn together a list of barriers, with somewhat different divisions than that reported by the Glass Ceiling Commission in the United States in that same year. Her main categories were cultural, organizational, individual, and governmental, with each category covering a long list of specific barriers. Each particular named barrier within one of those general categories could be understood as a potential cause of the outcome that women were not promoted, for, as with the American literature, any list of barriers functions as a list of possible causes of glass ceiling outcomes.

The Australian researchers also exhibited a greater worry about the proliferation of labels and about naming the phenomenon with a recognizable and accurate name. Still (1995), for example, spent some time considering the validity not just of the labels inherited from the American researchers—glass ceilings, glass walls, and sticky floors—but of others that had come into use in Australia in the early 1990s: "perspex ceiling," "sticky cobweb," and "greasy pole." In another chapter in the same volume as Still's 1995 chapter, Eleanor Ramsay (1995, 175), at this stage pro-vice-chancellor of the University of South Australia, railed against what she saw as the obfuscation that occurred because the labels used were "abstractions, metaphors and euphemisms which shroud in vague generalities the occurrences to which they allude." She argued that nothing would change until the behavior that created that ceiling was identified and not just labeled but named. These members of the "feminist establishment" were by no means united in their social scientific approaches, or in their stances on solutions to the problem, but they were united in their appreciation of labeling and naming. Indeed, the final chapter of this Australian volume (Porter 1995) urged a "spring clean" of all these labels.

The urge to get the name right echoed an earlier call in the Australian context. In 1993, a "national forum" meeting of the Affirmative Action Agency and the Institute of Public Administration of Australia had convened under the title "The Glass Ceiling: Illusory or Real?" One paper reported from the forum, by Andrew Hede (1994), had no doubt that there was a real phenomenon at issue but argued that there were too many labels and that they were mostly misleading. Hede preferred the name "sticky steps," because the barriers operated at many or every level (rather than the one level implied by the glass ceiling), and that such stickiness was, like the glass ceiling, equally unseen by either men or women, both in advance and afterward. But, as he recognized, the name "sticky steps" would not take, because it was not so glamorous for the popular literature (or so catchy for advocacy). Yet, there was a very serious point behind all this: an accurate label would encourage effective theorizing, and so "provide the basis for a theory about women's under-representation in management" (Hede 1994, 79).

The power to name rested on the social scientists' abilities to take note of something they initially did not understand but whose research had enabled them to identify, describe, and otherwise reveal its characteristics, as well as construct initial accounts of those characteristics and of the causal relations that create, underpin, or perpetuate them. And these activities in turn depended on being able to put an accurate label on—that is, name—the phenomenon. This is exactly the circular process—virtuous not vicious—that Australian feminist Dale Spender wrote about in *Man Made Language* (1980) when she discussed how a particular kind of women's "experience without a name" was given a name in the case of "sexual harassment."[23] Only when accurately named could the phenomenon be recognized by those who experienced it and addressed by those who took responsibility for preventing it. This power of naming has been much associated with feminist arguments, particularly in relation to the naming of experience. But this process of accurate naming must be equally the case for any scientist grappling with a previously unrecognized phenomenon—be they a social/human scientist or natural scientist. Why? Because naming a phenomenon is not just labeling it—it involves its recognition.

Facts and Values in the Social Science Domains

In a case such as the glass ceiling, it does not make sense to ask how facts and values might be separated, for the subject of study—a social scientific phenomenon—is a cultural one that embeds both facts and values.[24] It was a cultural transformation in values in certain societies that meant the facts of women's employment were no longer taken as natural but were reconstituted as social problems. And it was the result of social scientific research that these facts were transformed from facts about women to facts about institutions. At the same time, as we have seen, the phenomenon that social scientists investigated under the name of the glass ceiling varied in its characteristics across times, places, and occupations or industries, even while, as a general characteristic, many of these institutional elements proved common across locales. Whether the phenomenon was recognized, whether it was seen as normal or problematic, and the responses to it were all specifically local. The glass ceiling phenomenon manifested differently according to the community within which it was observed and was thought about differently within those different communities. Facts and values are constitutive of cultures, and the cultures of societies vary.

At the same time, the different knowledge communities who studied the phenomenon also exhibited considerable variety in the way they approached the topic, the knowledge they brought to its description, the modes of analysis used, and the facts that they drew out from the evidence. These knowledge communities might be broadly divided into two cultures: those that began from a scientific approach and those that began from a community's experience. But this division between scientific and community expertise does not provide much analytical grip, for in the social sciences there are not two completely separate and different knowledge cultures, but a network within which both shared and different elements occur. As we have seen, social science knowledge depended here upon a broad network of actors and organizations: both policy and academically oriented social scientists, those working in nonprofits and the activists of civil society, along with private-sector organizations and government departments that sponsored research in the field. These all contributed to the process of establishing the phenomenon by assembling a matrix of materials: journalistic pieces, commissioned reports, government reports, academic research papers, and books. Both the network of people and matrix of materials were cross-referring.[25]

This is not to suggest that all social science fits into this triangle of activist–government policy–social science (any more than we would place all twentieth-century natural sciences into a military–industrial–big science triangle). Rather, the point is that across a range of social science disciplines and projects of research, we can find many with policy orientations that attract social movements and create such networks of researchers and matrices of materials. And although the congruence of all these elements in the social sciences may appear to be a post-WWII development, we can find these kinds of cooperation stretching back in time—think of the poverty research of the late nineteenth century and the way that activists, government officials, and academics gathered around Charles Booth's poverty project and his massive survey of London or around Jane Addams and the Hull House initiatives in Chicago. The joining of social movements and philanthropic investigators with government action and social scientific work is certainly not "modern" in the experience of the social sciences.

Alongside this triangle of interests came a commitment to advocacy. These social scientists working on the glass ceiling did not believe in a supply-side theory of knowledge: they did not think that knowledge about barriers to women's and ethnic minorities' employment need only be supplied and would of its own accord trickle down into the wider community, let alone be acted upon.[26] So, contrary to the values often espoused for science and scientists that caution against advocacy overstepping the line between personal values and scientific activities, those involved in this project did not necessarily see any conflict between objectivity and advocacy—rather the opposite. But these social scientists believed that their advocacy, whether a duty or a permissible right, must be based on solid social science research as a guarantor of their objectivity. This too, of course, is an older position (see Furner 1975). Such advocacy was a feature of the public domain in the late nineteenth century, just as in the modern period we find social scientists' advocacy about the glass ceiling displayed through interviews, public lobbying, and so forth, regardless of where the social scientific activities had been conducted and under which umbrella organization, be it commissioned from a consultancy or carried out within academia.

Another important element of the research in the glass ceiling case, one that is characteristic of the historically constituted framework within which social science operates, is the status given to experiential knowledge. Social sciences had, from their disciplinary establishment in the late nineteenth century, relied on two modes of gathering evidence for analysis: One kind extracted personal experience in the form of extended interviews, life histories, social surveys, and so forth; the other kind constructed impersonal numbers or statistics. While the qualitative social sciences and the feminist movement shared a commitment to the value of experience-based knowledge, the discussion here shows that research on the glass ceiling phenomenon relied on both kinds of work: experiential accounts and statistical accounts. Crucial to the initial recognition of the phenomenon, and then in different forms and at different points, was the evidence of experience—in both social scientific and activist circles. Such experiential and personally articulated knowledge is considered valid in just the same way that patient groups are considered to have valid knowledge of particular kinds about the medical conditions that they (or their close relatives) experience, and that those who live with environmental problems are considered to understand certain aspects of those phenomena very well (see Wynne 1991, 1992). The social science triangular network of knowledgeable actors thus becomes a square comprising activist, government policy, and social science, with personal experience marking the final corner.

The importance of personal experience as an evidence base for the social sciences, not just in this glass ceiling case but in many fields and for many topics, brings us to issues in the public understanding of science and the nature of expertise: both topics need to begin from a somewhat different set of assumptions in dealing with social sciences. As I argued in Morgan (2010), in the social sciences those who make up "the public" have a considerable experience of social science subject matters for they all live in society, take part in the economy, experience bureaucracy, and so forth. These experiences are part of their knowledge about the world they live in, the same world that the social scientist investigates. In the case of medical knowledge, the literature on the public understanding of science has suggested that such experienced knowledge is often complementary to the scientifically based expert knowledge of the subject—it may be less lawlike, and less informed about hidden causes, but experiential knowledge is more detailed in certain respects, more accurate about the variations that occur within the same condition, because those phenomena are observed and experienced on a day-to-day basis. If this is the case in medicine, the same is likely to be the case in the social sciences, with the parallel implication that experience produces a particular kind of knowledge and expertise, which may well be organized and stratified differently from the scientist's knowledge gained through scientific investigation but is nevertheless reliably useful. It might also be fruitful to compare these issues of expertise and public engagement in social sciences with those of natural sciences where the experienced knowledge of "amateur" observers has remained important (e.g., in botany or astronomy) or where lay people have been active participants in the construction of scientific knowledge (as shown in the classic article on boundary objects by Susan Leigh Star and James R. Griesemer [1989]). Nevertheless, I suggest that there are still some differences between the social sciences and those natural sciences dependent upon lay participants that stem from the fact that social sciences are usually constructing knowledge not of other objects but of ourselves as subjects, just as we see in this case of the glass ceiling.

These reflections on the sources of social scientific knowledge suggest that a concept of "civil" or "community ontology" might prove a usable and useful complement to Sheila Jasanoff's (2005) "civic epistemology" which refers to the way different societies come to establish what counts as scientific knowledge in the public domain—it typically exhibits national variations according to the different ways in which expert knowledge is created, assessed, and used in public life. I use "civil" or "community ontology" to refer both to the role of civil society as a cooperative partner in the delivery of social scientific knowledge and equally—and perhaps more saliently—to the role of the community's experiential knowledge (acquired through experience) as opposed to scientific knowledge (acquired through scientific inquiry) in the construction of the content of that knowledge. The notion may also be relevant for some of the natural sciences. Methods of earthquake prediction in China (see chapter 12 in the present volume) might be framed with the notion of community ontology in mind as much as in the study of glass ceilings in contemporary society. For the social sciences, at any rate, civil or community ontology goes alongside civic epistemology—these are distinct but related characteristics of how a society comes to know things about itself.[27]

Historical, sociological, and philosophical studies of the natural sciences have taught us much. This study of the recognition and initial establishment of a social phenomenon suggests why the social sciences may require different accounts from those told about the natural sciences, but it is not intended to stake out hard-and-fast differences. Rather, it is designed to bring to light, or into better focus, the ways in which epistemology and culture intersect for the social sciences at two different levels—that of the knowledge communities, and that of the community within which those knowledge communities live and work. But this is not just about epistemology. Those wider communities raise questions and prompt what is found problematic and thus what is studied. And in doing so, they are closely involved in defining matters of content and ontology for the social sciences, whereas for those studying the natural sciences such community matters have more often been understood as defining matters of context.

We can see the implications of this by returning to the interdependence of values and facts that characterized the research on the phenomenon of the glass ceiling: in defining what it was, what characteristics it held, and what causes might be involved. The implication I draw is not that the intimacy between facts and values inherent in any culture negates the possibility of recognizing and establishing such a phenomenon as genuine for social scientists and for their subjects. Rather, it is the opposite—the values involved were critical in problematizing, inspiring, and developing the work that was done to establish the facts of the phenomenon, while those facts describing the phenomenon revolved back onto the way values were understood and expressed. That means not that the fact-value distinction collapses but that, by using the distinction, social scientists turned the subject of glass ceilings into an object: they turned what was previously understood as a subjective experience of individuals into an objectively recognized and described social scientific phenomenon.[28]

Notes

Thanks to Aashish Velkar for research assistance in tracing both Australian and American materials and to Linda Oppenheim, industrial relations librarian, for help in tracing Catalyst's reports at the Firestone Library, Princeton University (and see Oppenheim 1991). An earlier version of this chapter was presented as the Distinguished Lecture to the Forum for the History of Human Science, History of Science Society annual conference at Montreal, November 2010; at the workshop "Communicating across Cultures: Science, Medicine, and Society," University of South Carolina, November 2010; at the "Knowledge/Value" workshop, Anthropology Department, University of Chicago, June 2011; and at an Imperial College London seminar, October 2011. I thank participants at all presentations for their questions and comments, as well as participants at the Les Treilles workshop on "Culture without Culturalism" (and the Fondation Des Treilles that funded it) and the conveners, Evelyn Fox Keller and Karine Chemla. Thanks in addition to Sabina Leonelli, Claude Rosental, Evelyn Fox Keller, Jonathan Regier, and two reviewers for their written commentaries and to Leverhulme Trust Programme Grant F/04007/Z: "The Nature of Evidence: How Well do Facts Travel?" at the Department of Economic History, London School of Economics, which funded the original research.

1. For the mythology of the term's first usage and Bryant's usage, see Paige Churchman, "The Glass Ceiling: Who Said That?," *Glass Hammer*, April 9, 2009, http://www .theglasshammer.com/. At any rate, the term seems to have been in sufficient circulation for it to be used without explanation in Scott 1985.

2. For example, analysis was drawn from large-scale surveys undertaken by the American Management Association and reported in Sutton and Moore 1985 in the prestigious *Harvard Business Review*.

3. C. Hymowitz and T. S. Schellhardt, "The Glass Ceiling: Why Women Can't Seem to Break the Invisible Barrier that Blocks them from the Top Jobs," *Wall Street Journal*, March 24, 1986, 1D, 4D–5D; quote from 1D.

4. Catalyst, founded in 1962, played a critical role in the 1980s and 1990s in research on glass ceiling issues (statistical work, surveys, and case studies) and in working to change the situation. Their research publications pop up in almost every bibliography and their name in almost every press report.

5. Hymowitz and Schellhardt, "Glass Ceiling," 4D.

6. Hymowitz and Schellhardt, "Glass Ceiling," 5D.

7. This report was undertaken by the Hudson Institute, an offshoot of the secularizing activities of the Rand Corporation (with an address at Herman Kahn Center)—see Johnston and Packer 1987.

8. Legal statutes against discrimination in employment had been part of the Civil Rights Act of 1964 (amended in 1972 and 1978; pay had been the subject of the Equal Pay Act of 1963), but the focus here appeared to be on discrimination at the hiring stage. The executive orders of Lyndon Johnson that covered federal government contractors (on race discrimination in 1965 and sex discrimination in 1967) not only widened the discrimination net to include promotion, training, and so forth, as well as recruitment, but also introduced the "affirmative action requirement," namely, that contractors were required to take action to ensure equity.

9. A follow-up review returned to some of these companies to report on their progress in overcoming the barriers found—see U.S. Department of Labor 1992.

10. *Raleigh News and Observer*, August 25, 1991, quoted in Bullard and Wright 1993, 189.

11. An excellent account of this commission is given by Boyd Childress in the entry "Glass Ceiling" in *Encyclopedia of Business*, 2nd ed., accessed October 19, 2010, www .referenceforbusiness.com/encyclopedia/For-Gol/Glass-Ceiling.html.

12. Such activity was not just at the federal level—there were also state- and city-level glass ceiling initiatives in these years. For examples of their activities, see Johnson 1995 and Chicago Area Partnerships 1996.

13. J. A. Lopez, "Workplace: Study Says Women Face Glass Walls as Well as Ceilings," *Wall Street Journal*, March 3, 1992, B1, and see Catalyst 1992 for the original report.

14. Hymowitz and Schellhardt express this as men who are "quick to feel the woman who is tough isn't being womanly, while the woman who isn't tough isn't worth having around" ("Glass Ceiling," 1D). The term "glass escalators" may have been around earlier but was probably first used in a research paper by Christine Williams 1992.

15. Guy's 1994 account in places conflates the outcomes with the barriers that create those outcomes.

16. In other words, the argument had been between biological or social/cultural determinism; see chapter 4 in the present volume for a discussion of cultural essentialism.

17. See Dobbin 2009 (I am grateful to Claude Rosental for alerting me to this research).

18. I thank Evelyn Fox Keller for a discussion of this point.

19. One report, by the *Christian Science Monitor* (December 8, 2010), referred to work published in *Psychological Science*; the other, published online December 9, 2010, at the *Glass Hammer* (see note 1 for website address), reported work published in the *British Journal for Social Psychology*.

20. I thank Alison Booth for taking me through the works that were prompted by her paper, and the relationships of those authors involved to the set of authors of her paper.

21. On the role of models as knowledge carriers, see Morgan 2011. Morgan 2014 uses an example from industrial economics to make more generic claims about resituating knowledge. See also Mansnerus 2011, who shows how epidemiological models carry facts around public health and research sites.

22. Sorting out the relationships between pay differentials and promotions is a tricky problem both for analysis of the statistical evidence and for mathematical modeling; Booth et al.'s 1998 paper is by no means the only account. It functions here as an important example using the glass ceiling terminology, whereas earlier economic modeling work had been done under the label of employers having "a taste for discrimination."

23. See Spender 1980, 182–190. Ramsay had made direct reference to this argument in her claim to replace metaphors with names.

24. The interdependence of facts and values in social science and humanities research is not, of course, unusual and is explored in several chapters in Howlett and Morgan 2011.

25. So, for example, even in Bryant's first serious investigation of the glass ceiling phenomenon in Bryant et al. 1984 we find a discussion of both Kantor's 1977 "classic" study and research from the Wellesley College Center for Research on Women (another of the most active academic research centers working on this topic), as well as listing Catalyst among the major resource centers. A decade later Catalyst's 1994 report included a twenty-seven-page annotated bibliography of academic, public, and third-sector publications.

26. The reference is to the parallel assumption (held by some economists) that U.S. president Ronald Reagan's "supply-side" tax cuts would create "trickle-down" wealth

for the rest of the community. Naomi Oreskes (2011) picks up the term to suggest most climate scientists may have believed in a similar supply-side effect, that is, their findings on the role of human activity in prompting global warming would trickle down from the scientific experts leaving the way open to others—perhaps with less expertise—to take the arguments and evidence into the public field.

27. There is another sense in which they go together in the case of liberal democracies, where citizens have the right to express the knowledge of their experience; see Morgan 2010.

28. There is a nice irony here that a contemporaneous strand in science studies reinterpreted the detached objects of scientific research to become subjects—ones engaged, through resistance and compliance, in the research project of the scientists under study (as in Michel Callon's 1986 discussion of the scallops of Saint Brieuc Bay).

References

Berheide, C. W. 1992. "Women Still 'Stuck' in Low-Level Jobs." *Women in Public Service: A Bulletin of the Center for Women in Government*, no. 3 (Fall): 1–4.

Booth, A. L., M. Francesconi, and J. Frank. 1998. "Glass Ceilings or Sticky Floors." Discussion paper 1965. London: Centre for Economic Policy Research.

Booth, A., M. Francesconi, and J. Frank. 2003. "A Sticky Floors Model of Promotion, Pay, and Gender." *European Economic Review* 47: 295–322.

Bryant, G., and the editors of *Working Woman*. 1984. *The Working Woman Report: Succeeding in Business in the 80s*. New York: Simon and Schuster.

Bullard, A. M., and D. S. Wright. 1993. "Circumventing the Glass Ceiling: Women Executives in American State Governments." *Public Administration Review* 53(3): 189–202.

Callon, M. 1986. "Some Elements of a Sociology of Translation: Domestication of the Scallops and the Fishermen of St. Brieuc Bay." In *Power, Action and Belief: A New Sociology of Knowledge?*, ed. J. Law, 196–223. London: Routledge.

Catalyst. 1992. *On the Line: Women's Career Advancement*. New York: Catalyst.

Catalyst. 1994. *Cracking the Glass Ceiling: Strategies for Success*. Report commissioned by the Glass Ceiling Commission, U.S. Department of Labor. New York: Catalyst.

Chang, H. 2004. *Inventing Temperature: Measurement and Scientific Progress*. Oxford: Oxford University Press.

Chicago Area Partnerships. 1996. *Pathways and Progress: Best Practices to Shatter the Glass Ceiling*. Chicago: Chicago Area Partnerships.

Childress, Boyd. "Glass Ceiling." *Encyclopedia of Business*. 2nd edn. Accessed October 19, 2010. http://www.referenceforbusiness.com/encyclopedia/For-Gol/Glass-Ceiling .html.

Dobbin, F. 2009. *Inventing Equal Opportunity*. Princeton, NJ: Princeton University Press.

Furner, M. O. 1975. *Advocacy and Objectivity: A Crisis in the Professionalization of American Social Science, 1865–1905*. Lexington: University of Kentucky Press.

Guy, M. E. 1994. "Organizational Architecture: Gender and Women's Careers." *Review of Public Personnel Administration* 14: 77–90.

Hede, A. 1994. "The Glass Ceiling Metaphor: Towards a Theory of Managerial Inequity." *Canberra Bulletin of Public Administration* 76: 79–85.

Howlett, P., and M. S. Morgan, eds. 2011. *How Well Do Facts Travel?* Cambridge: Cambridge University Press.

Jasanoff, Sheila. 2005. *Designs on Nature: Science and Democracy in Europe and the United States.* Princeton, NJ: Princeton University Press.

Johnson, B. K. 1995. "Governor's Task Force Report on the Glass Ceiling," Minnesota Planning.

Johnston, W. B., and A. H. Packer. 1987. *Workforce 2000: Work and Workers for the Twenty-First Century.* Indianapolis: Hudson Institute.

Kantor, R. M. 1977. *Men and Women of the Corporation.* New York: Basic Books.

Laabs, J. J. 1993. "First Person: The Sticky Floor beneath the Glass Ceiling." (Interview with Catherine White Berheide.) *Personnel Journal* 72(5): 35–39.

Limerick, B., and B. Lingard, eds. 1995. *Gender and Changing Educational Management.* Second Yearbook of the Australian Council for Educational Administration. Sydney: Hodder Education.

Mansnerus, E. 2011. "Using Models to Keep Us Healthy: The Productive Journeys of Facts across Public Health Research Networks." In *How Well Do Facts Travel?*, ed. P. Howlett and M. S. Morgan, 376–402. Cambridge: Cambridge University Press.

Morgan, M. S. 2010. "'Voice' and the Facts and Observations of Experience." In *New Methodological Perspectives on Observation and Experiment,* ed. W. J. Gonzales, 51–69. La Coruña: Netbiblio.

Morgan, M. S. 2011. "Travelling Facts." In *How Well Do Facts Travel?*, ed. P. Howlett and M. S. Morgan, 3–39. Cambridge: Cambridge University Press.

Morgan, M. S. 2014. "Resituating Knowledge: Generic Strategies and Case Studies." *Philosophy of Science* 81: 1012–1024.

Morrison, A. M., R. P. White, E. Van Velsor, and Center for Creative Leadership. 1987. *Breaking the Glass Ceiling: Can Women Reach the Top of America's Largest Corporations?* Reading, MA: Addison Wesley.

Morrison, A. M., R. P. White, E. Van Velsor, and Center for Creative Leadership. 1992. *Breaking the Glass Ceiling: Can Women Reach the Top of America's Largest Corporations?* 2nd ed. Reading, MA: Addison Wesley.

Oppenheim, L. 1991. "Women and the Glass Ceiling." Selected References, Industrial Relations Section, Princeton University, January, no. 255. dataspace.princeton.edu/jspui/bitstream/88435/dsp01df65v7872/1/255.pdf.

Oreskes, N. 2011. "My Facts Are Better than Your Facts: Spreading Good News about Global Warming." In *How Well Do Facts Travel?*, ed. P. Howlett and M. S. Morgan, 136–166. Cambridge: Cambridge University Press.

Porter, P. 1995. "The Need for a Spring Clean: Gendered Educational Organisations and Their Glass Ceilings, Glass Walls, Sticky Floors, Sticky Cobwebs and Slippery Poles." In *Gender and Changing Educational Management,* ed. B. Limerick and B. Lingard, 234–243. Sydney: Hodder Education.

Ramsay, E. 1995. "Management, Gender and Language: Who Is Hiding behind the Glass Ceiling and Why Can't We See Them?" In *Gender and Changing Educational Management*, ed. B. Limerick and B. Lingard, 174–183. Sydney: Hodder Education.

Scott, A. F. 1985. Book review of *Mabel Walker Willebrandt: A Study of Power, Loyalty, and Law*, by Dorothy M. Brown. *Western Historical Quarterly* 16(2): 207.

Shapin, S., and S. Schaffer. 1985. *Leviathan and the Air-Pump: Hobbes, Boyle, and the Experimental Life*. Princeton, NJ: Princeton University Press.

Spender, D. 1980. *Man Made Language*. London: Routledge and Kegan Paul.

Star, S. L., and J. R. Griesemer. 1989. "Institutional Ecology, 'Translations' and Boundary Objects: Amateurs and Professionals in Berkeley's Museum of Vertebrate Zoology, 1907–1939." *Social Studies of Science* 19(3): 387–420.

Still, L. V. 1995. "Women in Management: Glass Ceilings or Slippery Poles." In *Gender and Changing Educational Management*, ed. B. Limerick and B. Lingard, 106–120. Sydney: Hodder Education.

Still, L. V. 1997. *Glass Ceilings and Sticky Floors: Barriers to the Careers of Women in the Australian Finance Industry* (A report prepared for the Human Rights and Equal Opportunity Commission and Westpac). Commonwealth of Australia.

Sutton, C. D., and K. K. Moore. 1985. "Probing Opinions: Executive Women—20 Years Later." *Harvard Business Review* 63(5): 42–66.

U.S. Department of Labor. 1991. *A Report on the Glass Ceiling Initiative*. Washington, DC: U.S. Government Printing Office.

U.S. Department of Labor. 1992. *Pipelines of Progress: A Status Report on the Glass Ceiling*. Washington, DC: U.S. Government Printing Office.

U.S. Department of Labor. 1995a. *Good for Business: Making Full Use of the Nation's Human Capital, The Environmental Scan*. Washington: U.S. Govt. Printing Office.

U.S. Department of Labor. 1995b. *A Solid Investment: Making Full Use of the Nation's Human Capital, Recommendations of the Glass Ceiling Commission*. Washington: U.S. Govt. Printing Office.

Williams, C. L. 1992. "The Glass Escalator: Hidden Advantages for Men in the 'Female' Professions." *Social Problems* 39(3): 253–267.

Wynne, B. 1991. "Knowledges in Context." *Science, Technology and Human Values* 16: 111–121.

Wynne, B. 1992. "Misunderstood Misunderstanding: Social Identities and Public Uptake of Science." *Public Understanding of Science* 1: 281–304.

CLAUDE ROSENTAL

7

Modes of Exchange
The Cultures and Politics of Public Demonstrations

To manage its research and development programs and to help define and implement European policies and politics, the European Commission (EC) has used various forms of public demonstrations in recent years. These include public demonstrations of technology that may be compared in some ways with Bill Gates's famous software demos.[1]

Here, I analyze the ins and outs of this peculiar phenomenon and the extent to which it can be described in terms of the politics and culture of public demonstrations. My argument is mainly based on sociological observations I have conducted on the running of a large European research and development (R&D) program sponsored by the EC, called Advanced Communications Technologies and Services (ACTS).

Starting from the analysis of this social process, I reflect on the social uses of public demonstrations, with the aim to contribute to the development of a systematic framework of analysis for such phenomena (see also Rosental 2004, 2005, 2013). Although the notion of public demonstration should sound familiar to most readers, the contours and stakes of my project here may not seem obvious at first sight. Indeed, the terms "demonstration" and "public demonstration" (i.e., demonstration conducted in public) are used in many social spaces. But the connections between the practices they refer to are not self-evident. Broadly speaking, it seems that "demonstration" implies an audiovisual development whose main intended or declared purposes are proving, convincing, or teaching, although its actual roles may be more diverse. For example, "demonstration" and "public demonstration" are commonly used to refer to experimental proofs or specific parts of physics lectures in the academic world, to performances of marketing presenters, and to street protests. It is thus difficult to think of all demonstrative practices as belonging to one and the same field of inquiry.

Besides, these practices are often perceived as isolated or anecdotal events of social life, not worth extensive theorizing. The stakes of these practices are more visible only on specific occasions. Such was the case in the PowerPoint demonstration Colin Powell gave at the United Nations on February 5, 2003, in support of a war against Iraq. This also applies to demonstrations of certain home products on television that have a major impact on sales. Public demonstrations have, in fact, major stakes in many domains, including economic life (e.g., as sales practices or tools for product design and launching), politics (e.g., as collective mobilizations or performances designed to test or persuade a large audience), and science and technology (as public proofs or teaching devices).

Here, I reveal some of the sociological, anthropological, and political stakes of demonstrative practices in general and highlight how these practices may be considered as part of a common domain of investigation despite their apparent diversity. Social science studies of demonstrative practices are often disconnected. A number of works, both in the history of science and technology and in scattered publications in sociology, anthropology, and other social sciences, have explored different aspects of these phenomena in more or less depth. Several authors have illustrated the roles of public demonstrations as persuasion tools and rhetorical devices in various settings (see, e.g., Latour 1983; Bloomfield and Vurdubakis 2002; Stark and Paravel 2008). The nature of and epistemological debates on public demonstrations of technology have also been studied by a number of authors. In particular, the ways these demonstrations have been likened or opposed to geometrical proofs, to experiments, to lectures, or to displays of virtuosity, as well as their uses as spectacles and entertainment, have been documented across history.[2]

Several studies have focused on whether and to what extent public demonstrations are or should be perceived as fiction or reality. Some authors have portrayed demonstrations as pure illusion or mutually agreed-upon fiction (see Wagner and Capucciati 1996; Lunenfeld 2000, 13–26) or as technological dramas that disable critical faculties (see Lampel 2001). Comparable to Gabriel Tarde's (1903) portrayal of society as being composed of insane hypnotists followed by sleepwalkers, certain public demonstrations have been described as performances of hypnotists influencing crowds (see Duval 1981). Other studies have analyzed public demonstrations as multiply framed experience combining fabrication and reality (see, e.g., Smith 2009). In the framework of these approaches, audiences may have multiple or fluctuating experiences and may be at least partly aware of the fiction taking place in front of their eyes. In particular, using a dramaturgical metaphor, Erving Goffman has depicted demonstrations as performances, with performers playing teaching or evidential roles.[3]

For the present day, scattered publications in the social sciences give an idea of some of the uses of public demonstrations in activities and domains of social life beyond the scientific field.[4]

However, the material features, grammar, and structure of public demonstrations appear to have attracted more attention than the interactions between demonstrators and their public or the actual effects of demonstrations on these audiences. Most studies look at demonstrations of final products, and far fewer at the uses of demos in the development of projects. More generally, it seems that many creative although barely visible uses of public demonstrations remain to be explored. The anthropological, sociological, and political stakes of these practices clearly call for more systematic analyses. My aim here is precisely to contribute to such an effort.

The ACTS program was managed by the Directorate General for Telecommunications, Information Market and Exploitation of Research (DG XIII) of the EC during 1994–1998; it was followed by other European programs: Information Society Technologies (1998–2006) and Information and Communication Technologies (2007–2013). Participants in the ACTS program included researchers, engineers, and executives from various European countries, many of whom were working for telecommunication and computing firms.

My investigations of ACTS activities drew on several sources and combined different methods. By the end of the 1990s, I had conducted a series of interviews among ACTS participants in Europe, made some ethnographical observations of a large ACTS meeting in Brussels, and collected various types of textual and multimedia documents. These include a series of ACTS and independent reports and CD-ROMs and brochures produced by ACTS participants and the ACTS program. The list of documents also includes electronic presentations of ACTS projects published on various European online databases, European newsletters and publications, newspaper articles, technical publications of ACTS participants, electronic exchanges of ACTS participants in a specialized forum, and video clips of public demonstrations of technology.

Demonstrating in a Competitive Context

One of the main purposes of ACTS was to help develop a very high-speed communication network in Europe in the image of the Internet 2 project in the United States. More specifically, ACTS was intended to contribute to the development of a physical network (named JAMES), of multimedia applications, and of telework experiments within major industrial firms and participating European institutions.

Officials in ACTS also had to demonstrate the achievements of their program to political and economic authorities and to the public.[5] Indeed, they routinely had to face questions and criticisms about the management of their colossal budget from both European Parliament members and various industrial lobbies.[6]

The process that led to the dismissal of the EC in 1999 illustrates the pressure that the European Parliament put on EC officials.[7] The dismissal was caused by a report charging the EC with fraud cases, bad management, and nepotism. This report was written at the request of the European Parliament (see MacMullen 1999; Georgakakis 2000; Meyer 2001). Officials of the EC also had to face contradictory demands of various lobbies such as the European Telecommunications Network Operators' Association (ETNO), the telecommunication operators' lobby.[8] As the plans for the EC Fifth Framework Programme were being finalized (1998–2002), ETNO was publicly criticizing European Community funding of short-term commercial projects focusing on information society applications such as telemedicine, tele-education, electronic commerce, and multimedia content applications (see Chappaz 1997). According to ETNO, such projects were already developed and tested outside different EC R&D programs. They duplicated private sector work and represented a waste of public European money. This position was said to be supported by an "independent" report headed by Etienne Davignon, a former EC vice president for research and industry. ETNO was therefore lobbying heavily to channel the funds of the Fifth Framework Programme toward telecommunication infrastructure R&D, arguing that a modern telecommunication infrastructure needed to be implemented before developing applications.

Other lobbies expressed different viewpoints. For example, Richard Sitruk, director of European Telecommunications Information Services in Brussels, argued that funding application projects would stimulate the development of telecommunication infrastructure. ACTS officials then had to show that they adopted a relevant and balanced position between the funding of telecommunication operators and the subsidies of smaller firms developing multimedia applications,[9] and that each competing group and view was supported by the program in an effective way. Showing the productivity of each aspect of the program helped to counter the claims of groups asking for a larger share of the subsidies on the basis of the supposed uselessness of competing projects.

To manage criticisms and forge a consensus, ACTS managers had to demonstrate the achievements of their program. The emphasis put on demonstration as well as on research and technology development (RTD) was explicit in the title of an ACTS call for proposals: "Third Call for Proposals for RTD Actions for the Specific Program for Research and Technology Development, including Demonstration, in the Field of Advanced Communications Technologies and Services."

The term "demonstrate" had several meanings. One of its main meanings, according to the context of competition I have just described, was to exhibit technological accomplishments to convince audiences, or provide a proof, of the feasibility of technical projects. Officials at ACTS were careful to show results that could appear tangible and convincing both to economic and political authorities (e.g., European Parliament members) and to the public. To this end, they set up a material economy of visibility of possibly convincing results that requires analysis.

A Large Set of Demonstrative Tools

Several devices were used in the effort to show convincing results in the framework of the ACTS program. One of them consisted in the regular production and distribution of summary reports in Brussels and beyond. Such summary reports displayed many statistics on the projects and used synoptic devices such as tables and figures.[10] They were more comprehensible for bureaucrats, industrialists, and political authorities than many technical documents produced by ACTS projects. Besides, ACTS funded a large set of projects—around 150—that produced masses of technical publications. These publi-

TABLE 7.1. Benefits of the Advanced Communications Technologies and Services Program as Perceived by Program Participants

Perceived Benefit	No. Projects	% Respondents
Improved corporate image	62	45
Increased number of R&D employees	61	44
New business or research areas	55	40
Improved scientific reputation	52	37
Improved scientific performance	47	34
Increased contract research	43	31
Increased number of technical employees	36	26

Numbers are based on answers to a closed questionnaire given to participants; the table reports only answers that were chosen by more than a quarter of ACTS participants.

Source: Reprinted from ACTS, "Results, Impact and Exploitation," interim report, DGXIII/B Ref: AC 1997/1339, May 15, 1997, 20.

cations could not all be brought together on readers' desks, whereas ACTS summary reports could circulate quite easily.

Summary reports from ACTS contained detailed lists of, or figures on, registered patents, contributions to standards, and experiments and publications produced by or attributed to the ACTS program. They often included short presentations of ACTS projects. They also displayed statistics summarizing the results of surveys conducted among ACTS participants on the basis of questionnaires. These statistics showed the structure of experiments and technological applications designed by ACTS participants, as well as projects' goals and the program's "benefits" for the participants. For instance, table 7.1 shows how the program's benefits for the participants were presented in one of the program's reports.

Tables such as this one possess notable assets to convince managers, members of the administration and parliament, and various politicians of the productivity of the program. The categories that were used here to assess the impact of the program fitted the language of business people concerned with marketing issues ("corporate image," "reputation"), employment ("number of employees"), market development ("new business areas," "contract research"), and performance in general ("scientific performance"). Unlike many scientific and technological descriptions of results, this vocabulary was also understandable to political authorities and nonspecialist audiences.

Officials at ACTS relied also on other devices to demonstrate the productivity of the program, as well as to facilitate what they called the "dissemination of information." Electronic databases were built to display information on the projects on the Internet—especially well-prepared abstracts. Journalists were hired to display exciting results to a large audience, especially in the framework of European publications.[11] Success stories of ACTS projects were also conveyed using different means of communication, including CD-ROMs.

For instance, a CD-ROM titled "ACTS Multimedia Success Stories" (Analysys Ltd. 1996) used a specific format to present selected ACTS projects as success stories. It described the nature of each "product," the aims and objectives of the project, the parties involved, the technological platform, the learning process, and the exploitation and success factors. Success was analyzed in four ways: internal factors, external factors, evidence of success, and reasons for success. For example, here are the evidence of and reasons for success invoked in the CD-ROM for a project named Mira-III Teleradiology, a multimedia conferencing system that supports the viewing and manipulation of medical image data over networks:

> *Evidence of success:* The market is showing signs of interest. In October 1995, Telenor exhibited Mira-III at Telecom 95 in Geneva. This attracted about 120 serious applications from potential customers, although Telenor was not then in a position to sell the system because it was undergoing an internal reorganization. Current international contacts include a potential client in Ireland, a dealership in Australia, and interest from Middle Eastern countries.
>
> *Reasons for success:* Mira-III is a good product, supported by scientific presentations at conferences visited by radiologists. It is important to demonstrate knowledge of the technology in a professional field.

This description illustrates ACTS officials' concern about producing "accessible" texts and, in particular, sufficiently short, well-calibrated documents. These documents were accompanied by images and videos to make the readers' task even easier. Using a common presentation format for all projects also helped suggest that ACTS managed all funded projects in a harmonious fashion. Besides, this presentation highlights how presentations at conferences and demos were seen as generators of "success."

Demos

Running demos represented the most important way to demonstrate the achievements of the program. What does the term "demo" refer to in general (see Rosental 2007)? *Demo* is an abbreviation of "demonstration" while in fact referring to one specific form of demonstration; "demonstration," by contrast, remains a generic term. A demo exhibits a technological device in action, such as some computer software. The exhibition frequently occurs in front of a selected audience, following a carefully elaborated scenario. A demonstrator may comment on the running of the technical device, perhaps linking its operation to general properties of a theory or methodology. Demos are commonly used by researchers, engineers, executives, sales representatives, and consultants in various fields to demonstrate the feasibility of a technological approach, the value of a specific theory, or the proper functioning of a prototype or product. The audience may include a mix of academics and representatives of economic and political power.

Generally, a "repertoire" (or stabilized narrative) is prepared in advance, prior to being deployed in the demo. This exercise is scripted in the sense that a scenario or script is used to organize the action but is not usually expressed in writing, or even orally, as is done in the movie industry (see Grimaud 2005).[12]

The preparation commonly takes a long time for demonstrators, who are concerned with anticipating objections, doubts, and questions, and more generally controlling the possible interpretations and meanings associated with the presentation. Demonstrators may be also anxious to avoid computer crashes and to distract attention from possible technical limitations, as well as to insist on achievements.

When demonstrators are present during a demo, they tend to make themselves a representative of the system (sometimes even a sales representative). Generally, a whole setting is created. Extreme, spectacular displays of the working of the device may be mounted to impress the audience. Members of the audience may be invited to exchange views or manipulate the device once demonstrators have finished their personal performance. The outcome of the demo then depends very much on the demonstrator's ability to control the interaction. If this outcome is favorable, the positive impact of the demo can then be extended, as the witnesses can vouch for the reality of the achievements to a wider circle of actors.

The exhibition can be performed live, but it may also be recorded and made into a video.[13] The audience and the demonstrator usually do not appear on the video, although the voice of the demonstrator may be retained. A video of the demo obviates the need to transport cumbersome, often fragile mechanisms when making presentations to sponsors or customers. It also allows the demonstrator to avoid the risk of failure involved in random replication of real-time demos.

Demos for ACTS consisted in particular in showing the workings and use-fulness of multimedia applications and high-speed exchanges of information that facilitate various forms of telework. These demos gathered executives and managers of telecommunication and computing firms, engineers, re-searchers, EC senior officials, representatives of lobbying organizations, jour-nalists, and politicians of various European countries. Examples of demos for ACTS projects may be viewed on the Internet. This applies to Mira-III, mentioned above, and Isabel, a project focused on the development of tools for electronic meetings.[14] Indeed, some ACTS demos include the actual running of teleconferences.

A teleconference called "Twenty-First Century: The Communications Age, a Conference on the Future of Advanced Communications" illustrates how such demos work (see Geiger 1997). Organized in Brussels on June 18, 1997, it was intended as a showcase for a number of projects funded by ACTS, including the high-capacity network prototype JAMES. In particular, it was designed to show how ACTS accomplishments had made a high-resolution world teleconference possible. It was well covered by the media.

The conference gathered participants from various European countries, as well as from Japan and Canada. The list of speakers included the president of the European Parliament's Committee on Research, Technological Develop-ment and Energy, the Portuguese minister of science and technology, the co-founder and CEO of Netscape Communications, the director general of DG XIII, and distinguished members of the Ministry of Posts and Telecommu-nications in Japan, the European Broadcasting Union, and BT.[15] Telepresen-tations discussed policy and technological aspects of communications devel-opment and were intended to influence the development of future European Community policies and especially the preparation of the Fifth Framework Programme's activities in the field of communication technologies.

Demos of multimedia projects were thus combined with the exhibition of presentations, high-status presenters, and high-quality images. Organizing a teleconference involving economic actors and political authorities was a powerful way for ACTS officials to demonstrate the projects' results to actors concerned with public spending policies. These actors did not need to assess ACTS results on the basis of experts' advice alone or lengthy and weighty technical reports and papers, or even their own reading skills (an especially important asset for audiences that prefer moving pictures to texts).

The limited time needed to attend demos also offered a unique opportu-nity for busy economic and political authorities to grasp—or to believe they grasped—submitted projects. Such confidence is a precious asset in a world dependent on evaluation on the basis of limited time and know-how (see Rosental 2008, 2010; Lamont 2009). Demos for ACTS projects were thus crucial tools for various authorities both to assess the program and its projects and to make decisions on their future.[16] Altogether, they built a large "demonstration of strength" of the ACTS program (see Mukerji 2009). Indeed, demos were not used simply to highlight the reliability of the technologies under development and of the participants. Their number also contributed to demonstrating the ACTS program's productive power and its great capacity as a collective enterprise to create the conditions for technical progress.

Demos on Clips and on Lists

Officials at ACTS also used video clips of selected demos to produce CD-ROMS advertising the program's achievements. Developed by researchers in communication sciences funded by ACTS to "disseminate" the program's results (see also Brine 2000),[17] these CDS were distributed to ACTS participants, industrialists, and political authorities. This represented a major way of creating wide access to ACTS demos for multiple audiences, spreading the visibility of the program, and advertising its usefulness altogether.

Representatives of ACTS were eager to set up a mass production and distribution of demos. From the start of the program, they asked participants to run regular demos. Some of them were planned according to a four-year schedule. Demos on CDS were part of this grand scheme of production and distribution of demos.

The program also produced large numbers of reports on demos. For instance, some ACTS reports listed demos of ACTS projects and systematically described when, where, and how they had been run. These reports enlarged the audience potentially reached by demos. Indeed, their circulation made the demos visible to actors who never left their office in Brussels.

For instance, one of the interim ACTS reports, "Results, Impact and Exploitation,"[18] offers a detailed table of "Public Demonstrations" carried out by ACTS projects over a specific time period. Each demonstration is presented in sequence, and three columns provide information under the following headings: "Date" of the public demonstration, "Nature of Demonstration," "Target Audience and Reaction." For example, some of the public demonstrations carried out by the "Manipulation of Images in Real-Time for the Creation of Artificially Generated Environments" (MIRAGE) project are reported as shown in table 7.2, which underscores both the breadth and diversity of the audiences of ACTS demos and the presence of the media. The table shows

TABLE 7.2. Summary of Public Demonstrations Carried Out by the MIRAGE Project, as Presented in the Interim Report

Date	Nature of Demonstration	Target Audience and Reaction
02.09.96	Showing of "Eye to Eye" on dual display and sequential TV receiver to national press and technical journals.	Everybody most impressed with the demonstration. Result was a number of published articles.
12–16.09.96	Showing of "Eye to Eye," demo videos of virtual production, live demo of Virtual Edit Suite and character animation.	Most considered this the best 3D TV demo they had seen. Much interest in the virtual studio systems and VES.
18–21.09.96	Showing of "Eye to Eye" on large-screen projection and sequential TV. Videos of mirage virtual production and virtual characters.	Politicians, academics, broadcasters, and manufacturers were highly impressed by the standards achieved and the practical systems demonstrated.

Source: Reprinted from ACTS, "Results, Impact and Exploitation," interim report, DGXIII/B Ref: AC 1997/1339, May 15, 1997, 40.

how frequently demos could be run—three times in one month in this case—and the spectacular settings the demonstrators sometimes created. It also illustrates how reactions were systematically screened and depicted so as to emphasize audience enthusiasm. Altogether, a table like this one reflects, but also contributes to, the demo "industry" that ACTS managers were helping set up.

Demos at the Crossroad between Coordination and Competition Dynamics

As indicated above, the advance of ACTS projects was structured by periodical staging of demonstrations. In fact, ACTS participants often took advantage of the program's requirement to run demos on a regular basis according to their own diverse goals. Mounting demos allowed them to consolidate or create social links, for example, by stimulating interest in their project among new actors, thereby helping to generate new contracts or new partnerships. Demos helped participants not only to promote interest in their projects but also to sustain confidence in their work. This, in turn, helped to justify the funding of their projects to administrative and political authorities, company managers, and the public.

Some demo versions allowed research engineers to show off their work to advantage within their firm. They also helped academics to gain credit in the eyes of their peers or to find new industrial partners in various arenas. Different versions of demos could be combined and reused in other frameworks and for other occasions. When planned to run in a number of strategic spaces, they became part of demonstrative campaigns. They were business as usual for ACTS participants, so much so that they became virtually de rigueur.

Running demos contributed to defining a project's content in a dialectical way. Indeed, engineers and researchers generally took seriously any criticism and suggestion expressed by the audience during a demo and adjusted the orientations and reorientations of the projects accordingly. Systematic observation of audiences' reactions could even be used as tools for project management in defining the content of the research.

Demos helped ACTS participants coordinate actions with their audiences—peers, partners, and customers—and with other participants. Indeed, the latter had to display collaborative work in the framework of the program in order to benefit from EC funding. Demos by ACTS participants were a perfect tool to exhibit, if not simulate, common achievements in this framework. In the competition among European telecommunication operators unleashed by the end of national monopolies, ACTS participants were often in a complicated position regarding collaborating with one another. Because they often had competing approaches and interests, preparing common demos represented a least common denominator for them.

At meetings in Brussels, representatives of computing and telecommunication firms were concerned about what could be said and shown and what should not be revealed. They were often asked by their superiors to conceal certain aspects of their work under cover of technological black boxes. Demonstrators were negotiating these borderlines during demo interactions. Gaining information was generally more rewarding for the demonstrators than a well-kept secret, especially as it was generally difficult to trace the sources of information leaks. As a result, gifts and countergifts of information were very much in play. In such tense situations, demos might more closely resemble exchange tools than proof procedures.

Demos were also a point of reference in guiding decisions concerning the program's reorientations. As indicated above, ACTS gathered together actors with different (even conflicting) interests, who were supported by a variety of more or less influential lobbyists. In particular, EU officials had to arbitrate on European subsidies between major telecommunication operators developing a physical network and small computing businesses developing software.[19] Demos were used as benchmarks in the corresponding negotiations and therefore to consolidate or redefine the legitimacy of groups in the program.

Public Demonstrations as a Bridge between Science, Technology, and Society

Altogether, demos played a central role in establishing and structuring relationships between a large number of actors. They structured the work of participants (especially when they were used as observatories of audience reaction, tools for project management, and exchange apparatus). They also structured the redistribution of credit allocated to individuals, teams, and institutions, as well as to scientific and technological objects. Their impact was enhanced by a host of peripheral tools, such as written and oral reports, brochures, and CD-ROMS that presented success stories. Demos were the flagship of a fleet of demonstrative devices.

Consequently, various strategies permeated the use of demos and various other devices. The setting up and running of demos suited the complementary interests of several types of actors—scholars, engineers, firm executives, managers, politicians, journalists, administrative officers—and constituted a rare opportunity for interaction, competition, coordination of action, and building of partnerships.[20] Those actors would probably have never met without demos bringing them together.[21] The regulation of their exchanges was marked by recourse to spectacular demonstrations somewhat analogous to those that attracted the presence of scholars, entrepreneurs, and representatives of political and religious powers in France and in England in the seventeenth century—although without the conventional courtesies that accompanied the latter (see Shapin and Schaffer 1985).[22] At a global level, demos served as a privileged bridge between science, technology, and society.

Public Demonstrations as a Constitutional Topic

The process I have described allows us to understand why demonstration activities were at the heart of the chapter devoted to science and technology in an unratified European constitution project.[23] In other words, it explains how demonstrations have become a constitutional topic in Europe. Indeed, the European constitution project indicates that "the Union shall carry out the following activities, complementing the activities carried out in the Member States: (a) implementation of research, technological development and demonstration programs, by promoting cooperation with and between undertakings, research centers and universities; (b) promotion of cooperation in the field of the Union's research, technological development and demonstration with third countries and international organizations; (c) dissemination and optimization of the results of activities in the Union's research, technological development and demonstration."[24] This statement illustrates how the elaborate know-how in managing European R&D programs like ACTS based on the use of public demonstrations has helped shape the details of a political project at the pan-European level. It helps in understanding how European politics and policies of science and technology have been defined in management terms in the twenty-first century and how demonstration activities have become part of the toolbox of European public management.[25] Surprising as it may seem, demos have become key tools for European construction.

The Many Roles of Demonstrations

The phenomena I have analyzed raise a major sociological issue. Should they be described in terms of the politics and culture of public demonstrations, whether at the level of the EC or at a larger level? To address this issue, we need first to ask what is meant by "demonstration," "politics," and "culture."

Various meanings have been attached to the term "demonstration" across different sociohistorical spaces since antiquity.[26] Generally speaking, it seems that a "demonstration" refers to an audiovisual development whose main intended or declared purposes are to prove, convince, or teach, although its actual roles may be more diverse. "Audiovisual" may refer to writings, such as the written proof of a mathematical theorem. It may also refer to live or videotaped demonstrations of technology. The intended or declared purposes may be exclusively to prove, convince, or teach or may mix several of these goals. As in the case under study, demonstrations may simultaneously play less overt roles, and be used, for example, as opportunities to make contact with other actors, as observatories of audience reaction, as ways to collect new ideas and build partnerships, as tools for project management, or as transactional devices.[27] Such roles should not simply be conceived in epistemological terms, because they may be of an anthropological, economic, and political nature and a common object of inquiry for social scientists interested in knowledge production, markets, organizations, and politics.

We have been accustomed since antiquity to view demonstrations in the limited terms of proof and persuasion, or *apodeixis* and *epideixis*.[28] However, the ontology of demonstrations cannot be reduced to proof and persuasion devices in general. Exploring the dimensions of demonstrations as spectacles is certainly valuable, but insufficient. In particular, demonstrations cannot be understood a priori as one-way communication forms in general. As in the case under study, public demonstrations of technology may be used by researchers and engineers to communicate but also to gain information from members of an audience. The information gained may be used systematically by the demonstrators to shape the technologies and the corresponding theories periodically, to coordinate their action with others, to compete with them, to attempt to take control of the possible sponsors and customers, to coconstruct the technologies and the users (see Woolgar 1991; Oudshoorn and Pinch 2005), and to create a market around the emerging objects. In addition, demos cannot be reduced to tools used to sell science and technology. Demos are not only run once scientific and technological contents are stabilized, in order to sell them, but may also be used at many stages to define the projects' contents in a dialectical way. Therefore, demos may be a key element in the processes that bind the making and marketing of science and technology.

In the case of ACTS, reports, CDs, and live demos produced by the program were first of all intended to convince audiences of the program's productivity, of the feasibility of different technical approaches, and of the soundness of various claims. In some cases, they were intended to deliver a pedagogical message or serve as proofs for specific statements. Facts, figures, lists, arguments, success stories, video clips, and onsite demos combined toward these goals. But they played other roles as well: they shaped interactions and helped manage an anthropological, economic, and political order. We may therefore compare demonstrations to Marcel Mauss's (1954) total social facts and be sensitive to their effect on transactions, material and symbolic goods, and fate of groups. Preparing and running public demonstrations may mobilize or generate as many exchanges, resources, tensions, (re)distributions of alliances, and intense moments of social life as, for example, the preparation and celebration of another grand anthropological event in many societies, the wedding.

Demonstrations also seem to be at the heart of a system one could call "scientific capitalism." I use this expression to denote the following dynamics. Demonstrations are used by demonstrators to obtain symbolic credit and various material resources. Demonstrators invest these resources to produce more demonstrations. In turn, these demonstrations are used by demonstrators to obtain further symbolic credit and resources, and so on. Demonstrations appear to play in scientific capitalism the role that commodities play in the Marxian theory of capital.[29]

Certainly, demonstrations may be perceived as "disinterested" by some of their producers. But many postures, such as "disinterestedness," have unintended consequences, including the creation of a large economic and political system like capitalism (see Weber 2004, 2009; Shapin 2008). Thus, it seems relevant here to talk not only about demonstration but also about the politics of demonstrations.

Demo-cracy

The EC appears to have developed a politics of public demonstrations, if we use "politics" to refer to the regulation of public affairs within a given space, for example, Aristotle's affairs of the city (see Aristotle 1962). Indeed, EC representatives have employed public demonstrations as tools to regulate European Community affairs. They have used these demonstrations as methods and tactics to define and implement R&D policies. They have placed them at the heart of the art and science of running European affairs and of making and enacting collective decisions in the field of science and technology.

Such phenomena are not unique. Indeed, various studies indicate that politics of public demonstrations has developed in other spaces and on various scales (see Jasanoff 2005). For instance, NASA has deployed politics of demos at different organizational levels and used this to manage the relationships between the space agency and the public (see Rosental 2002). As public demonstrations appear to be widely employed in different units of social life, demo-cracies—regimes that use public demonstrations for the management of public affairs—seem to have developed on a large scale, and a large demo-cracy may have developed in the industrial world (see also Rosental 2013). Demos may be no less, or even more, important for collective mobilizations than mass media and street protests of social movements, especially as they can be widely seen via electronic networks. Public demonstrations are visibly important sources of contests and deliberation in the contemporary period (see also Callon 2003; Stark and Paravel 2008). In particular, antagonistic demos seem to play a large role in the competition for resources and in the political game.

According to Alexis de Tocqueville (1981, 55), "the world is not led by long and learned demonstrations." This claim seems to apply to demo-cracies, and especially to European demo-cracy. European Union affairs are not led by long and learned demonstrations but, to a large extent, by short and well-calibrated public demonstrations and, more particularly, by demos. Demo-cracy does not appear to be identical to democracy. In the EU case, demo-cracy visibly benefits the masses, if only in providing them with specific forms of access to the closed world of laboratories and their technological production. But it gives less power to the ancient *demos*, or people, as in an ideal democracy, than to skillful demonstrators and their institutions.

This is not to say that public demonstrations, and demos in particular, should be seen merely as all-powerful devices and tools allowing certain people to manipulate or mystify the masses (see Rosental 2009). Certainly, the resources of demonstrators and those of their audiences are often not equal, especially in terms of expertise (see also Collins 1988). But public demonstrations may fail to produce intended effects in some cases. They may be subject to variable interpretations, or be credited with different meanings, and produce mitigated and heterogeneous reactions (see also Rosental 2007). A given demo may even be judged to have "failed" by some members of an audience and to have "succeeded" by other members of the same audience. Besides, audiences are not a priori composed of credulous victims or enthusiastic idiots: spectators may remain skeptical and keep their critical sense in relation to a demonstration. Also, as in the EC case, demonstrations may be confronted with counterdemonstrations.

Moreover, public demonstrations represent opportunities for various forms of participation, intervention, and mobilization for actors who would be more removed from the management of public affairs otherwise. Such was the case for many ACTS participants. Public demonstrations help create spaces for politics aside from the main loci of political decisions, such as in the case of ACTS teleconferences (see also Barry 2001; Girard and Stark 2007; Lemieux 2008). Finally, there is no irrevocably fixed, stable divide between a group of demonstrators and a mass of nondemonstrators: demonstrators can take turns to a certain extent.

A Culture of Public Demonstrations

The case under study highlights also how a culture of public demonstrations has developed at the level of the EC. By this I mean that certain behavior has been cultivated, based on the accumulated and socially transmitted experience of individuals and groups in a particular space. More precisely, the use of public demonstrations for the management of R&D programs corresponds to a know-how that has been progressively tested, circulated, and cultivated within the EC to the point of being finally inscribed in a European constitution project.

As a matter of fact, EC administrative officers as well as successive and recurring participants to R&D programs have contributed to the development of specific ways of organizing and carrying out the production of public demonstrations. The EC not only has encouraged the running of public demonstrations in the framework of its R&D programs but also has positively promoted some practices over others, introducing distinctive formats and norms regarding the production of demonstrations. Demos have been privileged, while other forms of demonstration deployed in reports and multimedia documents have taken specific shapes.

In this way, the EC has contributed to creating a relatively homogeneous and distinct environment, rich in the periodical performance of spectacular demonstrations of technology, that creates "ruptures in any uniformities of practice" in the world.[30] It has shaped a space that is characterized by specific constraints and possibilities in terms of worldviews, communication, and action. In the framework of the ACTS program, demos were not only de rigueur; as we have seen, they represented the principal means for actors to depict and view technological achievements, coordinate action, and compete with one another. This environment contributed to defining what was or was not possible for the participating individuals.[31] In it, individuals were prepared for particular (and exclusive) ways of seeing and acting.[32]

This whole set of practices, values, and norms, corresponding to particular ways of communicating and behaving, defines by itself a distinct culture. It seems important to note that this culture cannot be described solely as "scientific." Indeed, in the case under study, both scientists and nonscientists—researchers, engineers, executives, political authorities, and EC administrative officers in particular—participated in and shaped a common environment. Besides, some actors encountered in this milieu had such rich professional trajectories and identities that it is particularly difficult to label them pure scientists or nonscientists. Using a general notion of culture as opposed to scientific culture helps draw our attention to the ways in which specific practices have migrated, mutated, and articulated with one another at the level of the EC. (I refer here to demos that are performed not only by scholars but also by consultants and other actors.)

Finally, this study suggests that a culture of demos has developed on a large scale. Demos are used not only by participants in European R&D programs: investigations I have conducted in Silicon Valley, at NASA, and in the field of artificial intelligence converge with various other studies in showing that demos are used on a large scale by engineers, scientists, consultants, marketing, and sales executives in the industrial world,[33] and that many similarities may be observed in terms of practices, norms, and values associated with their running. For example, some U.S. engineers give more than hundred demos a year. Nicholas Negroponte, who created the Media Lab at MIT, rewrote the old adage "publish or perish" as "demo or die," and Guy Kawasaki's "official" title at Apple Computers was "software evangelist."[34]

A number of artificial intelligence researchers I have encountered in Silicon Valley also value demos as ways to get in touch with peers and industrial partners, as opportunities to get feedback on their projects, or as tools for project management. They are no less constrained in using this form of demonstration in their working environment than were ACTS participants. Their comments on the working of technologies conform to narrative standards that can be compared with those developed by ACTS participants. As they place demos at the heart of their activity, they also appear closer to ACTS participants, in terms of professional practices and identities, than to other scientists who produce essentially written proofs, such as mathematicians conducting "pure" research.

Certainly, large and in-depth investigations would be required to confirm the existence and define the exact geography of this culture of demos. However, a working assumption of that kind already seems useful for understanding why and how we have ended up living in a world where demos have become so ubiquitous, that is, in a demo world.

Notes

I thank all the participants in the workshop "Cultures and Styles of Scientific Practices," especially the organizers, Karine Chemla and Evelyn Fox Keller, and my commentators, Fan Fa-ti and Nancy Nersessian, for their suggestions and criticisms related to previous versions of this chapter.

1. A demo consists of exhibiting a technological device, such as a piece of computer software, in action. Bill Gates's software demos explain how to use Microsoft products.

2. See esp. Schaffer 1983, 1994; Shapin 1988; Collins 1988; Hankins and Silverman 1995, 37–71; Dolza and Vérin 2003; and Thébaud-Sorger 2009.

3. See Goffman's (1974, 66–68) definitions of demonstrations as "technical re-doings," "performances of a tasklike activity out of its usual functional context in order to allow someone who is not the performer to obtain a close picture of the doing of the activity," or as an "ideal running through of an activity for learning or evidential purposes."

4. Some of the uses of public demonstrations beyond the scientific field include business practices such as market pitching (see Clark and Pinch 1992, 1995; Sherry 1998; Le Velly 2007) and product launching (Bloomfield and Vurdubakis 2002; Simakova

2010), economic experiments (Callon and Muniesa 2007), architecture (Houdart 2005; Yaneva 2009), the film industry (Grimaud 2005), the health sector (Winthereik, Johannsen, and Strand 2008), activities of hackers (Auray 1997), and forms of collective mobilizations and political participation (Barry 2001; Brian 2001; Girard and Stark 2007; Stark and Paravel 2008).

5. On the work and views of European commissioners, see Joana and Smith 2000 and 2002.

6. On the tensions of European governance, see Christiansen 1997.

7. On the European Parliament's control over the European administration, see Chauchat 1989. On the history of the relationships between the Parliament and the EC, see Schwed 1978. On the power of the European Parliament, see Tsebelis 1994.

8. On how large firms came to lobby in the European Union, see Coen 1997. On lobbying strategies in the EC and effects on professional groups, see Neale 1994.

9. On the history of EC telecommunication policies, see Sandholtz 1993.

10. On the use of European statistics and measurements by European officials to create a European consciousness, see Shore 1995.

11. On communication politics in Europe, see also Meyer 2002. On showing practices of journalists, see Dayan 2009.

12. On the embodiment of scripts in technological devices, see Akrich 1992. The development of the uses and contents of demos may be part of, and articulated with, the evolution of audiovisual practices and of the movie industry in particular, and especially of the nature of, and roles devoted to, trailers.

13. Many videos of demos are available on the Internet. For example, one of Bill Gates's demos may be viewed at www.youtube.com/watch?v=KqKC5A9JWTg (accessed July 25, 2016).

14. See cordis.europa.eu/infowin/acts/analysys/products/thematic/multimed /document/mira/mira01.htm, ftp.cordis.europa.eu/pub/infowin/docs/visabe11.mov, ftp.cordis.europa.eu/pub/infowin/docs/visabe12.mov (accessed July 8, 2013).

15. BT is a British multinational telecommunication services company with head offices in London, United Kingdom.

16. On EC initiatives in evaluation, see Levy 1997.

17. On the mobilization of experts for the management of European programs, see Ant 1997.

18. ACTS, "Results, Impact and Exploitation," interim report DGXIII/B Ref: AC 1997/1339, May 15, 1997.

19. On EC arbitration processes, see From 2002.

20. On the role of prototypes in aligning multiple and discontinuous social worlds, see also Trigg, Bødker, and Grønbæk 1991 and Suchman, Trigg, and Blomberg 2002.

21. For the effect of spatial arrangements on the administration of public affairs, see Domahidy and Gilsinan 1992.

22. For a typology of models of patronage across history, see Turner 1990.

23. See "Treaty Establishing a Constitution for Europe," *Official Journal of the European Union* 47 (December 16, 2004): 109–111.

24. "Treaty Establishing a Constitution for Europe," 109–110.

25. On the evolution of the EC, see Dimitrakopoulos 2004.

26. See esp. Serene 1982; Lloyd 1990, 1996; Jardine 1991; Hankins and Silverman 1995, 37–71; Netz 1999; Chemla 2009; and Rosental 2009.

27. Similarly, it might be reductionist to describe the action of an individual buying a newspaper at the same shop every day simply in terms of monetary transaction: this ritual may have other functions, such as providing an opportunity for the buyer to exchange views with the shopkeeper, to create a social link, and to avoid loneliness. These functions may be as important for the buyer as the purchase of the newspaper per se.

28. For insightful analyses of these concepts in ancient Greek science, see esp. Von Staden 1994 and Cassin 2004.

29. Note that demonstrations versus "inscriptions" (Latour 1993, 100–129) are at the heart of the economic cycles of scientific capitalism—"resources-demonstrations-resources."

30. Such phenomena may be perceived as contributing to define a culture; see Knorr Cetina 1999, 10.

31. On this important type of cultural feature, see Keller 2004, 15.

32. Ludwik Fleck would talk here about a "thought style"; see Fleck 1979, 64.

33. References to some of these studies appear in the reference list of this chapter.

34. See J. Markoff, "Nothing Up Their Sleeves? Masters of the High-Tech Demo Spin Their Magic," *New York Times*, March 11, 1996.

References

Akrich, M. 1992. "The De-scription of Technical Objects." In *Shaping Technology/ Building Society: Studies in Socio-technical Change*, ed. W. Bijker and J. Law, 205–224. Cambridge, MA: MIT Press.

Analysys Ltd. 1996. *ACTS Multimedia Success Stories*. Stuttgart: Infowin Project.

Ant, M. 1997. "The Role of Expertise in the Decision Chain of the European Commission Illustrated with the Aid of the Leonardo da Vinci Program." *Revue de l'Institut de Sociologie* 1: 191–201.

Aristotle. 1962. *The Politics*. Translated by T. A. Sinclair. Baltimore: Penguin Books.

Auray, N. 1997. "Ironie et solidarité dans un milieu technicisé: Les défis contre les protections dans les collectifs de 'hackers.'" In *Cognition et information en société*, ed. B. Conein and L. Thévenot, 177–201. Paris: Editions de l'École des Hautes Études en Sciences Sociales.

Barry, A. 2001. *Political Machines: Governing a Technological Society*. London: Athlone Press.

Bloomfield, B., and T. Vurdubakis. 2002. "The Vision Thing: Constructing Technology and the Future in Management Advice." In *Critical Consulting: New Perspectives on the Management Advice Industry*, ed. T. Clark and R. Fincham, 115–129. Oxford: Blackwell.

Brian, E. 2001. "Citoyens américains, encore un effort si vous voulez être républicains." *Actes de la recherche en sciences sociales* 138: 47–55.

Brine, J. 2000 "TSER and the Epistemic Community of European Social Researchers." *Journal of European Social Policy* 10: 267–282.

Callon, M. 2003. "Quel espace public pour la démocratie technique?" In *Les sens du public: Publics politiques, publics médiatiques,* ed. D. Cefai and D. Pasquier, 197–221. Paris: Presses Universitaires de France.

Callon, M., and F. Muniesa. 2007. "Economic Experiments and the Construction of Markets." In *Do Economists Make Markets? On the Performativity of Economics,* ed. D. MacKenzie, F. Muniesa, and L. Siu, 163–189. Princeton, NJ: Princeton University Press.

Cassin, B. 2004. "Managing Evidence." Paper presented at the Society for Social Studies of Science conference, Paris, August 26.

Chappaz, V. 1997. "Telecoms Are Looking towards Fifth Framework R&D." *ACTS NewsClips* 24, June 1. Accessed July 25, 2016. https://cordis.europa.eu/infowin/acts/ienm /newsclips/nc9724.htm.

Chauchat, M. 1989. *Le contrôle politique du Parlement européen sur les exécutifs communautaires.* Paris: LGDJ.

Chemla, K. 2009. "Apprendre à lire: La démonstration comme élément de pratique mathématique." *Communications* 84: 85–101.

Christiansen, T. 1997. "Tensions of European Governance: Politicized Bureaucracy and Multiple Accountability in the European Commission." *Journal of European Public Policy* 4: 73–90.

Clark, C., and T. Pinch. 1992. "The Anatomy of a Deception: Fraud and Finesse in the Mock Auction Sales 'Con.'" *Qualitative Sociology* 15: 151–175.

Clark, C., and T. Pinch. 1995. *The Hard Sell: The Language and Lessons of Street-wise Marketing.* London: HarperCollins.

Coen, D. 1997. "The Evolution of the Large Firm as a Political Actor in the European Union." *Journal of European Public Policy* 4: 91–108.

Collins, H. M. 1988. "Public Experiments and Displays of Virtuosity: The Core-Set Revisited." *Social Studies of Science* 18: 725–748.

Dayan, D. 2009. "Quand Montrer c'est Faire." *Divinatio* 29: 155–178.

Dimitrakopoulos, D. G., ed. 2004. *The Changing European Commission.* Manchester: Manchester University Press.

Dolza, L., and H. Vérin. 2003. "Une mise en scène de la technique: Les théâtres de machines." *Alliage: Culture, Science, Technique* 50–51: 8–20.

Domahidy, M. R., and J. F. Gilsinan. 1992. "The Back Stage Is Not the Back Room: How Spatial Arrangements Affect the Administration of Public Affairs." *Public Administration Review* 52: 588–593.

Duval, M. 1981. "Les camelots." *Ethnologie française* 2: 145–170.

Fleck, L. 1979. *Genesis and Development of a Scientific Fact.* Chicago: University of Chicago Press.

From, J. 2002. "Decision-Making in a Complex Environment: A Sociological Institutionalist Analysis of Competition Policy Decision-Making in the European Commission." *Journal of European Public Policy,* 9: 219–237.

Geiger, R. 1997. "21st Century: The Communications Age." *ACTS NewsClips* 26, July 1. Accessed July 25, 2016. https://cordis.europa.eu/infowin/acts/ienm/newsclips /nc9726.htm.

Georgakakis, D. 2000. "The Resignation of the European Commission: Scandal and Institutional Turning Point (October 1998–March 1999)." *Cultures et Conflits* 38–39: 39–71.

Girard, M., and D. Stark. 2007. "Socio-technologies of Assembly: Sense Making and Demonstration in Rebuilding Lower Manhattan." In *Governance and Information: The Rewiring of Governing and Deliberation in the 21st Century*, ed. D. Lazer and V. Mayer-Schoenberger, 145–176. Oxford: Oxford University Press.

Goffman, E. 1974. *Frame Analysis: An Essay on the Organization of Experience.* Cambridge, MA: Harvard University Press.

Grimaud, E. 2005. "Démonstrations d'automates et effets spéciaux à Bombay." Paper presented at the workshop Sociologie de la démonstration, École des Hautes Études en Sciences Sociales, Marseille, May 19–20.

Hankins, T. L., and R. J. Silverman. 1995. *Instruments and the Imagination.* Princeton, NJ: Princeton University Press.

Houdart, S. 2005. " 'Ceci est une persienne': Montrer/démontrer dans un projet architectural au Japon." Paper presented at the workshop Sociologie de la démonstration, École des Hautes Études en Sciences Sociales, Marseille, May 19–20.

Jardine, N. 1991 "Demonstration, Dialectic and Rhetoric in Galileo's 'Dialogue.'" In *The Shapes of Knowledge from the Renaissance to the Enlightenment*, ed. D. R. Kelley and R. H. Popkin, 101–121 Kluwer: Dordrecht.

Jasanoff, S. 2005. *Designs on Nature: Science and Democracy in Europe and the United States.* Princeton, NJ: Princeton University Press.

Joana, J., and A. Smith. 2000. "Le mariage de la carpe et du lapin? Une sociologie politique de la Commission européenne en chantier." *Cultures et Conflits* 38–39: 73–100.

Joana, J., and A. Smith. 2002. *Les commissaires européens: Technocrates, diplomates ou politiques?* Paris: Presses de Sciences Po.

Keller, E. F. 2004. *Expliquer la vie: Modèles, métaphores et machines en biologie du développement.* Paris: Gallimard.

Knorr Cetina, K. 1999. *Epistemic Cultures: How the Sciences Make Knowledge.* Cambridge, MA: Harvard University Press.

Lamont, M. 2009. *How Professors Think: Inside the Curious World of Academic Judgment.* Cambridge, MA: Harvard University Press.

Lampel, J. 2001. "Show-and-Tell: Product Demonstrations and Path Creation of Technological Change." In *Path Dependence and Creation*, ed. R. L. Garud and P. Karnoe, 303–328. Mahwah, NJ: Erlbaum.

Latour, B. 1983. "Give Me a Laboratory and I Will Raise the World." In *Science Observed*, ed. K. D. Knorr Cetina and M. Mulkay, 141–170. Beverly Hills, CA: Sage.

Latour, B. 1993. *Petites leçons de sociologie des sciences.* Paris: La Découverte.

Lemieux, C. 2008. "Rendre visibles les dangers du nucléaire: Une contribution à la sociologie de la mobilisation." In *La Cognition au prisme des sciences sociales*, ed. B. Lahire and C. Rosental, 131–159. Paris: Éditions des Archives Contemporaines.

Le Velly, R. 2007. "Les démonstrateurs de foires: Des professionnels de l'interaction symbolique." *Ethnologie française* 37: 143–151.

Levy, R. P. 1997. "Evaluating Evaluation in the European Commission." *Canadian Journal of Program Evaluation* 12: 1–18.

Lloyd, G. 1990. "The Theories and Practices of Demonstration in Aristotle." *Proceedings of the Boston Area Colloquium in Ancient Philosophy* 6: 371–401.

Lloyd, G. 1996. "Theories and Practices of Demonstration in Galen." In *Rationality in Greek Thought*, ed. M. Frede and G. Striker, 255–277. Oxford: Clarendon Press.

Lunenfeld, P. 2000. *Snap To Grid: A User's Guide to Digital Arts, Media, and Cultures.* Cambridge, MA: MIT Press.

MacMullen, A. 1999. "Fraud, Mismanagement and Nepotism: The Committee of Independent Experts and the Fall of the European Commission 1999." *Crime, Law and Social Change* 31: 193–208.

Mauss, M. 1954. *The Gift: Forms and Functions of Exchange in Archaic Societies.* Translated by I. Cunnison. Glencoe, IL: Free Press.

Meyer, C. O. 2001. "European Public Sphere as Watchdog—Transnational Journalism and the Dismissal of the European Commission." *Forschungsjournal Neue Soziale Bewegungen* 14: 42–52.

Meyer, C. O. 2002. *Europäische Öffentlichkeit als Kontrollsphäre: Die Europäische Kommission, die Medien und Politische Verantwortung.* Berlin: Vistas.

Mukerji, C. 2009. *Impossible Engineering: Technology and Territoriality on the Canal du Midi.* Princeton, NJ: Princeton University Press.

Neale, P. 1994. "Expert Interest Groups and the European Commission: Professional Influence on EC Legislation." *International Journal of Sociology and Social Policy* 14: 1–24.

Netz, R. 1999. *The Shaping of Deduction in Greek Mathematics: A Study in Cognitive History.* Cambridge: Cambridge University Press.

Oudshoorn, N., and T. Pinch, eds. 2005. *How Users Matter: The Co-construction of Users and Technology.* Cambridge, MA: MIT Press.

Rosental, C. 2002. "De la démo-cratie en Amérique: Formes actuelles de la démonstration en intelligence artificielle." *Actes de la recherche en sciences sociales* 141–142: 110–120.

Rosental, C. 2004. "Fuzzyfying the World: Social Practices of Showing the Properties of Fuzzy Logic." In *Growing Explanations: Historical Perspectives on Recent Science*, ed. M. N. Wise, 159–178. Durham, NC: Duke University Press.

Rosental, C. 2005. "Making Science and Technology Results Public: A Sociology of Demos." In *Making Things Public: Atmospheres of Democracy*, ed. B. Latour and P. Weibel, 346–349. Cambridge, MA: MIT Press.

Rosental, C. 2007. *Les capitalistes de la science: Enquête sur les démonstrateurs de la Silicon Valley et de la NASA.* Paris: CNRS Editions.

Rosental, C. 2008. *Weaving Self-Evidence: A Sociology of Logic.* Princeton, NJ: Princeton University Press.

Rosental, C. 2009. "Anthropologie de la démonstration." *Revue d'Anthropologie des Connaissances* 3: 233–252.

Rosental, C. 2010. "Social Studies of Evaluation." *Social Studies of Science* 40: 481–484.

Rosental, C. 2013. "Toward a Sociology of Public Demonstrations." *Sociological Theory* 31: 343–365.

Sandholtz, W. 1993. "Institutions and Collective Action: The New Telecommunications in Western Europe." *World Politics* 45: 242–270.

Schaffer, S. 1983. "Natural Philosophy and Public Spectacle in the Eighteenth Century." *History of Science* 21: 1–43.

Schaffer, S. 1994. "Machine Philosophy: Demonstration Devices in Georgian Mechanics." *Osiris* 9: 157–182.

Schwed, J. J. 1978. "The Parliament and the Commission." *Annals of the American Academy of Political and Social Science* 440: 33–41.

Serene, E. 1982. "Demonstrative Science." In *The Cambridge History of Later Medieval Philosophy*, ed. N. Kretzmann, A. Kenny, J. Pinborg, and E. Stump, 1:496–517. Cambridge: Cambridge University Press.

Shapin, S. 1988. "The House of Experiment in Seventeenth-Century England." *Isis* 79: 373–404.

Shapin, S. 2008. *The Scientific Life: A Moral History of a Late Modern Vocation*. Chicago: University of Chicago Press.

Shapin, S., and S. Schaffer. 1985. *Leviathan and the Air-Pump: Hobbes, Boyle, and the Experimental Life*. Princeton, NJ: Princeton University Press.

Sherry, J. F. 1998. "Market Pitching and the Ethnography of Speaking." *Advances in Consumer Research* 15: 543–547.

Shore, C. 1995. "Usurpers or Pioneers? European Commission Bureaucrats and the Question of 'European Consciousness.'" In *Questions of Consciousness*, ed. A. P. Cohen and N. Rapport, 217–236. London: Routledge.

Simakova, E. 2010. "RFID 'Theatre of the Proof': Product Launch and Technology Demonstration as Corporate Practices." *Social Studies of Science* 40: 549–576.

Smith, W. 2009. "Theatre of Use: A Frame Analysis of IT Demonstrations." *Social Studies of Science* 39: 449–480.

Stark, D., and V. Paravel. 2008. "PowerPoint in Public: Digital Technologies and the New Morphology of Demonstration." *Theory, Culture and Society* 25: 30–55.

Suchman, L., R. Trigg, and J. Blomberg. 2002. "Working Artefacts: Ethnomethods of the Prototype." *British Journal of Sociology* 53: 163–179.

Tarde, G. 1903. *The Laws of Imitation*. New York: Henry Holt.

Thébaud-Sorger, M. 2009. *L'Aérostation au temps des Lumières*. Rennes: Presses Universitaires de Rennes.

Tocqueville, A. de. 1981 *De la démocratie en Amérique*, vol. 2. Paris: Garnier-Flammarion.

Trigg, R. H., S. Bødker, and K. Grønbæk. 1991. "Open-Ended Interaction in Cooperative Prototyping: A Video-Based Analysis." *Scandinavian Journal of Information Systems* 3: 63–86.

Tsebelis, G. 1994. "The Power of the European Parliament as a Conditional Agenda Setter." *American Political Science Review* 88: 128–142.

Turner, S. P. 1990. "Forms of Patronage." In *Theories of Science in Society*, ed. S. E. Cozzens and T. F. Gieryn, 185–211. Bloomington: Indiana University Press.

Von Staden, H. 1994. "Anatomy as Rhetoric: Galen on Dissection and Persuasion." *Journal of the History of Medicine and Allied Sciences* 50: 47–66.

Wagner, A., and M. Capucciati. 1996. "Demo or Die: User Interface as Marketing Theatre." Paper presented at the Conference on Human Factors in Computing Systems, Vancouver, April 13–18.

Weber, M. 2004. *The Vocation Lectures: "Science as a Vocation"; "Politics as a Vocation."* Edited by D. Owen and T. B. Strong. Translated by R. Livingstone. Indianapolis: Hackett.

Weber, M. 2009. *The Protestant Ethic and the Spirit of Capitalism with Other Writings on the Rise of the West,* 4th ed. Oxford: Oxford University Press.

Winthereik, B. R., N. Johannsen, and D. L. Strand. 2008. "Making Technology Public: Challenging the Notion of Script through an E-Health Demonstration Video." *Information, Technology and People* 21: 116–132.

Woolgar, S. 1991. "Configuring the User: The Case of Usability Trials." In *A Sociology of Monsters: Essays on Power, Technology and Domination,* ed. J. Law, 58–99. London: Routledge.

Yaneva, A. 2009. "The Architectural Presentation." In *Networks of Design,* ed. F. Hackney, J. Glynne, and V. Minton, 212–219. Boca Raton, FL: Universal.

DAVID RABOUIN

8
———

Styles in Mathematical Practice

Although the use of the category of style in the history and philosophy of science has become fashionable in recent years, it has not appreciably affected the description of mathematical activity.[1] Among the very few works devoted to the issue of style in mathematics is a paper by the mathematician Claude Chevalley, published in 1935 in the *Revue de Métaphysique et de Morale* titled "Variations du style mathématique." Chevalley pointed to ways of writing in mathematics as an indication of "general tendencies," as he called them, much as one can identify styles in literature. These styles, he argued, circulated so widely through different places at particular times that they could characterize these periods. Chevalley's article presented an idea that deserves particular attention, which I pursue in this chapter: by focusing on ways of writing, it put particular emphasis on external manifestations as opposed to other traditional internal ways of characterizing styles in terms of concepts, theories, reasoning, or objects. In addition, as I aim to show, Chevalley's article departed from other familiar ways of characterizing styles in terms of cultural contexts (whether schools, nations, or traditions). One might even argue that his proposal, given its immediate historical context, was conceived as a reaction to a cultural interpretation of style (see the next section).

Chevalley's characterization of style seems to fall between two dominant uses of this category in the history and philosophy of science: the cultural one (sometimes called local) and a more general one represented by styles of reasoning à la Alistair Crombie and Ian Hacking.[2] Paolo Mancosu (2009, sec. 5), in his survey on mathematical style, credited Chevalley for accurately delineating the level of inquiry, at least on a programmatic level:

> In the case of mathematics there is good evidence that the most natural locus for "styles" falls, so to speak, in between these two categories. Indeed, by and large, mathematical styles go beyond any local community defined in simpler sociological terms (nationality, direct membership in a school etc.) and are such that the support group can only be characterized by the specific method of enquiry pursued. On the other hand, the method is not so universal as to be identifiable as one of the six methods described by Crombie or in the extended list given by Hacking.

My wish to expand on Chevalley's notion of style, however, goes beyond the desire to find a better integration of local and global; my main intention is to address the epistemological challenge formulated by Mancosu at the end of his survey, which I believe has not been answered satisfactorily: "The problem of an epistemology of style can perhaps be roughly put as follows. Are the stylistic elements present in mathematical discourse devoid of cognitive value and so only part of the coloring of mathematical discourse or can they be seen as more intimately related to its cognitive content?" (sec. 6). In more basic terms, does focusing on styles lead us to essential or to accidental characteristics of mathematical knowledge? Chevalley's reflections on this question are interesting not just because they provide information on how one mathematician viewed the importance of style in his own activity. They also point to an aspect that is sometimes considered as merely accidental or, at best, purely expressive, which he presents as essential: the cognitive role played by ways of writing in mathematical knowledge. Symmetrically, by considering these concrete aspects of mathematical activity, Chevalley reveals transmissions that are not tied to what is often considered the core of the cognitive content of mathematics. Following Gert Schubring (1996, 363), I will call them "shared epistemologies," in the sense of common understanding about the broad definition of the theories in play and the characterization of their objects.[3]

Chevalley's proposal reverses traditional ways of looking at mathematical knowledge by pointing to cases in which what is stable is the expressive part and what is potentially variable is its correlative content. This approach contradicts many characterizations of style in science and may help explain some of their limitations. Of the few epistemological discussions of style in mathematics that do exist, most rely on a classical approach in which theories, concepts, or objects are essential ingredients of mathematical knowledge. For example, Gilles Gaston Granger (1968, 20), author of one of the most elaborate philosophical analyses, states that "style appears to us . . . on the one hand as a way of introducing the concepts of a theory, of connecting them, of unifying them; and on the other hand, as a way of delimiting what intuition contributes to the determination of these concepts" (as translated in Mancosu 2009, sec. 6).[4] Hacking (2002, 181), for his part, insists that new styles accompany the introduction of new domains of objects or, better, in a clear reference to Kant, new conditions of possibility for objectivity: "My styles of reasoning, eminently public, are part of what we need to understand what we mean by objectivity. This is not because styles are objective (i.e., that we have found the best impartial ways to get at the truth), but because they have settled what it is to be objective (truths of certain sorts are what we obtain by conducting certain sorts of investigations, answering to certain standards)."[5] In both these characterizations, styles are inseparable from conceptual domains, which, to quote Hacking's Foucaldian terminology, allow for a certain "regime of truth."

Chevalley's styles are free of such presuppositions. They do not ask for identification of concepts, objects, structures, regimes of truth, or theories expressed through a certain style. Using the example of the modern axiomatic style, Chevalley remarks that in his sense, very different theories can share a common style (general topology, abstract algebra, measure theory). In this case, it even helps us to circulate among theories and discover new conceptual architectures.[6] On the other hand, as Chevalley indicates by the example of Karl Weierstrass's "ε-style," the conceptual content of a style can be seen as fully captured by the way of writing in and of itself (rather than *expressed* through it)—as if, in Leibnizian terms, writing is reasoning for us. I return to this issue at the end of this chapter, as it implies displacing attention from concepts and theories to inferences attached to material proxies (see "Other examples," below).

One of my aims in this chapter is to expand upon two features: (1) the relative independence of styles of writing with regard to their regimes of interpretation, particularly in relation to the delineation of theories, objects, and concepts involved; and (2) the cognitive value of the ways of writing in and of themselves (as external demonstrations carrying specific kinds of inferences). The first part of this chapter explores this program on a methodological level. I first describe Chevalley's styles in greater detail, focusing on the context in which they were expounded. I then confront this notion critically with the two main descriptions of style (local and general) mentioned above. Finally, I indicate other examples for which this category of style might apply and try to sketch a corresponding epistemological framework.

Chevalley's Styles

Chevalley's Characterization of Style

Commentators have often remarked that the notion of style, although quite popular in recent literature in history and philosophy of science, is rarely defined, and its epistemological foundations are rarely made explicit (Gayon 1996; Mancosu 2009). Even when authors try to provide a definition, it is striking that this is usually intelligible only through a list of examples (see Granger 1968, 20–21).[7] At first glance, Chevalley is no exception to this: he merely states that one can identify general tendencies in ways of writing mathematics and limits himself to focusing on two important examples:[8] (a) the ε-style developed by Weierstrass and his school to express basic properties of functions in real and complex analysis (in contrast to the style of algebraic analysis dominant in the previous period),[9] and (b) the axiomatic style initiated by David Hilbert for plane geometry in his famous *Grundlagen der Geometrie* (1899) and developed in various fields of research in the nineteenth and the beginning of the twentieth century: measure theory (Henri Lebesgue), general topology and functional analysis (Maurice Fréchet), abstract algebra (Richard Dedekind, Ernst Steinitz, Emmy Noether), and finally probability theory (for which Chevalley does not specify any names but clearly alludes to Andreï Kolmogorov).

This rather general description is interesting as soon as one realizes that it is expressed by one of the founders of the Bourbaki group, which had just been established at that time.[10] To acknowledge that there are "revolutions that inflect writing, and thus thought" (Chevalley 1935, 375; see n. 12) certainly seems opposed to a particular kind of naive mathematical Platonism sometimes ascribed to structuralist ideology (as if structure were some fixed abstract entity living in a Platonic realm). On this regard, one could jump directly to the conclusion of the paper, in which Chevalley insists on the fact that the axiomatic style, which expounds the notion of structure, is a style *like any other*. As a consequence, structure is presented not as a way to unravel the real ontology suited to mathematics in general but as a temporary practice loaded with a fruitful heuristic value: "As a result, contemporary mathematics tends to define mathematical objects in comprehension, i.e. by their characteristic properties, rather than in extension, i.e. through a construction. This feature has presumably nothing definite. But it is difficult to foresee right now in which direction it would develop. In any case, the actual trend seems far from having exhausted its internal dynamic" (Chevalley 1935, 384).[11]

There is another reason that Chevalley's proposal is not as naive as it may seem. Publishing a paper on the notion of style hardly seems innocent for a European mathematician writing in 1935. Only the previous year, an essay on the same topic was published: "Forms of Styles in Mathematical Creation" ("Stilarten mathematischen Schaffens"), written by Ludwig Bieberbach, a famous mathematician enrolled in the Nazi movement.[12] In this paper, which followed up other articles and lectures on the role of personality structures in mathematics, Bieberbach, in line with Erich Rudolf Jaensch's racial psychology and the *Blut und Boden* doctrine,[13] advocated the existence of very different styles of mathematical creation, namely, the J-style (characteristic of the German mind, therefore seen as closer to intuition) and the S-style (characteristic of the Jewish mind, said to be more abstract and analytic). This provoked harsh polemics in European mathematical journals, involving important contemporary mathematicians such as Harald Bohr and Geoffrey Hardy (see Segal 2003). Considering this context, the choice of the category of style, and of the examples given in the paper, seems significant. To begin with, we could note not only that the three great German mathematicians mentioned by Chevalley (Weierstrass, Hilbert, and Dedekind) are those who were most difficult (to say the least) for Bieberbach to integrate in his classification of styles (in which abstraction is correlated with the S-style), but also that Chevalley associated them closely with mathematicians of Jewish origin, such as Steinitz and Noether. One should also notice that the very purpose of the paper is to designate two styles of writing, that of the ε-style and that of the axiomatic style, which circulated between Germany and France. In this context Chevalley mentions two important French names: Lebesgue and Fréchet.

This reminds us, if necessary, that our analytical categories have a history and that, because of this history, their use may be far from neutral. In Chevalley's period, there were good strategic reasons to rescue the notion of style. However, these reasons are still valid nowadays: not only do we still have to combat religious or nationalist appropriations of the history of mathematics,[14] but less dramatic, the danger of cultural essentialism often lurks behind the use of the category of style in the history of science. It is important not only to emphasize the existence of styles circulating across different cultural determinations but also to invest this category positively by anchoring it in the ways of writing, that is, the more material aspect of this circulation (as opposed to what is attached to interpretations and meanings shared by a community of actors or, by contrast, to certain platonic ideas underlying the supposed universality of the supposed conceptual background).

The story Chevalley tells about the emergence of the ε-style presented as a consequence of a need for rigorous foundations in mathematics, and the subsequent development of the axiomatic style, may appear commonplace today, or as an oversimplifying myth. But one should recall that it was also a way to demonstrate to people such as Bieberbach the absurdity of an opposition between French and German mathematics (which was very common at the time). Whatever the historical reasons for the emergence of the ε-style, it would remain an interesting historical fact that this way of writing spread rapidly from the Berlin School in Germany and in France (to the point that it is still one of the standard ways to write mathematics in real and complex analysis). To give just one example of a French use, to which I shall return: Henri Poincaré, although opposed in many ways to Weierstrass's program of arithmetization, had no trouble in adopting the ε-style when dealing with complex functions (see, e.g., Poincaré 1883). It should also be noted here that in his paper Chevalley was not denying the existence of individual styles, which might involve cultural and biographical specificities (recall Poincaré and Weierstrass). His point was that these styles coexist with some general tendencies marked by circulation between individuals and cultures, and particularly between national traditions and schools.[15]

Epistemological Issues

Given these elements of context, Chevalley's paper is less naive than might appear, although its epistemological grounds are hardly rendered explicit. At first glance, the core of the paper seems a fairly commonsense reaction: there are general trends in the ways of writing mathematics, and these "styles" are not fixed but vary through cultures and history. However, here is an example of why I think one should take a distance from Chevalley's position: he states quite clearly at the beginning of the paper that styles are due to the influence of strong personalities in mathematics: "This style, under the influence of powerful mathematical personalities, is subject every once in a while to revolutions" (Chevalley 1935, 375). This is indeed the way he describes his two examples, citing Weierstrass and Hilbert. But he gives no convincing reason (in fact, no reason at all) that this should be so. Yet it is not difficult to find examples of ways of writing that show parallel developments for several authors who came from different and independent traditions. As just one example, contemporary to Chevalley's examples, Frederic Brechenmacher (2010) gives a nice case study on the development of "matrix calculus" from 1850 to 1938; it appears that this style of writing with tables of numbers was developed by several authors independently and in quite different contexts before converging in a unified theory. There seems to be no clear difference between this example and that of the ε-style: it was a new way of writing, which inflected mathematical thought, and it coincided with the development of a new way not only of doing mathematics (later called linear algebra) but also of conceiving some of its fundamental objects. It is also the case, for example, for the Cartesian style of geometry. This new algebraic style, although associated with the name of René Descartes, is completely unintelligible if one does not see first that it became stabilized rapidly as a mixture of (at least) two independent origins, one from Descartes and the other from Pierre de Fermat. Moreover, what was understood by the name "Descartes" was a direct reference not to a "strong personality" but primarily to a book (the larger Latin edition of *La Géométrie* in 1659–1661), which has a history of its own, and particularly an editor, Frans van Schooten, who played a pivotal role in shaping its content.

Another aspect on which Chevalley is unclear is whether or not something like a shared epistemology has to be attached to any given style. As I have pointed out, he certainly claims that changes in the ways of writing inflect thought. This becomes explicit through an analysis of the inflexion produced by the different styles. In the case of the ε-style, the main feature is the emphasis on inequalities, whereas the style of algebraic analysis focused on equalities (Chevalley 1935, 378–379). In the case of the axiomatic style, the emphasis is on comparison between theories and the way a detailed conceptual analysis of the role played by each axiom forces us to reshape them. What remains unclear is whether or not these features are part of more general shared epistemologies, for example, the trend toward rigorous foundations, which Chevalley also invoked. If this were the case, it would not be difficult to find counterexamples. Poincaré, as we saw, adheres when necessary to the ε-style without encompassing Weierstrass's general epistemology in any way.

The same could be said about the "Modern algebra" promoted by Dedekind, Steinitz, and Noether, and mentioned by Chevalley, as a prototypical example of the new axiomatic style. As Leo Corry (2004) pointed out in his seminal study on this tradition, the fact that Dedekind, Steinitz, and Noether shared the same epistemology (often described as a kind of structuralism) is far from obvious and can probably be ascribed to a myth created retrospectively by Noether herself to enhance the image of the Göttingen School.[16]

Since Chevalley is not explicit about this, it is not easy to know exactly what he had in mind when claiming that there are "revolutions that inflect writing, and thus thought." Considering the previous counterexamples, I would limit myself to a narrow interpretation: the epistemological content of a style of writing does not have to involve elements of a shared epistemology (although it may, in fact, do so). When Poincaré used the ε-style, it was not because he shared a certain conception with Weierstrass of what the objects (real numbers, functions, etc.) involved in this manipulation were and about the good (in this case "rigorous") delineation of the theories, but because this way of writing allowed some powerful inferences that were not possible in the previous style.

Comparison with Other Notions of Mathematical Style

Hacking's Program

The two main uses of the category of style identified in survey essays such as those of Jean Gayon (1996 and 1999), Ian Hacking (2002), and Paolo Mancosu (2009) are mainly local (also called "cultural") and general (also called "methodological"). Because of their philosophical background, these authors have a strong tendency to conceive the local usage as a merely descriptive category used by historians and intended to characterize the epistemic orientation of a given community or even a single person.[17] Even if they do not deny that this use of style is interesting in itself, they tend to present it as unhelpful in the resolution of epistemological problems, precisely because it leaves us with local descriptions and no way (or even no desire) to reconnect them from a larger epistemological perspective.

The view of these authors, strongly expressed by Hacking, seems to be that the local studies in history of science that have flourished in recent years offer source material that could be used to build a new kind of epistemology. This could facilitate harmonious cooperation between "styles for historians" and "styles for philosophers," and reestablish the dialog among three kinds of inquiries, which Hacking (1992, 130) points out "have almost ceased to speak to each other":

A. Newly gained analyses of, and case-studies of, the fleeting "micro-social" interactions of knowers and discoverers, their "macrosocial" relationship to larger communities, and the material conditions and objects in which the discoveries are made and which they are about. At this level, the relevant events last a week or at most a few decades.
B. Current philosophical conceptions of truth, being, logic, meaning, and knowledge.
C. Models of relatively permanent, growing, self-modulating, revisable features of science. . . . The result of their persistence is a body of what is counted as objective ways of determining truth, of settling belief, of understanding meaning, a body of nothing less than logic itself. (131)

One could certainly agree that these fields of research developed largely independently in recent years. This development is problematic if one considers that they all aim at giving a description of what scientific activity is. One could also agree that the traditional use of the category of style by historians was mainly incorporated into the first tradition.[18] The exception here was Crombie (1994), who was one of the rare adherents of Hacking's third position—thus, Hacking was able to present his own program as a way to match A (e.g., social history of science) and C (Crombie's long-term history) via a detour through B (his own post-Putnam philosophical agenda).[19]

However, this presentation takes for granted what should first be put into question: do we really need a philosophical framework to match A and C via the welcoming connection of B? Many historians would think not. The difficulty appears clearly when Hacking (1992, 131) mentions the constructionalist option: "They study the first shift at the factory of facts. Quitting work early in the day, they leave us in the lurch with a feeling of absolute contingency." But what is wrong with this feeling of "absolute contingency"? What if there were no metaphysical framework to ground historical contingencies—or, better, what if absolute contingency were, in fact, the only true metaphysical framework? What if, to borrow Clifford Geertz's beautiful expression, there were nothing but "local knowledges"—the history of scientific knowledge being nothing else, to mimic a famous dictum of Jacques Lacan's (1966, 47), than the integral of the equivoques left by local language games (see Geertz 1983)?[20]

All of these questions should be taken into account to develop a unifying notion of style compatible with "analyses of, and case-studies of, the fleeting 'microsocial' interactions of knowers and discoverers." Hacking (1992, 131) has something to say regarding this when he briefly mentions that the constructionalists "give little sense of what holds the constructions together beyond the networks of the moment, abetted by human complacency." As a consequence, he emphasizes the fact that we need to give an account of a phenomenon that seems to be as important as the variability of representations in the history of science: stabilization.[21] Hence the following program: "A proposed account of self-stabilizing techniques begins by observing that a style becomes autonomous of the local microsocial incidents that brought it into being. Then there is the detailed account of how each style does stabilize itself" (Hacking 2002, 196). No one would deny the intense variability occurring on a micro (social/historical) level, and these many accidents were certainly hidden by the traditional approach of the history of science expressed in terms of grand teleologies leading to the actual state of the theories in play. However, Hacking is right that it leaves us with a new challenge: how to understand that, notwithstanding the local variability, something stabilizes in science.

Hacking's own proposal is nonetheless very disappointing when it comes to mathematics. The *longue durée* that he proposes to consider as a process of stabilization is nothing less than the whole history of deductive mathematics! The first element of Crombie's "style of scientific thinking" was indeed "the simple method of postulation exemplified by the Greek mathematical sciences" (Hacking 2002, 182). Even if Hacking claims that "styles do not determine a content, a specific science," in this case he makes a startling exception: "We do tend to restrict 'mathematics' to what we establish by mathematical reasoning, but *aside from that*, there is only a very modest correlation between items (a) through (f) and a possible list of fields of knowledge."[22] In other words, mathematics is the only case in which a style of reasoning is correlated by Crombie and Hacking with a science and in which the longue durée is that of a whole field of knowledge!

Exploring the list of Crombie's examples more closely, Hacking (2002, 185) finally acknowledged another mathematical style that he named "the algorismic style of reasoning," which he attributed to the "Indo-Arabic style of applied mathematics."[23] But that did not really influence his first characterization. Styles, according to Hacking, are strongly connected with the introduction of new objects,[24] but he did not make explicit what kind of objects—if any—were associated with the algorismic style. We are left with the preceding characterization in which mathematical objects in general (as nonempirical entities) were supposed to be introduced with the postulational style. This may be the reason that Hacking designates the "algebrizing of geometry" (he does not say much more about it than calling it an Arabicizing of the Greek) not as a third independent style but as a combination of the two others (as if it were an evolution of the type of objects supposedly introduced in the first style and characteristic of pure—as opposed to applied—mathematics).[25]

However, the fact that Hacking finally acknowledges the existence of another style of reasoning in mathematics and that he claims it can combine with the postulational style to form a new style (algebrized geometry) open the door to a multiplication of styles with no clear limit to their possible proliferation. If one accepts the algebrizing of geometry as a new style, why not the style of indivisibles, which developed so prominently at the same time in Europe? Or the synthetic style of geometry, advocated as a reaction to the algebrized style at the end of the eighteenth century? Or the modern axiomatic style (as distinct from the ancient axiomatic style)?[26] What would be the reasons for rejecting these styles and accepting the algebraico-geometrical one? And if we were to accept these more finely differentiated descriptions, why not then distinguish Simon Stevin's style of symbolical algebra (centered on polynomials) from that of François Viète (centered on equations)? Lazare Carnot's style of synthetic geometry from Gaspard Monge's? Jean-Victor Poncelet's sketch of projective geometry from Karl Von Staudt's, and so on? What would then be the limit in the precision and delineation of mathematical styles of reasoning?

One can certainly follow Mancosu's suggestion that we need a middle-range description—or, rather, that we should at least be able to determine clearly to what extent we want our category of style to be efficient (a point on which Hacking is ultimately less clear than he claims, except in pointing to the introduction of new domains of objects). The emphasis on the processes of stabilization is certainly a good start, but as we have seen, Hacking is not clear about the kind of stabilization he has in mind. This brings us back to our first question. The discontent arose through a local study of styles. What exactly is the problem with these local styles? Why should styles necessarily be part of a larger (philosophical?) point of view? At a descriptive level, we have a possible answer through the focus on the process of stabilization, especially when it occurs across different social or cultural contexts. It is true that, in this sense, we should at least try to bridge the gap between local determinations and longue durée descriptions, but as was pointed out by Catherine Goldstein (1995), this articulation can certainly be done without recourse to epistemology. One just needs to follow the long-term trajectories (what Goldstein calls the "readings") of mathematical documents (see also chapter 13 in this volume). So, once again, why should we need a larger metaphysical framework?

The Role of Philosophy

If we take "metaphysical" to mean something as general as Hacking's (1992, 131) designation of "philosophical conceptions of truth, being, logic, meaning, and knowledge," with explicit reference to the type of work done by Hilary Putnam, in my opinion it would be very difficult to escape such a framework. But this has nothing to do with the idea that philosophy would be a welcoming haven where historians would cease fighting each other, the place where local descriptions would find a way to reconnect with each other. At some point, we have to recognize in our historical documents where actors' knowledge arises. Historians, by not explaining what they consider as the proper characteristics of these facts of knowledge, take the risk of projecting their own prejudices onto the documents. To borrow from the case made by George Smith (see n. 26), the persistence of Newtonian results in modern physics and the cumulative aspect of science were simply rendered invisible to the Kuhnian generation of historians of science raised in a conception of scientific knowledge as a noncumulative process. The perception of a (historical) fact, like any other fact, is dependent on an epistemic background that renders this fact visible. To detect continuity or discontinuity, while identifying the context pertinent to a fact of knowledge, one has to rely on this background, which does not amount to a simple reflexivity (epistemology of history) but belongs to a certain theory of knowledge (epistemology tout court).

Even radical constructionists have in mind a certain conception of how meaning and knowledge are supposed to function. The repeated references to the Wittgensteinian language games in this literature make it quite clear.[27] Consequently, the necessity of an epistemological dimension has nothing to do with a normative position of philosophy. It should rather resemble a kind of hygiene: a way to render the individual conception of where knowledge occurs and how it functions as explicit as possible. Symmetrically, as rightly emphasized by Hacking, this epistemological move is a good way to improve our conception of what knowledge is by confronting it with historical occurrences. The interaction with philosophy should certainly not be considered as one-way here.

I mention this methodological issue because this is precisely where an important problem with the notion of style occurs. In the literature on style, especially on local styles, it is often assumed that the identification of knowledge and its correlation with a certain locus (be it a nation, a school, a laboratory, or an individual) can be expressed in terms of a system of beliefs and understanding and, more generally, in terms of the meaning of the entities in play. Here, for example, is one of the rare explications of this notion by a historian: "While the notion of style is intuitively appealing, there have been few attempts to clarify and apply it convincingly to comparative cases. My preference is to analyze *communication processes*, to identify basic units of communication with *shared epistemologies of science* and to study interaction between such units; which may either lead to a *negotiation* for a common understanding, or to misunderstandings" (Schubring 1996, 363). The idea that styles have to do with shared epistemologies of science and common understanding or misunderstanding is indeed very widespread—whether explicit or not.[28] It is also a good example of the danger of hidden epistemological assumptions. Indeed, it is clear from Gert Schubring's (2005) analysis that the system of beliefs he would favor is based on agreement/disagreement on the nature of some fundamental concepts and/or values.

I do not intend to deny that this type of dis/agreement plays a very important role in the history of science, but I would like to emphasize the fact that the contrary is also true. More often than not, actors share certain aspects of their scientific activity even if they strongly disagree on the nature of the fundamental objects/concepts/theories and values involved. More interesting, as was pointed out by Putnam (one of Hacking's metaphysical heroes), it has to be the case if we want any scientific theory to have a history. If we want representations of a scientific entity (e.g., the atom) to change through history in such a way that one version (e.g., Ernest Rutherford's) is incompatible with the other (e.g., Erwin Schrödinger's), but without abandoning them to totally incommensurable theories, we have to give up the idea that scientific activity is based on transparent concepts or coherent systems of beliefs about the description of objects and theories—what Putnam (1975, 275–276) called the "principle of the benefit of doubt." To give an account of the historical evolution of concepts and/or theories, we need to delineate what is invariant under this evolution; for obvious reasons, concepts and theories themselves, at least as classically understood, cannot be good candidates here. In any case, and this is my main point, this discussion is a typical area where epistemological issues seem inescapable.

Other Examples and an Epistemological Program

To convince the reader that the examples mentioned by Chevalley are not isolated, I would like to devote the last part of this chapter to briefly indicating other examples. I start by summarizing the orientations of a joint program with my colleague Sébastien Maronne on the development of the Cartesian style of geometry from 1650 to 1750 and then turn briefly to other cases studied by historians of mathematics.

The Cartesian Style of Geometry

By "Cartesian style of geometry" I refer, to start in a rather simplistic fashion, to the mix of algebra and geometry that Descartes and other mathematicians developed in the first half of the seventeenth century and that stabilized itself at the beginning of the next century, finally to become, after a long period of time, what would later be called analytic geometry or, better, algebraic geometry. The Cartesian style of geometry was already one of the examples chosen by Granger (1968), and we saw that Hacking also alludes to it by mentioning the algebrizing of geometry. It will therefore provide a convenient way to indicate the divergence between their approaches and the one I advocate in this chapter.

According to Granger (1968, 47–55), Descartes's style has first to be characterized by the philosophical idea that spatial extension can be reduced to what is clearly and distinctly knowable in it, that is, what in Descartes's parlance can be accessed through a process of "exact" measurement. The control over these geometrical entities is ensured by the fact that they conform at the same time to a kind of regular construction and to a form of computation.[29] On this ground, geometry can be rigorously based on algebraic calculations, and the acceptable geometrical objects can be redefined as those accessible through the algebraic manipulations. This allows an extension of the realm of ancient geometry (what would nowadays be called algebraic curves of any degree), a complete classification of these geometrical objects, and a clear exclusion of what is not accessible through these exact determinations (what Descartes designates as "mechanical" curves). One can see that, considered as a way of reasoning, it could be described (as proposed by Hacking) as a combination between the ancient postulational style and the new algorismic one. Regarding the introduction of new objects, the Cartesian style appears more like an extension than like a creation in the strict sense (it extends the realm of acceptable curves in geometry).

This overall description could sound like a quite neutral characterization, if based on a purely conceptual analysis. In fact, it corresponds to the most common way to characterize this new geometry, especially among philosophers. From the point of view of the historian, however, it raises several difficulties, summarized as three general debates.

First, there is a lively debate among historians of mathematics about whether or not the core of Descartes's *Géométrie* consists of treating curves through algebraic equations. According to commentators like Henk Bos, this feature, although very important, has to be interpreted in a larger setting: the program of the geometrical construction of equations. According to Bos (1984, 2001), this long-forgotten context is the only way to understand the very puzzling structure of the treatise. Other important arguments support of this claim; for example, Descartes never treats straight lines and circles by their equations (whereas circles play a very important role, e.g., in his construction of normals to curves). According to other commentators, such as Enrico Giusti (2000), the core of Descartes's project is the method for finding normals demonstrated in the second book of *La Géométrie*. In this case, the identification of curves with their equation seems central and constitutes the core of the Cartesian modernity.

Second, if we take Cartesian geometry as related to the belief that any acceptable curve in geometry has to be definable in terms of an algebraic characterization, it appears that this is precisely the point that was not accepted by Descartes's heirs. Although there is no doubt that Descartes defined only geometrical curves in this way, and although he clearly grounded his overall project on a kind of calculus of segments (presented at the beginning of book I), which limits the realm of acceptable (constructible) objects to algebraic ones, it is a very striking historical fact that he was not followed by his first readers, even the most faithful ones, regarding these two aspects. This supposed shared epistemology (the redefinition of the very nature of mathematical objects, in this case geometrical entities) was therefore not the core of the real change that occurred in mathematics in the middle of the seventeenth century, and of the acceptance of a new style of doing mathematics.

Third, any attempt to base a change of style solely on an internal conceptual analysis, as Granger proposes, risks ignoring some documents attesting to the same kind of conceptual change occurring in previous authors with less success or simply without sufficient transmission. This is precisely the case with the algebrizing of geometry, which as expressed by Hacking is first an "Arabicizing of the Greeks." A better knowledge of the Arabic sources has shown that there were important predecessors of Descartes as early as the ninth century. We therefore need a better way to assess the proper originality of Descartes's achievement (if there is one) and the real beginning of a new style (as opposed to the extension of an already existing algebraic style in geometry; see Rashed 2005).

What is interesting in the characterization of styles as ways of writing is that it allows us to shortcut all of these difficulties. As this description of the three main debates surrounding the interpretation of Descartes's *Géométrie* shows, such a characterization is of major historiographic importance. Regarding the first debate, it is clear that in our view both interpretations can be true at the same time, since *La Géométrie* is situated exactly at the juncture of different lines of historical development. More than that, these lines, developed in one and the same book, might run in opposite conceptual directions (especially regarding the role of geometric intuition in mathematics and the identification of curves with algebraic counterparts). History, in contrast to systematic philosophy, has no problem with contradiction, since the same fact can be interpreted over time in quite opposite ways.

Even if Bos is right that the treatise is governed by a strange composition, which has to be explained historically by reference to a particular context (that of the construction of equations), and even if it is true that Descartes was not systematically treating curves through equations, and more generally—as he still explained to Princess Elisabeth in 1643—still heavily relied on some diagrammatic elements in his analysis of mathematical problems,[30] Giusti is also right that some important readers did not pay much attention to that general structure and focused on the method of normals expounded in the second book. For this method, the identification of curves with equations and the use of proper algebraic tools are pivotal. Although it was still linked in Descartes's practice with some diagrammatic elements, it is undeniable that authors such as Johan Hudde or Isaac Newton increasingly gave it a purely algorithmic nature (see Panza 2005).

Finally, and related to the second and third debates, the attention paid to ways of writing appears a good way to escape difficulties created by the conflation of conceptual and stylistic changes. Indeed, one can claim that Descartes's ideas are located simultaneously in the continuity of important conceptual changes, which already occurred with the algebrization of geometry in Arabic mathematics, and at the origin of a new way of writing mathematics. The best way to indicate this difference between the conceptual and the stylistic level is expressed through the second debate: there was a circulation of Descartes's style made explicit by the actors themselves but based on profound disagreements regarding the basic concepts involved in this new kind of geometry. What these authors took up, I claim, was not some Cartesian conceptions but a certain way of doing mathematics, which was embodied in a new way of writing it. What struck them was not a redefinition of mathematical objects but powerful techniques for solving geometrical problems, which could be interpreted in various ontological frameworks.

One might object that styles as ways of writing designate here simply what can be better characterized as the circulation of a certain method. But this is not the case: many Cartesian mathematicians, such as John Wallis or James Gregory, did not follow Descartes's method for normals and preferred Fermat's. In the latter, the core procedure is not the identification of the coefficients of two equations, as in Descartes, but another technique relying on the introduction of a quantity e, which acts as an increment of the variable and which, after identification of some points on the curve, can be considered as becoming zero. At that stage it can therefore be eliminated from the equations to produce the expected result. The question is then, what did our authors consider as being new and equally powerful in *both* these methods (if they did)?

One possible answer to this question is the striking feature of these two methods that they display inferences that are simultaneously coupled to a geometrical setting (more precisely, to the resolution of geometrical problems) but have no geometrical interpretation whatsoever.[31] Both Descartes's and Fermat's methods rely on a kind of inferential black box coupled with geometrical reasoning. This allows us to give a more precise characterization of the Cartesian style (at least for one important aspect): its core is not the use of algebra in and of itself (which existed long before Descartes and Fermat) but the coupling of specific kinds of computational inferences with geometrical ones. In this sense, one can say that the Cartesian style of geometry, even if it did not suddenly disappear, took a dramatic turn around 1750 with the first formulations, which, as later emphasized by Joseph-Louis Lagrange, were free from any diagrammatic inferences—Leonhard Euler (1748) can be considered a starting point here. It would give rise to a new style that Chevalley (along with many other actors and commentators) described in his paper as algebraic analysis (and was later replaced by that of Augustin-Louis Cauchy and Weierstrass; see Fraser 1989; Jahnke 1993; Panza 2007).

The Leibnizian Style of Differential Calculus

The next example is that of Gottfried Wilhelm Leibniz's new style of (transcendental) geometry, which shows many analogies with the Cartesian style. It is intriguing that the Leibnizian way of writing differential calculus developed so quickly at the beginning of the eighteenth century, with such deep disagreements on the nature of its fundamental concepts and objects. To establish this in detail would require a separate study, but let me just cite one beautiful passage by Leibniz:

> When our friends were disputing in France with the Abbé Gallois, Father Gouye and others, I told them that I did not believe at all that there were actually infinite or actually infinitesimal quantities; the latter, like the imaginary roots of algebra, were only fictions, which however could be used for the sake of brevity or in order to speak universally. . . . But as the Marquis de l'Hôpital thought that by this I should betray the cause, they asked me to say nothing about it, except what I already had said in the Leipzig *Acta*. (Leibniz to Pierre Dangicourt, September 11, 1716, quoted in translation in Jesseph 2008, 230)

In other words, when the main supporters of the Leibnizian calculus in the *Académie des Sciences* asked Leibniz to take a clear public position on the nature of the fundamental entities involved in his calculus, they realized that he disagreed with them (and asked him politely to keep silent so as not to ruin their whole strategy!). This phenomenon is not limited to the first generation of supporters of the differential calculus. A profound evolution of the concept involved in differential calculus, already emphasized by Jean le Rond d'Alembert, was the progressive introduction of a new interpretation of its basic constituents in terms of limit processes—an interpretation that is put forward in the Newtonian tradition.[32] The Leibnizian calculus developed quite rapidly as a mixture of features derived from Leibniz's and Newton's techniques, whereas the two men were involved at the end of their lives in an important controversy about their respective originality and launched two apparently incompatible traditions (culturally speaking).[33]

Set Theory

A similar example, closer to modern day, could be the development of set theory. It is well known that set theory was involved in important debates and disagreement about conceptual issues. The distinction between set theory as a way of writing and as an axiomatic theory is a very important historical fact for understanding the incredible success of the style, despite the quarrel over what was involved in the theory. In his study of this history, José Ferreiros (1999, xix) places great emphasis on the distinction "between set theory as an autonomous branch of mathematics—as in *transfinite* set theory or *abstract* set theory—and set theory as a basic tool or language for mathematics: the set-theoretical *approach* or the *language* of sets. As indicated above, abstract set theory came about after the set-theoretical approach began to develop, not the other way around."

Although Ferreiros does not dwell upon this aspect in his book, the distinction was also of importance in the subsequent developments. Hélène Gispert (1995) provides a very interesting study on these issues in the context of the French school of analysis (René-Louis Baire, Emile Borel, and Henri Lebesgue). She shows, in particular, that there were strong disagreements over the interpretation of set theory before the so-called foundational crisis. Moreover, one could be a strong supporter of Georg Cantor's way of writing, as was Borel, and find his theory of transfinite numbers (for which set theory was built as a theory) "totally useless" (*pas la moindre application*, according to Baire).[34] The distinction between the theory and the way of writing is also an important philosophical fact. Indeed, one may wonder, in common with logicians such as Solomon Feferman, whether the adoption of a set-theoretical way of writing implies a real commitment to the so-called theory axiomatized by mathematicians such as Ernst Zermelo and Abraham Fraenkel in the first half of the twentieth century. Feferman and others seem to give a negative answer, which reinforces the impression of discrepancy between the two aspects (see Feferman 1998, 229–248).

The Euclidean Style of Geometry

To underline the epistemological issues at stake in this research program, I present as a last example the characterization of the Euclidean practice of geometry proposed by Ken Manders in his article "The Euclidean Diagram" (2008). Modern mathematicians such as Mortiz Pasch and David Hilbert heavily criticized the use of geometric intuition in ancient geometry because it involved hidden (and sometimes erroneous) assumptions. Without denying the strength of these criticisms, Manders claims that, paradoxically, they underscore the urgent need to understand the *success* of Euclidean geometry. Indeed, no modern mathematician would claim that Euclid's theorems are false. In fact, it is very rare to find false statements in ancient Greek geometry, although some of them may lack the generality we might expect. Hence, it should seem strange, especially for modern mathematicians claiming that mathematics needs rigorous demonstrations, that a geometry relying on misleading intuitions did not lead to false results.[35] But there is another mystery: since Euclid relies heavily on *reductio ad absurdum*, his geometry is full of impossible representations. It is therefore impossible to engage in a semantic reading of them (Manders 2008, 84–87).

Manders's proposal is that both mysteries can be solved at the same time by looking at the way in which diagrams are used not as representing particular entities (what he calls the "semantic" view on diagrams) but as proxies for carrying specific inferences. His interpretation is that Greek mathematicians lacked the resources to express some relations (which he calls "co-exact") through discourse, whereas other ("exact") relations, such as equality, could not be expressed by diagrams. Typical examples of co-exact relations given by the diagram (and only by the diagram) would be intersections, inclusion of one region into another, and contiguity of two regions (Manders 2008, 93). Greek geometers elaborated a style of discourse coupling these two types of complementary resources.

Even if this is not the terminology employed by Manders, what he described in terms of correlation between different kinds of inferences built into the discourse and diagrams (exact and co-exact) corresponds well to what I have proposed to designate as a style—even if, following Reviel Netz (1999), more emphasis could be put on a precise study of the material display involved in this correlation. This approach allows us to recount a very important historical fact: as testified by ancient commentators such as Pappus and Proclus, disagreements about the nature of geometrical objects were common in antiquity. These disagreements went far beyond the philosophical debate mentioned by Hacking on the ontological status of mathematical entities: they also involved mathematical questions about how to manipulate these objects and what standards should be set for the demonstrations. The debates continued among the Arab authors, assuming ever greater importance—to the extent that they contributed to introducing new concepts in geometry, such as geometrical transformations (Crozet 2010).

Material Anchors and Conceptual Blends

To conclude this section, I sketch a kind of epistemological framework that could be adapted to this description of mathematical styles and clarify what I have alluded to on several occasions concerning their material aspect. My main inspiration here comes from a study by the anthropologist Edwin Hutchins about the phenomenon of stabilization in human reasoning. According to Hutchins, ethnological studies have shown at least two very different strategies regarding stabilization. The first is linked to the way in which we stabilize reasoning by interpreting its components (e.g., when we interpret a process of reasoning in a familiar context, which is easier to handle)—this could be called semantic.[36] The other strategy does not appeal to interpretations and meanings but uses (perceived or imagined) material settings, which, so to speak, reason for us. This is what happens, for example, when someone computes with a slide rule or even with fingers (Hutchins 2005, 1565, 1571). In this case, parts of the inferences are tied not to the interpretation of the operations involved but to the very functioning of the material setting. To describe this kind of cognitive strategy, Hutchins coined the term "material anchor" of reasoning. He has provided a series of examples ranging from simple perception (seeing people waiting in line) to much more sophisticated reasoning, such as that used in Micronesian navigation (see Hutchins 2005 for details). The important point for us is that the material anchor functions as a proxy to carry inferences (e.g., first come, first served in the case of the queue) and that this mixture of a material disposition and conceptual elements is a kind of blend that can be treated as a whole (and then completed, extended, or transformed).[37]

As Hutchins noted, this strategy of material anchoring has not yet been thoroughly studied. It opens the way to a very interesting program not only for cognitive anthropology but also for history and philosophy of mathematics. There seems, indeed, to be a form of continuity between this rather widespread use of material anchors in everyday reasoning and the use of symbolic artifacts in mathematics, a continuity that remained hidden because of the modern myth claiming that formal languages are not material, and resemble rather a transparent picture of concepts:

> In some cases it is possible to do the cognitive work by imagining the manipulation of a physical structure. In others, the case of the slide rule for example, it is necessary to manipulate the physical device itself because it is not possible to imagine it accurately enough to be of use. The cultural process of crystallizing conceptual models in material structure and saving those up through time puts modern humans in a world where thinking depends in significant measure on the availability of a set of physical structures that can be manipulated in this way.
>
> A final turn on this path is that when a material structure becomes very familiar, it may be possible to imagine the material structure when it is not present in the environment. It is even possible to imagine systematic transformations applied to such a representation. This happened historically with the development of mathematical and logical symbol systems in our own cultural tradition. Beginning as external representations physically embodied and operated upon with manual skills, we learn to imagine them and to operate on the imagined structures. . . .

Unfortunately, much of cognitive science is based on the mistaken view that this relatively recent cultural invention is the fundamental architecture of cognition. (Hutchins 2005, 1575)

This picture conforms to what I have tried to develop in this chapter: style as ways of writing can indeed be considered as a kind of blend in which a material anchor is coupled with a conceptual space.

Conclusion

My claim in this chapter is certainly not that ways of writing offer the one and only way to understand the phenomenon of circulation and stabilization in history of mathematics, or that this is the only legitimate use of the category of style. I claim, however, that this use is a suitable and fruitful way to understand some phenomena of stabilization found in the history of mathematics—precisely the case in which conceptual variability is important. Moreover, I claim that it offers a very important view on the nature of mathematical knowledge, which places objects, concepts, and theory in the background and brings material displays and the kind of inferences they carry to the fore. The material anchors approach, because it places the representational aspects in the background, is also valuable for understanding how these mathematical theories, concepts, and objects can stabilize in history without projecting a teleological view onto their development as already constituted (consequently denying the very idea of their stabilization by a historical process). A study of mathematical styles could therefore help in sketching an answer to one of the main challenges currently posed in the integration of history and philosophy of science: how to give an account of the constitution of theories and domains of objectivity as a historical process.

Notes

Unless otherwise indicated, all translations are my own.

1. On the epidemic character of style, see the survey Gayon 1996 or its English translation, Gayon 1999. For a complementary survey of the (rather sparse) literature on mathematical style, see Mancosu 2009.

2. On this typology, see Gayon 1996.

3. My use of this notion is opposite to Schubring's, as according to him shared epistemologies could be used as a good characterization of style.

4. On the characterization in terms of form and content, see also Granger 1968, 6–7, as well as Otte 1991, 238: "Style does not mean the form or the relation between form

and content but the meaning of a unity of the two. This unity however is not absolute and static but is a unity of process."

5. See also n. 34 about the fact that a style accompanies the introduction of new objects. Hacking (2012, 606) takes the same stance in his recent article on the subject: "Styles of scientific thinking and doing were said to be characterized by their objects and their methods of reasoning. Early on, the styles project maintained that each style of scientific thinking & doing introduces a new class of objects."

6. This first aspect would be accepted by Crombie and Hacking, but on a much larger scale—way too large, in fact. According to these two authors, axiomatic style is indeed a style of reasoning that has remained stable since the Greeks! As we shall see, this hinders them from seeing the dynamic aspect of modern axiomatic style that distinguishes it very clearly from the ancient one. Chevalley (1935, 384) places great emphasis on this dynamic aspect (see n. 15).

7. Hacking (2002, 181–182) also follows this strategy, remarking that "Crombie does not expressly define 'style of scientific thinking in the European tradition.' He explains it ostensively by pointing to six styles that he then describes in painstaking detail." Hacking describes his own contribution as a way to isolate "at least a necessary condition for there being such a 'style.'"

8. "Mathematical style, just like literary style, is subject to important fluctuations in passing from one historical age to another. Without doubt, every author possesses an individual style; but one can also notice in each historical age a general tendency that is quite well recognizable. This style, under the influence of powerful mathematical personalities, is subject every once in a while to revolutions that inflect writing, and thus thought, for the following periods" (Chevalley 1935, 375; translation in Mancosu 2009, sec. 4).

9. The canonical example of this style, still familiar to undergraduate students, is the modern definition of "continuity." For a real function (i.e., a function mapping some element x of the set of the real numbers, $\mathbb{R}$, to other elements $f(x)$ of $\mathbb{R}$), it would take the following form:

"Let f be a real function, x_0 a point in its domain of definition, and ε, δ, some positive quantities ($\in \mathbb{R}^+$).

The function f would be said to be continuous in x_0 if:

For any ε, there exist a δ so that: $|x - x_0| \le \delta \Rightarrow |f(x) - f(x_0)| \le \varepsilon$."

I intentionally do not write symbols for quantification, since they were invented *after* the development of the Weierstrass style. What is important in this style is not only that it provides rigorous definitions for some intuitive notions, such as continuity, but also that it commands a repertoire of classical techniques of demonstration, which were developed by Weierstrass and his school.

10. "Nicolas Bourbaki" was the pseudonym of a group of French mathematicians created at the end of 1934, which played a central role in the development of contemporary mathematics. Claude Chevalley was a member of this group from the very beginning, alongside with Jean Dieudonné, André Delsarte, Charles Ehresmann, and André Weil.

11. He continues: "The various theories, which were until now kept separated, did not presumably reach yet their final form. Many of them could certainly be analyzed as

superimpositions of more general theories; others are felt to be equivalent or to derive from a same source. The structural analysis of already known facts is hence far from being achieved, not to speak of new facts which manifest themselves from time to time" (Chevalley 1935, 384). Among the Bourbaki group this opinion does not seem to have been unique to Chevalley, and it is noteworthy that the same kind of claim is made at the end of "L'Architecture des Mathématiques," the famous manifesto published by the group in 1948 (for an English translation, see Bourbaki 1950).

12. I am grateful to Moritz Epple for pointing out this historical context to me when I first presented my ideas on Chevalley's notion of style in 2008.

13. *Blut und Boden* (Blood and Soil) is a racial ideology, promoted by the Nazi movement, which emphasizes the relationship between population and territory. It supported a description of psychology types according to races (like that of Jaensch).

14. See Chemla and Keller 2008, a collection of papers presented at the 2006 workshop "When Nations Shape History of Science," and subsequent work by Agathe Keller on Vedic mathematics (Keller 2003 and www.reseau-asie.com/colloque/thematiques -du-1er-congres-2003/atelier-2-histoire-des-sciences-en-asie-history-of-sciences-in-asia/, accessed August 1, 2016); see also Raina 1999.

15. "Without doubt, every author possesses an individual style; but one can also notice in each historical age a general tendency that is quite well recognizable" (Chevalley 1935, 375).

16. One could also note that Fréchet (another name mentioned by Chevalley) developed a kind of "empiricist" view of geometry quite different from Hilbert's "formalism" and Göttingen proto-structuralism; see Arboleda and Recalde 2003.

17. This second distinction (between individual and community), not equivalent to the first one (between methodological and cultural), is also often referred to as that between an "individualizing" and a "generalizing" concept of style (e.g., the style of Shakespeare vs. Shakespearean style). Philosophers usually insist on the fact that they consider only the generalizing notion.

18. This is especially true through the notion of national style (see Daston and Otte 1991) or styles attached to given schools.

19. In the introduction to his essay Hacking (1992, 130) emphasizes this important "philosophical task in our times" as a way to reconnect: "(1) Social studies of knowledge, of the sort pioneered by David Bloor and Harry Barnes in Edinburgh. . . . (2) Metaphysics, particularly the debates that resulted from Hilary Putnam's series of revised positions. . . . (3) The Braudelian aspect of science, that is, the long-term slow-moving, persistent, and accumulating aspects of the growth of knowledge."

20. This approach is not yet well developed in the history of mathematics, but some very interesting indications can be found in the notion of "readings" introduced in Goldstein 1995. According to this study (on one of Fermat's theorems), the history of a mathematical result can be described through the series of its interpretations, whether they are made by historians (*lectures historiennes*) or by mathematicians (*lectures mathématiciennes*). No particular epistemological background seems to be needed to reconcile long-term history and local determinations.

21. Hacking (2002, 192) talks about "quasi-stability of science." For a very interesting case study on the persistence of Newtonian physics into the contemporary field, see Smith 2010.

22. Hacking 2002, 182. The letters refer here to the six different "styles of reasoning" isolated by Crombie and summarized by Hacking.

(a) The simple method of postulation exemplified by the Greek mathematical sciences

(b) The deployment of experiment both to control postulation and to explore by observation and measurement

(c) Hypothetical construction of analogical models

(d) Ordering of variety by comparison and taxonomy

(e) Statistical analysis of regularities of populations, and the calculus of probabilities

(f) The historical derivation of genetic development

23. Note the important—and very debatable—qualification of "applied" mathematics.

24. "Every style of reasoning introduces a great many novelties including new types of: objects / evidence / sentences, new ways of being a candidate for truth or falsehood, laws, or at any rate modalities / possibilities. . . . Each style, I say, introduces a number of novel types of entities, as just listed. Take objects. Every style of reasoning is associated with an ontological debate about a new type of object. Do the abstract objects of mathematics exist? That is the problem of Platonism in mathematics" (Hacking 2002, 189).

25. "The historian will want to distinguish several types of events. There is the extinction of a style, perhaps exemplified by reasoning in similitudes. There is the insertion of a new style that may then be integrated with another, as has happened with algorismic reasoning, combined with geometrical and postulational thought" (Hacking 2002, 194).

26. These styles are the ones proposed in De Lorenzo 1971, summarized in Mancosu 2009. Granger 1968 also proposed contrasting Descartes's style of geometry with that of Girard Desargues (see also Otte 1991). He used the example of the "vectorial style" that flourished at the beginning of the nineteenth century (in a completely different way of algebrizing geometry than that of analytical geometry).

27. The evolution of Bruno Latour's conception toward a fully fledged metaphysical framework is quite significant in this regard; see Latour 2012.

28. It is very pregnant, for example, in Ludwik Fleck's (1935) characterization of the *Denkstil* (thought style) associated with a *Denkkollektiv* (thought collective).

29. Descartes called this kind of regular construction "continuous movement." The form of computation is related to the way in which the constructions involved can be described as combinations of geometrical proportions. This geometrical calculus is technically related to two important mathematical advances: the abandonment of the homogeneity constraint when dealing with geometrical magnitudes and, along with it, a proper definition of multiplication between them, able to bypass the traditional problems of dimensionality.

30. René Descartes to Elisabeth, Princess of Bohemia, November 1643 (Descartes 1996, 38).

31. By contrast, one main goal of authors who brought algebra into geometry since the Arab mathematicians was to secure algebraic inferences by grounding them either in substitutions operating on basic equalities or in geometry.

32. See the seminal paper by Bos 1974.

33. On the tradition of the Newtonian calculus, see Guicciardini 1989.

34. As Baire expressed in a letter to Borel in 1905 (quoted by Gispert 1995, 58).

35. Note that exactly the same point applies to the first century of differential calculus.

36. Interestingly, Hutchins (2005, 1558) describes this type of stabilization as "culturally based."

37. Note that a material anchor can enter into quite complex processes and involve different layers of conceptual mappings; see Hutchins 2005, 1563–1565.

References

Arboleda, L. C., and L. C. Recalde. 2003. "Fréchet and the Logic of the Constitution of Abstract Spaces from Concrete Reality." *Synthese* 134(1–2): 245–272.

Bieberbach, L. 1934. "Stilarten mathematischen Schaffens." *Sitzungsbericht der preußischen Akademie der Wissenschaften*, 351–360. Berlin: Akad. d. Wissenschaften.

Bos, H. J. M. 1974. "Differentials, Higher Order Differentials and the Derivative in the Leibnizian Calculus." *Archive for History of Exact Sciences* 14: 1–90.

Bos, H. J. M. 1984. "Arguments on Motivation in the Rise and Decline of a Mathematical Theory: The 'Construction of Equations,' 1637–ca. 1750." *Archive for History of Exact Sciences* 30: 331–80.

Bos, H. J. M. 2001. *Redefining Geometrical Exactness: Descartes' Transformation of the Early Modern Concept of Construction*. New York: Springer.

Bourbaki, N. 1950. "The Architecture of Mathematics." *American Mathematical Monthly* 57(4): 221–232.

Brechenmacher, F. 2010. "Une histoire de l'universalité des matrices mathématiques." *Revue de synthèse* 131(4): 569–603.

Chemla, K., and A. Keller, eds. 2008. "When Nations Shape History of Science." *Disquisitions on the Past and Present* 18: 169–210.

Chevalley, C. 1935. "Variations du style mathématique." *Revue de Métaphysique et de Morale* 3: 375–384.

Corry, L. 2004. *Modern Algebra and the Rise of Mathematical Structure*. 2nd ed. Basel: Birkhäuser.

Crombie, A. 1994. *Styles of Scientific Thinking in the European Tradition*. London: Duckworth.

Crozet, P. 2010. "De l'usage des transformations géométriques à la notion d'invariant: La contribution d'Al-Sijzī." *Arabic Sciences and Philosophy* 20(1): 53–91.

Daston, L., and M. Otte, eds. 1991. "Style in Science." Special issue of *Science in Context*, 4(2).

De Lorenzo, J. 1971. *Introducción al estilo matematico*. Madrid: Editorial Tecnos.

Descartes, R. 1659–1661. *Geometria à Renato Descartes anno 1637 Gallice edita; postea autem Unà cum Notis Florimondi De Beaune . . . Operâ atque Studio Francisci a Schooten. . . .* Amsterdam: Apud Ludovicum and Danielem Elzevirios.

Descartes, R. 1996. *Œuvres de Descartes*. Edited by C. Adam and P. Tannery. Vol. 4. Paris: Vrin.

Euler, L. 1748. *Introductio in analysin infinitorum.* Lausanne: Bousquet.

Feferman, S. 1998. *In the Light of Logic.* Oxford: Oxford University Press.

Ferreiros, J. 1999. *Labyrinth of Thought: A History of Set Theory and Its Role in Modern Mathematics.* Basel: Birkhäuser.

Fleck, L. 1935. *Entstehung und Entwicklung einer wissenschaftlichen Tatsache: Einführung in die Lehre vom Denkstil und Denkkollektiv.* Basel: Schwabe.

Fraser, C. 1989. "The Calculus as Algebraic Analysis: Some Observations on Mathematical Analysis in the 18th Century." *Archive for History of Exact Sciences* 39: 317–331.

Gayon, J. 1996. "De la catégorie de style en histoire des sciences." *Alliage* 26: 13–25.

Gayon, J. 1999. "On the Uses of the Category of Style in the History of Science." *Philosophy and Rhetoric* 32(3): 233–246

Geertz, C. 1983. *Local Knowledge: Further Essays in Interpretive Anthropology.* New York: Basic Books.

Gispert, H. 1995. "La théorie des ensembles en France avant la crise de 1905: Baire, Borel, Lebesgue et tous les autres." *Revue d'histoire des mathématiques* 1(1): 39–81.

Giusti, E. 2000. *La naissance des objets mathématiques.* Paris: Ellipses.

Goldstein, C. 1995. *Un théorème de Fermat et ses lecteurs.* Saint-Denis: PUV.

Granger, G. G. 1968. *Essai d'une philosophie du style.* Paris: Armand Colin.

Guicciardini, N. 1989. *The Development of Newtonian Calculus in Britain, 1700–1800.* Cambridge: Cambridge University Press.

Hacking, I. 1992. "Statistical Language, Statistical Truth and Statistical Reason: The Self-Authentication of a Style of Scientific Reasoning." In *Social Dimensions of Sciences,* ed. E. McMullin, 130–157. Notre Dame, IN: University of Notre Dame Press.

Hacking, I. 2002. *Historical Ontology.* Cambridge, MA: Harvard University Press.

Hacking, I. 2012. "'Language, Truth and Reason' Thirty Years Later." *Studies in History and Philosophy of Science* 43(4): 599–609.

Hilbert, D. 1899. *Grundlagen der Geometrie.* Leipzig: Teubner.

Hutchins, E. 2005. "Material Anchors for Conceptual Blends." *Journal of Pragmatics* 37: 1555–1577.

Jahnke, H.-N. 1993. "Algebraic Analysis in Germany, 1780–1840: Some Mathematical and Philosophical Issues." *Historia Mathematica* 20(3): 265–284.

Jesseph, D. 2008. "Truth in Fiction: Origins and Consequences of Leibniz's Doctrine of Infinitesimal Magnitudes." In *Infinitesimal Differences: Controversies between Leibniz and His Contemporaries,* ed. U. Goldenbaum and D. Jesseph, 215–233. New York: De Gruyter,

Keller, A. 2003. "Vedic Mathematics: Enjeux multiples de l'histoire des sciences en Inde." halshs-00069276. Accessed June 24, 2016. halshs.archives-ouvertes.fr/docs/00/06/92/76/PDF/VM_enjeux_multiples.pdf.

Lacan, J. 1973. "L'étourdit." *Scilicet* 4: 5–52.

Latour, B. 2012. *Enquête sur les modes d'existence: Une anthropologie des modernes.* Paris: La Découverte.

Mancosu, P. 2009. "Mathematical Style." In *The Stanford Encyclopedia of Philosophy,* ed. E. N. Zalta. plato.stanford.edu/archives/spr2010/entries/mathematical-style.

Manders, K. 2008. "The Euclidean Diagram." In *The Philosophy of Mathematical Practice*, ed. P. Mancosu, 80–133. Oxford: Oxford University Press.

Netz, R. 1999. *The Shaping of Deduction in Greek Mathematics: A Study in Cognitive History*. Cambridge: Cambridge University Press.

Otte, M. 1991. "Style as a Historical Category." In "Style in Science," ed. L. Daston and M. Otte. Special issue of *Science in Context* 4(2): 233–264.

Panza, M. 2005. *Newton et les Origines de l'analyse, 1664–1666*. Paris: Blanchard.

Panza, M. 2007. "Euler's *Introductio in analysin infinitorum* and the Program of Algebraic Analysis: Quantities, Functions and Numerical Partitions." In *Euler Reconsidered: Tercentenary Essays*, ed. R. Backer, 119–166. Heber City, UT: Kendrick Press.

Poincaré, H. 1883. "Sur un théorème de la théorie générale des fonctions." *Bulletin de la Société Mathématique de France* 11: 112–125.

Putnam, H. 1975. "Language and Reality." In *Mind, Language and Reality*, Vol. 2 of *Philosophical Papers*, 272–290. Cambridge: Cambridge University Press.

Raina, D. 1999. "Nationalism, Institutional Science and the Politics of Knowledge." PhD diss., Göteborgs Universitet.

Rashed, R. 2005. "La modernité mathématique: Descartes et Fermat." In *Philosophie des mathématiques et théorie de la connaissance: L'œuvre de Jules Vuillemin*, ed. R. Rashed and P. Pellegrin, 239–252. Paris: Albert Blanchard.

Schubring, G. 1996. "Changing Cultural and Epistemological Views on Mathematics and Different Institutional Contexts in Nineteenth-Century Europe." In *Mathematical Europe: Myth, History, Identity*, ed. C. Goldstein, J. Gray, and J. Ritter, 363–388. Paris: Editions de la Maison des Sciences de l'Homme.

Schubring, G. 2005. *Conflicts between Generalization, Rigor and Intuition*. New York: Springer.

Segal, S. 2003. *Mathematicians under the Nazis*. Princeton, NJ: Princeton University Press.

Smith, G. E. 2010. "Revisiting Accepted Science: The Indispensability of the History of Science." *Monist* 93(4): 545–579.

III. THE MAKING OF SCIENTIFIC CULTURES

9

Historicizing Culture

A Revaluation of Early Modern Science and Culture

The construction of cultural difference and conflict has been a long-standing practice, not only in politics, the media, or the arts but also in the sciences. In one prominent example, Eastern and Western scientific cultures have supposedly been separated by a great divide in mentalities, ways of reasoning, or social organization.[1] This view is still being reinforced, for instance by a recent opinion piece arguing that "scientific culture in China is quite different from Europe and the U.S."[2] Some hold traditional Chinese culture responsible for impeding China's scientific progress (Gong 2012) and for recent scientific fraud scandals.[3] As a reaction to the often generalizing, essentialist, determinist, and discriminatory tendencies of much discourse about culture,[4] some scholars have proposed that we refer to "culture" in the plural only, indicating the internal diversities and instabilities of cultures.[5] Others have preferred to discard the concept of culture altogether or even to write against it (see Abu-Lughod 2006).[6]

On the one hand, such linguistic purification may seem ridiculous or, at best, window dressing. Indeed, common language use often resists linguistic engineering, especially if it is not accompanied by social action.[7] On the other hand, words and concepts not only reflect but can also initiate changes in social perceptions and realities. The ways in which words, concepts, and social realities intertwine and interact are best visible in historical accounts that cover key moments of change. In this article, I take a historicizing approach to the concept of culture. I study a key moment in the early modern period, when the Jesuit polemicist Antonio Possevino for the first time conceptualized and used extensively the notion of culture to describe activities beyond agriculture. This notion of culture reflected new social realities, brought about by the Reformation and Counterreformation, but more important, it was coined to promote and accompany controversial social and scientific change. As I will show, the early modern notion of culture was developed by Possevino to spearhead scientific and educational reform, promoting a political as well as an epistemic program.

Historicizing culture not only describes how historical uses and abuses of the concept relate to changing realities, power relations, and ideologies in the past; it also contributes to a more refined and reflexive application of the concept today. Historicization can uncover multiple layers of meaning that may be hidden within contemporary words and concepts and open up older and newer meanings for current use. It counteracts essentializing tendencies, elucidating the fractured nature of concepts, giving insight in how current meanings have been generated in sometimes very different conditions and contexts. Striking resonances with the past may also help us take a fresh perspective on current debates. Historicization thus brings more nuance and differentiation to current uses of the term "culture" without abolishing the concept. This may in turn lead to new shifts in social perceptions and may even inspire social change.

The Historiography of Science and Culture

Since the rise of cultural anthropology, the concept of culture has long been popular in academia, including history of science and science studies. Often, the concept is used as a kind of stopgap, a useful word when one cannot immediately find a more precise alternative (this is evidenced by most of the books that have "culture" in the title). In contrast, some authors have introduced culture as a well-delineated analytical category, as in epistemic culture (Knorr Cetina 1999), epistemological culture (Keller 2002), or cultures of experimentation (Rheinberger 1995, 1997, 2011), to gain more insight into scientific practices.[8] A different but well-established approach in the history of science is to focus on actors' categories. Historians of science have been remarkably reflexive about key concepts, including the notion of science itself.[9] This has not been the case for the concept of culture, especially for the earlier periods. In this chapter, I study culture as an actor's category in the early modern period, focusing on concrete uses of the word, especially in the context of the traditional sciences, epistemic institutions, and educational reform.

The recognized reference for a historical study of keywords, and of the word "culture" in particular, is the work of Raymond Williams.[10] He contrasts his approach to philology and etymology because he connects actors' categories to the social world to which they belonged (Williams 1983, 19). The keyword "culture" was special, Williams believed, because it provides unique access for understanding these social realities. A common objection to this approach was succinctly formulated by Quentin Skinner (1980, 562): "What can we hope to learn about our culture and society through studying the changing meanings of words?" If we want to understand how people in the past saw the world, what distinctions they made, and which classifications they accepted, we should focus on the concepts they possessed, not the words they used. Nevertheless, Skinner concludes that there is a lot we can learn from the history of keywords: "The surest sign that a group or society has entered into the self-conscious possession of a new concept is that a corresponding vocabulary will be developed, a vocabulary which can then be used to pick out and discuss the concept with consistency" (564). A history of words is important, even if they do not track concepts perfectly. Sensibility to actors' categories not only gives a more precise understanding of past thought, but discovering significant shifts in past meanings of words also uncovers key moments of conceptual innovation often occurring in tandem with social change.

Raymond Williams located the origin of the modern concept of culture in the late eighteenth century. Inspired by Williams's work, scholars such as Andrew Cunningham and Perry Williams (1993, 409–410) located the origin of science during 1760–1848, interpreting science as an intrinsically modern invention. This is not a coincidence, as the development of the concepts of science and culture is intertwined, and important changes have occurred concurrently. For Cunningham and Williams, identifying the origin of science is predicated on similarities with what we today mean by the word "science," that is, science "in the modern sense" (420–421). But what Cunningham and Williams construct as relevant similarities (e.g., the formation of new disciplines, or values such as objectivity) can surely be displaced by a focus on differences. The contrast between the tabletop science of nineteenth-century gentlemen practitioners and post-WWII big science, defined by the military-industrial-scientific complex, can seem more striking than the similarities.[11] Something similar holds for Raymond Williams's claims. Instead of looking for the origins of the concept of culture in the late eighteenth century, one could focus on a different set of similarities, for instance, with the 1940s or with the early modern period.[12] Searching for origins has lost some of its splendor recently, but presupposing a well-defined modern sense of concepts may also be misguided because it obscures the variety of meanings today. A pertinent outcome of reflection on these debates, in my eyes, would be a clearer awareness that it is possible to distinguish different similarities and dissimilarities among the variegated usages of a concept in different historical periods and geographies. It is important to pay close attention to the changing uses of words and concepts in their own terms and contexts, and from there it will be possible to find relevant similarities and resonances, as for instance between the early modern and current notions of science and culture.[13]

A focus on actors' categories also implies taking into account different languages. The European vernacular words denoting culture (culture, *coltura, cultura, Kultur, cultuur, kultur,* etc.) all derive from the Latin substantive "cultura."[14] The close etymological relations among these words and a shared core of meaning make it reasonable to refer to one concept of culture for this linguistic area, even if this concept may have different layers of meaning that have their own complex history, geography, and linguistic expression.[15] The importance of the Latin "cultura" for the historicization of culture goes beyond etymology, of course, as Latin continued to be the main language for intellectual discourse and international communication until the eighteenth century. The classical word "cultura" has its roots in the active verb "colere," which has a range of meanings: to inhabit, to take care of, to cultivate, to adorn, to praise, to esteem, and to venerate. "Cultura" commonly referred to agriculture, with exceptional metaphorical extensions such as "improving the soul" (*cultura animi*) or "religious worship" (*cultura dei* or *cultura deorum*). The term was applied to something specific, "cultivation of," and implied an activity. Other derivatives of "colere" were *cultus* and *excolere* (the "ex" adds the idea of improving), alternative forms that have long been preferred to the relatively rare "cultura." It is only in the early modern period that the use of "cultura" expanded, changed its meaning and assumed new importance. In the following pages, I study in detail one crucial episode of this momentous shift.

Antonio Possevino's *Cultura Ingeniorum*

The first sustained use of the word "culture" in the West, in the form of the Latin "cultura" and the Italian "coltura," can be found in the *Cultura Ingeniorum* (*The Culture of Talents*) by the Jesuit Antonio Possevino (1533–1611).[16] This text originally comprised the introductory chapters to the huge two-volume *Bibliotheca Selecta* (1593) and became known as the *Cultura Ingeniorum* when it was published separately in Latin (1604) and in Italian (1598). These books were widely distributed in the whole of Europe, even in Protestant regions (see Balsamo 2001). The work marks an increasing use of, and initiates the increasing importance of, the concept of culture in the early modern period. The *Cultura Ingeniorum* presents a Jesuit pedagogical program to culture Christian adolescents in religion and the sciences. The text should be understood in the context of the *Bibliotheca Selecta*, which in many ways presented a mirror image of the *Index Librorum Prohibitorum*. Instead of prohibiting books, the *Bibliotheca Selecta* was a compendium of orthodox texts deemed proper for Christian students, approved by a special high committee and by the pope himself. The *Cultura Ingeniorum* should be read in the context of Possevino's militant work for the Society of Jesus. Read as such, it presents a textbook for an epistemic Counterreformation.

Possevino was a second-generation Jesuit, one of the Society's foremost activists and polemicists. In his first years as a Jesuit he was sent to religiously and politically sensitive areas, where he debated about religion in contests with Protestant leaders and educated noblemen. He innovated religious mass propaganda in many ways, including by posting leaflets and posters everywhere, printing and disseminating catechisms, publishing tracts, and touring towns to barter good Catholic books for heretical ones (which were later destroyed). He even arranged for the books he mentioned from the pulpit to be sold at the entrance of the church when the parishioners left after mass.[17] During his travels as a militant Catholic, Possevino founded many Jesuit colleges in the regions that are now France, Italy, the Baltic states, Poland, the Czech Republic, and Romania. Later, after service also as secretary of the Jesuit Superior General, he became one of the most important diplomats for the Holy See, traveling throughout Europe. After his diplomatic missions to Sweden, he was in Poland and Russia between 1580 and 1585, negotiating with the king and the tsar in an unsuccessful attempt to bring the Russian Orthodox Church into the Catholic fold. After these personal experiences in diplomacy, he concluded that such political strategies were doomed to fail.[18] This reinforced his belief that education and propaganda were the best way to win hearts and minds for the Catholic Church.

Possevino wrote the *Bibliotheca Selecta* after his retirement from diplomacy, but the idea of creating a standard-setting Catholic compendium had long been simmering in his mind ("Causae et idea operis," in Possevino 1593, 1:1–11). As the subtitle of the *Bibliotheca Selecta* indicated, the book provided a method of study (*ratio studiorum*), focusing on history, the sciences (*diciplinis*), and the salvation of all people. Possevino attached great importance to the battle of the books between Protestants and Catholics, and he wanted to respond to Konrad Gesner's influential *Bibliotheca Universalis* (1545), an enormous humanist bibliography, the pinnacle of Erasmian and Reformed scholarship. In the eye of Catholic censors, however, Gesner's work was replete with heretical references, and it was soon put on the *Index*.[19] Possevino's *Bibliotheca Selecta* is more than a Catholic bibliography, however. Created concurrently with the *Ratio Studiorum*, the official plan for Jesuit education published in 1599, the *Bibliotheca Selecta* was a seminal work for the Jesuit educational mission, spelling out its program, its methods, and its means. In particular, it laid down an orthodox canon of traditional scientific disciplines and presented a method for censuring and expunging heretical, evil, and false ideas. It is in this context that the concept of culture became revaluated and acquired prominence for the first time in Western Europe. In Possevino's work, culture came to stand for the Jesuit educational agenda, constructed in direct opposition to the humanists' educational program and their ideal of *humanitas*. Possevino appropriates some of the humanist themes and ideas but accommodates them to the Counterreformation project. To accomplish this transformation, he makes specific use of the agricultural dimension of the word "cultura," making the concept of culture into a slogan for the Jesuit ideology (see Possevino 1610, 29, 40, 71, 110, 133–134).[20]

We can trace the increasing importance of the concept of culture for Possevino by looking at subsequent editions of the *Bibliotheca Selecta* and the *Cultura Ingeniorum*. In the first edition of the *Bibliotheca Selecta* (1593), the first book does not bear a title (except for "Liber Primus"), but the term "cultura," together with its cognates "colere," "excolere," and "cultus," is used many times in the introductory chapters.[21] In a period when these words were relatively rare, it is striking to see them used recurrently in the first sixty-five folio pages of the book. However, only one chapter heading refers explicitly to culture ("The culture of talents that is more concise, easy and fruitful"; Possevino 1593, 1:33).[22] In 1598, Possevino published a separate Italian edition of the twelve introductory chapters (less than half of the original "Liber Primus"), with revised contents and structure, in a translation by Mariano Lauretti. This 115-page book, remarkably titled *Coltura degl'ingegni*, was the first book-length treatment exclusively devoted to culture. An inexpensive quarto edition in the vernacular, it was aimed at being disseminated as widely as possible. What is especially striking in this new edition is the new chapter structure, which was completely revised. Not only did the title of the book refer to culture but also many of the new chapter headings. From chapter 20 to chapter 35, "coltura" (and exceptionally also *coltivare*) appeared in almost every title. In the second revised edition, in 1603, of the complete *Bibliotheca Selecta*, the overhauled chapter structure was retained. The first book was limited to the original first twelve chapters on the cultivation of talents (now divided in fifty-three headings), giving them special importance, while the other chapters on history were moved to a different section. The first book was now named *De Cultura Ingeniorum*,[23] and as in the 1598 Italian edition, "cultura" appears in many of the chapter titles. From then onward, Possevino would publish many separate inexpensive octavo editions, in Latin, of the first book only, under the title *Cvltvra Ingeniorum; examen ingeniorum Io. Huartis expenditur* (*The Culture of Talents; [in Which] the Examination of Talents by Juan Huarte Is Judged*), which were distributed throughout the entire world.[24] Possevino had clearly recognized the potential of the concept of culture and had decided to prioritize "coltura" and "cultura."

It is instructive to look at Possevino's concrete uses of the word "cultura." Most of the time, "cultura" is used in juxtaposition with "ingenium," signifying the cultivation of a talent.[25] "Cultura" is thus an active word that is applied to an object, conforming to the traditional usage in Latin. In a number of cases, however, Possevino uses "cultura" on itself, without an object, even if "ingenium" is still implied. In the course of the book, "cultura" is progressively used in the more abstract sense of education. It is not just the practice of an individual to learn and improve his skills; it becomes a method, a program of study, and finally, an institution. In elaborating this notion of cultura, Possevino draws on the complex network of earlier meanings of the word. Possevino refers to the Hippocratic analogy, for instance, which explicates the relation between science and the mind as the way seeds grow in the soil.[26] The analogy points to the necessary qualities of the mind/soil for the sciences/seeds to take hold. It also highlights the importance of cultivation, of following the right schedule, and of the right fit between sciences/seeds and mind/soil. Possevino goes much further than mentioning this analogy, however—he presents an elaboration of more than one hundred pages, constantly returning to the agricultural metaphor. The analysis is brought to a new level when he presents the details of cultivation, including the selection of the seeds and soil, as well as different ways of manuring, weeding, and pruning, but applied to education. Furthermore, because for him religion and knowledge were inseparable, he is able to interconnect different core meanings of the word "culture"—agriculture, religion, improving the mind—into an indissoluble whole, making the various meanings resonate.

Possevino starts the *Cultura Ingeniorum* with a chapter on the dignity of man, a traditional humanist theme. For the humanists, the ideal was humanitas, becoming fully human by forming oneself. There are many ways to be human, and this freedom exhilarated them. As Pico della Mirandola wrote in his *Oration on the Dignity of Man* (1486): "In man, at the moment of his birth, God has planted omnifarious seeds and the germs of every form of life. Cultivated, they will grow and bear their fruit in him."[27] Man was like a chameleon and could take on all kinds of shapes and identities (see Pico della Mirandola 1557, 315). Although Possevino had earlier been considered a humanist scholar, in the 1590s he presented a very different, Jesuit perspective on the matter.[28] Too much freedom precludes cultivation, he argues, and freedom should be curbed (Possevino 1610, 97–100). For Possevino, the dignity of man lies not in his freedom but in his relation to God as his origin and endpoint. God created man with an inborn aspiration to knowledge and religion, and cultura implied developing the destiny given to you by God (11–12). Yet Possevino argued that trying to cultivate the whole person is a waste of time and energy. Instead, one should cultivate only one's main talent, the seed planted in man by God. He writes: "For knowledge is cultivated [*colitur*] together with religion, when every life and action is directed at one goal and at one ideal" (33).[29] One should avoid fragmenting one's mental powers and losing one's mental sharpness by taking on too much. In Possevino's method, the student has to cultivate his mental faculties and direct them toward one attainable goal. In this way, the whole person would be engaged, and real results ("fruits") could be achieved (67–68). Possevino's approach is characterized by a desire for efficiency and concrete fruits, in service of the student's life goal and, of course, in service of the Catholic world. Cultivating talents will lead to the development of the sciences as well as to individual salvation, but it also assumes that Christians take up their role in a larger divine plan.

Possevino was an excellent communicator and propagandist. As the subtitle of later editions indicates, he also took advantage of the hype around the Spanish physician Juan Huarte's 1575 book, *Examen de ingenios para las ciencias* (*Examination of Talents for the Sciences*).[30] Huarte had developed a physiological theory of man's natural capacities or talents (*ingenium*), based on Galen's theory of qualities and humors. Huarte argued that one should learn to recognize one's own abilities in order to choose a science, art, or profession that corresponds best to one's talents. This would be most satisfactory for oneself, but it would also be efficient and beneficial to society. This inspired Possevino, but what the physician Huarte saw as a natural talent the Jesuit reinterpreted as a gift from God. (He also suggested that child prodigies, usually seen as autodidact miracles of nature, were theodidacts instead; Possevino 1610, 38).[31] It was one's duty to God to develop this given talent, and this was really the central goal of one's life. It was therefore crucial to find out what this talent was. Possevino did not follow Huarte's naturalist suggestions but insisted on a selection mechanism controlled by the Church. To decide about one's future, a young man should make a general confession to an ordained priest and present all the arguments for and against his possible calling. The priest would then present the adolescent's case to God, who would enlighten him and help student and priest to make the right choice. After this life choice, further selection mechanisms were available in universities and cathedral schools. All said and done, the selection of talents and the vocation of students were in the hands of the Church hierarchy.

Possevino's discussion of Huarte was only a preliminary to his real topic. The most important question for him was not how to identify but how to cultivate a young man's ingenium.[32] He writes: "Thanks to culture [*cultura*], the human talent becomes the most divine (said Plato); but if this [culture] is lacking, it [the human talent] becomes diabolic. For the more fertile the fields are, the more evil weeds they produce, except if they are cultivated [*colantur*] and purged" (Possevino 1610, 71).[33] Many people fail in their lives because of lack of talent, illness, a contrary attitude, divine intervention, or other reasons. However, a major cause of failure is lack of education (*culturae defectu*). Prayer, purity of body and soul, discipline, and motivation were all important for making cultivation successful, but for Possevino the key element was the *ordo* or the method of study. In chapter 35, Possevino gives an example of how a model student should not indiscriminately attend lectures by professors. Instead, he should practice Jesuit spiritual exercises to focus his mind, combined with bodily exercise, a healthy regimen, and enough sleep (102–109). Students should study selected works, but instead of just taking notes in lecture classes, they should take care to thoroughly digest the material. After reading, a student should meditate on the material for one hour, after which he should make notes on his thoughts and conclusions (see also Possevino 1610, chaps. 32, 32, 34). As Possevino writes:

Furthermore, it always damages all kinds of talents greatly when teachers neglect the duty of demanding (home)work from their students, when they neglect the repetition, the group discussion, the response (whenever they are questioned) and the disputations.[34] Indeed, just as the field will not be perfectly cultivated if one merely plows or sows it without covering the seeds with earth, in the same way, talents will not be properly cultivated if after public preparatory lectures, that which I have mentioned [i.e., demanding homework, repetition, etc.], is not employed. (93)[35]

Only then will their studies bear fruit (*fructu*), and only then will they, in turn, be able to sow the seeds (*semen*) of their studies in other places for the benefit of society (109).

The *Bibliotheca Selecta* provided a compendium of the sciences buttressing Possevino's method as described in the *Cultura Ingeniorum*. The first part deals with religious disciplines and institutions, including theology, divine history, and pedagogical institutions. It provides extensive refutations of those he considers to be schismatists, heretics, and atheists. The second part deals with worldly sciences such as jurisprudence, natural and moral philosophy, medicine, mathematics (including astronomy, geography, cosmography, etc.), history, art, and literature, and it has a final chapter touching on diverse subjects. The book on philosophy focuses mainly on Plato and Aristotle, stressing the precautionary measures one should take before Christian minds can use their philosophy fruitfully and without detriment ("fructuosius uti absque noxa"; Possevino 1593, 2:94).[36]

Possevino presented the sciences from a religious perspective. Knowledge and religion always went together, and this meant censuring, pruning, and correcting the sciences. Dangerous philosophical systems were censured, along with improper artistic expression.[37] It was necessary to be vigilant, because aside from errors in opinion, students needed to be made aware of the interference of evil ideas and sciences. "But he who conceals the faults that have improperly interfered in the good sciences [*scientias*] also abandons culture [*culturae*; i.e., the right pedagogy]," wrote Possevino (1610, 85).[38] It was therefore important to promote the "good" sciences, such as the natural sciences (*scientiam naturalium*), geometry, astronomy, theology, and medicine, and to weed out the germs of evil that have been injected into them, such as demonic magic, geomancy, astrology, physiognomy, and alchemy (86).

As the chapters of Possevino's book progress, the meaning and scope of culture expand. From a discussion of the talents and dignity of man, the cultivation of the mind, and the cultivation of the sciences, the meaning broadens to the Jesuit pedagogical system. Possevino writes, for instance: "Then, he who has disregarded this [pedagogical method], . . . has abandoned this culture [i.e., the Jesuit pedagogical system], which I have indicated above" (Possevino 1610, 80).[39] Not following his method is abandoning culture itself, that is, the right (Jesuit) way of studying and teaching. In later chapters, Possevino also discusses educational institutions in detail, exploring "the places where the same culture [*cultura*; i.e., the Jesuit pedagogical system] can be more rightly implemented" (109).[40] He discusses private schools, universities, religious colleges, and seminaries, again employing agricultural metaphors. All these places have particular characteristics, just like farming fields, but for education the results depend less on the terrain, the air, or social and political conditions and more on the aptitude of the students, the quality of the teaching staff, and especially on the method of study, the rules and statutes of the educational institutions (110). He adds: "In the clerical seminars, . . . there is nothing worthy of greater attention and culture [*cultura*]" (111).[41]

In chapter 40, Possevino addresses the education at the Jesuit colleges, with special focus on the Collegio Romano. Significantly, the title reads "Quae cultura ingeniis excolendis adhibeatur in Collegiis Societatis IESU," or "which educational system [*cultura*] has to be adopted to cultivate [*excolere*] talents in the Colleges of the Society of Jesus." Here, the verb "excolere" is used to denote the active cultivation of talents. In contrast, "cultura" has become something to be adopted or introduced in order to cultivate. Given the content of the chapter, it makes sense to translate it as educational system, or even educational institution. "Cultura" in this chapter has come to comprise buildings, the number of teachers, the curriculum, the organization of the classes, how students can move to the next grade, the course schedule, how discipline is maintained, selection mechanisms, special rules for foreign pupils, and so forth, all of which is discussed under the general title of "cultura," in a book on *Cultura Ingeniorum*. The agricultural association is still present, yet the fruits are not so much on an individual level as they are institutional; that is, they contribute to the development of the sciences. Possevino writes, "Which culture [*cultura*] to implement in the religious colleges of this kind, and which fruits they yield in the sciences [*disciplinis*], will be evident . . . when we have sketched the model of the Collegio Romano of our Society" (Possevino 1610, 132).[42] He proceeds to discuss in detail institutional matters such as financing, administration, offices, teaching schedules, working conditions, and teaching methods. Thus, finally for Possevino, culture comes to stand for the whole institution of Jesuit education.

Possevino's work details the Jesuit epistemic culture, to use Karin Knorr Cetina's (1999) phrase, that is, the establishment of the conditions of possibility for knowledge creation. This was exactly the project Possevino was engaged in. He tried to construct the amalgam of arrangements, mechanisms, and institutions that made up the backbone of the Jesuit way of knowing. He did not produce new knowledge himself. He was not even involved in forming students and upcoming knowledge producers. He established the institutions in which a particular way of knowledge creation and transmission could become possible. Indeed, the *Bibliotheca Selecta* was a manual for Jesuit teachers, explaining how they should form the new epistemic and religious elite. During his long career, Possevino also founded core epistemic institutions, such as Jesuit colleges and seminaries. Most important, he devised the educational system and its pedagogical method, and he selected the material that counted as knowledge, which made a particular Jesuit kind of knowledge creation and transmission possible. This new epistemic culture was what Possevino, in the end, called culture. The product of these mechanisms and institutions was a new cultured epistemic elite.[43]

From Possevino's point of view, culture was a crucial positive force, elevating man and bringing him closer to God. One could also take another, more ideological perspective, however, from which things would have looked rather different. Placed in its social and political context, it becomes clear that Possevino's notion of culture reflected confessional tensions and was developed in support of the Jesuit apparatus and the Counterreformation project. In contrast to Gesner's *Bibliotheca Universalis*, Possevino's vision of knowledge was selective, and he aimed at separating the wheat (Catholic truth) from the chaff (heretical ideas). Possevino hammered on the importance of cultura and specialization, promoting the development of useful competences so that adolescents could become workers in the Lord's vineyard (Possevino 1610, 133–134). Adolescents were not just prepared for a religious vocation: the secular colleges were supposed to cultivate men for an important active role in Catholic society. One such secular model was the Christian soldier who transformed the world to the glory of God (*Il Soldato Cristiano* is the title of another work by Possevino). The culture of an individual Christian would imply growing the seeds planted in him by God and extirpating sins and errors like weeds. However, the agricultural metaphor not only allowed for the pruning of sins or books; it also applied to the social sphere, legitimizing the persecution of heretics for instance (see Höpfl 2004, 133–134).

Possevino was mainly engaged in educational struggles, while his secular counterpart, the Christian prince, could resort to more direct techniques for cutting off the bad branches of Christianity and for building a bulwark to defend the Catholic faith.[44] Nevertheless, the battle of the books did not have just a literary impact. Possevino's words should be understood as deeds that attempted to change reality, and they made a difference in the all too real battles between Catholics and Protestants. "Culture" was therefore operative and performative. His appropriation of the word "culture" helped to bring into being the Jesuit pedagogical and epistemic vision, a project he was also engaged in implementing in practice. His texts, such as the *Bibliotheca Selecta* and the *Cultura Ingeniorum*, and his activities in educating princes, founding colleges, printing catechisms, and other proselytizing actions are two sides of the same coin. These activities corresponded to the dissemination of so many seeds that would bear fruit one day. In the end, he envisioned that this would lead to a counterreformed world, unified under the aegis of the Holy See. The central aim of Possevino's work was the glory of God, and culture became a newly minted battle cry in the fight against heresy and error.

Possevino's concept of culture is multivalent. The still active meaning of culture as a process of cultivation goes against essentializing tendencies. For Possevino, an individual does not automatically belong to a culture. Everyone has a choice of culture and the potential to improve through culture. Despite Possevino's best intentions to educate and save souls, his notion of culture was also deeply antagonistic, which plays out most clearly on the social level. The methods of education (or indoctrination, as some would have it) that his concept of culture stood for were used to create epistemic and religious identities. Jesuit activism played an important role in enhancing confessionalization during the seventeenth century, which to some extent reified different forms of Christianity.[45] Possevino's culture was thus not only a nondeterministic, active and transformative concept; it could also contribute to the essentialization of differences, which were not restricted to religion, as for many, Protestant or Catholic culture became essential to defining their life.

Conclusion

Raymond Williams located the origin of the modern notion of culture at the end of the eighteenth century. Culture, he argued, was coined as a refuge from the new scientific, technical, and industrial spirit that defined late eighteenth-century modernity. Williams (1995, ix) saw it as a "holistic protest against the fragmenting effects of industrialism (the society science had helped produce)." Williams also uncovered this notion of culture for ideological reasons. He fought against the received version of British culture, revaluated working-class culture, and constructed an alternative conception of culture for the New Left (Williams 1960; 1989, 1–38; 1981, 97–98). For him, this was inextricably connected to his historical work: "To understand several urgent contemporary problems—problems quite literally of understanding my immediate world—achieved a particular shape in trying to understand a tradition" (Williams 1983, 13). In this article, I have uncovered earlier significant notions of culture, which indicate a more complex tradition and genealogy. Late sixteenth-century culture, as propounded by the Jesuit Antonio Possevino, was a programmatic notion that did not reject but promoted the sciences. This notion of culture still had its roots firmly in earlier agricultural and religious meanings of culture, but it was also radically transformed. Possevino's elaborate description of how to manure, prune, and weed the sciences took the metaphor to a new level and transcended simple analogies.

Such metaphors are still prevalent in current-day discourse about culture and science. When arguing that China's cultural history is the main factor impeding its scientific progress, Peng Gong (2012, 411) writes: "Two cultural genes [the thought of Confucius and Zhuangzi] have passed through generations of Chinese intellectuals for more than 2,000 years. . . . They helped to produce a scientific void in Chinese society that persisted for millennia. And they continue to be relevant today." Gong uses the notion of cultural genes (a naturalizing version of Richard Dawkins's memes?) to indicate how strongly ingrained specific traits are and how directly they determine the actions and beliefs of the Chinese as a unified group.[46] On one hand, Gong's cultural determinism and the contradistinction between culture and science are key differences with Possevino's view. On the other, the use of biological metaphors bears a striking resemblance to the early modern period. Furthermore, these cultural genes can be overcome, Gong thinks, by inculcating children with the scientific spirit early in their education. The Chinese educational system, which still perpetrates traditional Chinese values, should be overhauled by a new kind of culture (to use Possevino's term) in order to revolutionize Chinese science.

Culture has always been a value-laden concept. In the early modern period, the core meaning of the word "culture" referred to improvement, but the question of how to improve, and in which direction, remained open for interpretation and appropriation. Possevino constructed his notion of culture as an alternative to humanitas: it represented a strict, goal-oriented, and utilitarian education according to the Jesuit mold. This revalued concept did not reflect an existing situation but was meant—performatively—to help create a new epistemic reality. Possevino's notion of culture (resonating with Knorr Cetina's epistemic culture) stood for a new knowledge regime, encompassing ways of teaching, pedagogical methods, and even epistemic institutions such as colleges and universities. This notion of culture lies at the basis of the burgeoning Jesuit educational institutions, setting into motion its wide-ranging epistemic and social machineries. Its goals were admirable, providing efficient instruction for adolescents and setting new standards of education. But educational policy cannot be separated from politics. In this case, culture became a rallying cry for the Counterreformation, promoting confessionalization and resulting in the reification and ossification of religious identities. Possevino's diplomatic and political activities can be interpreted (to invert Carl von Clausewitz's famous principle) as religious war conducted by other means.[47] When Possevino became frustrated with politics, he came to see the Jesuit educational project as a superior way to solve the same problems. Consequently, to extend the metaphor, culture arose as politics conducted by other means.

Notes

I thank the editors of this volume for their ongoing support and encouragement. I received helpful advice from four anonymous referees, as well as from many colleagues and friends. All translations are my own.

1. On the great divide, see Hart 1999. G. E. R. Lloyd opposes the idea of hypothesized Chinese mentality and replaces it by less reified "styles" (Lloyd 1990, 12). For orientalism, see, e.g., Said 1978 and 1993. There are many other examples, some of which are explicitly framed in terms of conflict between "cultures." The term "culture wars" was used in Germany at the end of the nineteenth century to describe the clash between Protestants and Catholics. Today, the term also refers to ideological tensions between Left and Right (e.g., Kuper 1999, 1–20). For the divide in East-West ideologies during the Cold War as culture wars, see Judt 2005, 197–225. See also the discussion in chapters 2 and 4 in this volume.

2. Deliang Chen, quoted in C. Wickham, "Insight: China Rises in Science, but Equation May Have Flaws," Reuters, 2012, http://www.reuters.com/article/us-science-china -idUSBRE84R06G20120528 [accessed August 31, 2013].

3. L. Lim, "Plagiarism Plague Hinders China's Scientific Ambition," National Public Radio, 2011, http://www.npr.org/2011/08/03/138937778/plagiarism-plague-hinders -chinas-scientific-ambition [accessed August 31, 2013].

4. See the introduction to this volume, as well as, e.g., Kuper 1999, Barth 1969, Gupta and Ferguson 1992, and Keesing 1994.

5. According to the literature, Johann Gottfried Herder had already proposed to refer to "cultures" in the plural, stressing human diversity, in the late eighteenth century (see,

e.g., Stocking 1968, 201–214; but see Denby 2005 for a counterargument). See Kuper 1999 for the use of "cultures" by anthropologists. See chapter 4 in this volume for recent strategies of pluralization. For civilizations in the plural, see, e.g., Avlami and Remaud 2008.

6. See Brumann 2005 for a collection of citations of recent anthropologists criticizing the essentializing tendencies in the use of the concept of culture, as well as his pragmatic defense of the concept.

7. After some time, newly introduced politically correct words often acquire the negative connotations of the words they were meant to replace. For the debate on politically correct terminology and the attack on political correctness by conservative critics, see Wilson 1995, Lea 2009, and Hughes 2010, 289. For examples of linguistic engineering, which works best in totalitarian societies, see Collins and Glover 2002, Klemperer 2000, and Ji 2004.

8. See also the other contributions in this volume. Some but not all of these concepts can be fruitfully applied to describe pre-nineteenth-century science.

9. See, e.g., the articles collected in Cunningham 2012. For the methodological background, see esp. Skinner 1969.

10. I focus here on Williams because of his role in the historiography of culture and his stature as a founder of the discipline of cultural studies. For different methodologies in the historiography of concepts, see some of the classics, e.g., Skinner 1969; Koselleck 1989a, 1989b, 2002, 2006; and Williams 1960, 1983. For comparisons and debates between these models, see, e.g., Sheehan 1978, Palonen 2002, Richter 1990, and Skinner 1980. Evidently, these methodological discussions continue to this day.

11. This is not to say that Cunningham and Williams's proposals do not make sense. Their work should also be historicized and understood in its proper academic and historiographical context. Cunningham and Williams sensibly reacted against the dominant narrative of the scientific revolution, in which important differences between the nineteenth- and seventeenth-century "science" were neglected. As a result of their work, however, the dominant narrative about science has changed; it now obscures similarities and exaggerates differences with earlier periods, and is in need of revision.

12. Raymond Williams himself remarked on an important shift and a considerable expansion of the meaning of "culture" in the 1940s: before the war, "culture" had been used to denote either social superiority or the arts, but after 1945, it referred to some central formation of values (especially in literature and criticism), or it had taken on the anthropological sense of a particular way of life (as in "Chinese culture"). See Williams 1983, 11–12.

13. For more methodological reflections, where I put more emphasis on genealogical connections, see Vermeir 2013.

14. For histories of the word "culture" and civilization in different languages, see esp. Febvre et al. 1930, Niedermann 1941, Benveniste 1966, Rocher 1969, and Bénéton 1975. For *Kultur*, see Knobloch et al. 1967 and Fisch 1992, and for the related *Bildung*, see Vierhaus 1972.

15. These shared roots are especially important because we do not have a reliable way to individuate concepts. Should we categorize the premodern and modern concept of culture as one and the same? Are the English and Chinese concepts of culture different concepts or variations of the same? Does the current word "culture" refer to one or many

concepts (cf. the contention that there are two—or more—concepts of liberty; Berlin 2002; Skinner 2006, 2008)? These are complex questions that are very difficult to solve. For the linguistics area I cover in this article, it makes sense to refer to one concept of culture. When it is important to highlight differences, I will refer to different *notions* of culture, in an attempt to recognize the unity of a concept, as well as its internal variety.

16. There is a good bibliography for Possevino's political activities (especially as a diplomat). On the *Bibliotheca Selecta* see, e.g., Balsamo 2001, 2006; Zedelmaier 1992; and Biondi 1981. There is much literature on separate sections in the *Bibliotheca Selecta* (e.g., on specific geographical areas, the arts, or history), but the *Cultura Ingeniorum* remains understudied.

17. On Possevino's bibliographical practice, see Balsamo 2001 and 2006.

18. They also led to personal disappointment, not only because of political setbacks abroad but also because his proselytizing work had become so controversial and politicized that the pope was forced to ban him from Rome.

19. For the relation between the *Bibliotheca Selecta* and the *Bibliotheca Universalis*, see Zedelmaier 1992 and Balsamo 2001.

20. The term "ideology" is arguably anachronistic, but it has been usefully applied to the Jesuits by Steven Harris. I refer here to the elaboration of the notion of ideology in Harris 1989, 45–48.

21. Where Possevino uses these words as synonyms, I take them to stand for the same concept, that is, the concept of culture. Nevertheless, I indicate slight variations in meaning, signaling, for instance, that "excolere" is used in a more active sense than is "cultura."

22. "Cultura ingeniorum quaenam breuior, facilior, fructuosior."

23. In the table of contents, the subtitle "quaeve cuique Disciplinae sint indonea; Ioannis autem Huartis Examen Ingeniorum expenditur," which does not appear on the first page of the book, is added.

24. The important early editions are Possevino 1593, 1598, 1603, 1604a, 1605, and 1610. There were also less important editions in 1606 and 1607. Most references in this article are to the most complete 1610 edition of the *Cultura Ingeniorum*.

25. In general, "ingenium" means talents. Possevino defines "ingenium" as the capacity to learn ("ingeniù ea indoles est, qua facile, vel difficulter res, aut artes addiscuntur"; Possevino 1610, 34–35). The juxtaposition with "ingenium" is even more pronounced for the Italian "coltura."

26. Possevino probably draws on the Spanish physician Juan Huarte de San Juan (Possevino 1610, 40), as discussed below. See also the passage in Huarte de San Juan 1593, 5v: "Y assi dize Hypp. que el ingenio del hombre tiene la mesma proporcion con la sciencia que la tierra con la semilla: la qual aunque sea de suyo fecunda y paniega, pero es menester cultivarla, y mirar paraque genero de simiente tiene mas disposicion natural." (And so Hippocrates said that the talent [*ingenium*] of man has the same proportion to science as the earth has to the seed; even if the soil is fertile and can produce wheat, we must cultivate it and look for what kind of seed has the best natural disposition [for the soil].)

27. The title "On the Dignity of Man" was coined in later editions by later sixteenth-century humanist scholars and attests to the interest in the theme of human dignity;

see, e.g., the Basle humanist edition, Pico della Mirandola 1557, 315: "Nascenti homini omnifaria semina et omnigenae vitae germina indidit Pater. Quae quisque excoluerit illa adolescent, et fructus suos ferent in illo."

28. In general, for a preliminary assessment of the relation between Jesuit pedagogy and humanism, see O'Malley 2000.

29. "Sapientia enim simul cum Religione colitur, vbi vita, & actus omnis ad vnum caput, & ad vnum summum refertur."

30. After his death, a second, revised version expurgated by the Inquisition was published: Huarte de San Juan 1593. During the early modern period, this popular book also appeared in French, Italian, English, Latin, German and Dutch, in many editions. The book sparked a controversy, with diverse appraisals and rebuttals, e.g., Persio 1576; Zara 1615; and Possevino 1593, 1598. For recent new archival material on Huarte, see Virues-Ortega et al. 2011.

31. Possevino also criticized Huarte, arguing, for instance, that talents, human faculties, and professions could not be as neatly categorized as Huarte thought (Possevino 1610, 45–58).

32. In stark contrast to Possevino's work, there are only three references to "culture" in Huarte de San Juan 1593, notably in paraphrases of classical authors such as Cicero and Hippocrates.

33. "Humanum ingenium (aiebat Plato) cuItura fit diuinissimum; at haec si defit, fit diabolicum. Agri enim quò foecundiores, eò peiores germinant herbas, nisi colantur & purgentur."

34. For the Latin technical terminology of student and teaching practices, see, e.g., Hamesse 1988.

35. "Porrò qualibuscumque ingenijs magnoperè semper obfuit, cùm defuere praeceptores officio exigendi à discipulis pensum, repetitione, collatione, responsione (quotiescumque interrogantur), ac disputationibus. Quemadmodum enim perfecta agri cultio non est, si quis aret tantummodo, aut serat, sed tellure tegendum est semen; sic non rectè excoluntur ingenia si post praelectiones publicas, ea, quae dixi, non adhibeantur."

36. For warnings against Plato, see Possevino 1593, 2:78; for Aristotle, 2:94–96.

37. For Possevino's censorship of the arts, see Dekoninck 2008.

38. "At & culturae deest qui reticet, quae falsae sese inter veras scientias improba immiscuerunt."

39. "HAEC igitur qui omisere . . . , culturam eam deseruere, quam innuimus supra." Possevino is explicitly referring here to a part of his method that rejects the dictating of lectures, which is counterproductive and oppresses the students' minds.

40. "Loca, ubi eadem cultura rectius adhibetur."

41. "In Seminariis Clericorum, quae afflatu Spiritus Sancti Synodus Tridentina iussit iustitui, vt in singulis Dioecesibus efformarentur operarii, nihil est quod maiore animaduersione ac cultura dignius sit."

42. "At & quae cultura in Religiosis istiusmodi Collegijs adhibeatur, quemve fructum in disciplinis ferant, liquebit adhuc planius, si quemadmodum è publicis Academijs ea, quae Salmanticae est, proposita fuit, sic ex ijsdem Collegijs, Romani Collegij Societatis nostrae quasi idea quaedam (quod ad res nostras spectat) delineetur."

43. Carella 1993 argues that the *Bibliotheca Selecta* was mainly directed at the education of the higher nobility.

44. Possevino (1604b) writes to his princely patron: "So I would hope you to be a mutual example to them, so that in a few years, all their majesties, and your highnesses, with one heart and a single value, will be united for every mighty league, and form a bulwark against the enemies of Christianity." (3r–v: "Che onde speri di essere à loro di scambieuole essempio, accioche fra pochi anni, tutte le Maestà loro, & l'Altezze Vostre con vn core, & con une istesso ualore sieno strette insieme per ogni potente lega, & antimurale contra i nemici della Christianità.") For the way that the Christian prince should exercise censorship and eliminate errors (more by encouragement than punishment) to create a cohesive Christian community in his domains, see Possevino 1610, 149–154.

45. The study of confessionalization has been an important paradigm in the historiography of religion since the 1970s, but its strong claims have recently been tempered. In the context of this chapter, it is clear that Jesuit education formed students according to a certain mold, and Possevino's work is one of the clearest expressions of this. But this education did not fix the (religious) identity of the students, and many profited from the excellent Jesuit educational system without becoming militant Catholics, or even without becoming Catholic at all.

46. Although in principle a cultural gene is not an essence (it can mutate), in this context it serves to explain the extraordinary stability and resistance to change of Chinese culture, as if it were endowed with an essence.

47. For Clausewitz's principle, see Clausewitz 1832, 1:28: "Der Krieg ist eine bloße Fortsetzung der Politik mit anderen Mitteln" ("War is merely a continuation of politics by other means").

References

Abu-Lughod, L. 2006. "Writing against Culture." In *Anthropology in Theory: Issues in Epistemology*, ed. H. L. Moore and T. Sanders, 466–479. Oxford: Blackwell.

Avlami, C., and O. Remaud. 2008. "Civilisations: Retour sur les mots et les idées." *Revue de synthèse* 129: 1–8.

Balsamo, L. 2001. "How to Doctor a Bibliography: Antonio Possevino's Practice." In *Church, Censorship, and Culture*, ed. G. Fragnito, 50–78. Cambridge: Cambridge University Press.

Balsamo, L. 2006. *Antonio Possevino S.I. bibliografo della Controriforma e diffusione della sua opera in area anglicana*. Firenze: Olschki.

Barth, F. 1969. *Ethnic Groups and Boundaries: The Social Organization of Culture Difference*. Oslo: Universitetsforlaget.

Bénéton, P. 1975. *Histoire de mots: Culture et civilisation*. Paris: Presses de la Fondation Nationale des Sciences Politiques.

Benveniste, E. 1966. "Civilisation, contribution à l'histoire d'un mot." In *Problèmes de linguistique générale*, 336–345. Paris: Gallimard.

Berlin, I. 2002. "Two Concepts of Liberty." In *Liberty*, 166–217. Oxford: Oxford University Press.

Biondi, A. 1981. "La biblioteca selecta di Antonio Possevino: Un progetto di egemonia culturale." In *La "Ratio studiorum": Modelli culturali e Pratiche educative dei Gesuiti in Italia tra Cinque e Seicento*, ed. G. P. Brizzi, 43–75. Roma: Bulzoni.

Brumann, C. 2005. "Writing for Culture: Why a Successful Concept Should Not Be Discarded." In *Concepts of Culture. Art, Politics and Society*, ed. A. Muller, 43–77. Calgary: University of Calgary Press.

Carella, C. 1993. "Antonio Possevino e la biblioteca 'selecta' del principe cristiano." In *Bibliothecae selectae: Da Cusano a Leopardi*, ed. E. Canone, 507–516. Florence: Olschki.

Clausewitz, C. von. 1832. *Vom Kriege*. 3 vols. Berlin: Dümmlers.

Collins, J., and R. Glover, eds. 2002. *Collateral Language*. New York: New York University Press.

Cunningham, A. 2012. *The Identity of the History of Science and Medicine*. Farnham, UK: Ashgate.

Cunningham, A., and P. Williams. 1993. "De-centring the 'Big Picture': The Origins of Modern Science and the Modern Origins of Science." *British Journal for the History of Science* 26: 407–432.

Dekoninck, R. 2008. "Une bibliothèque très sélective: Possevino et les arts." *Littératures classiques* 66: 71–80.

Denby, D. 2005. "Herder: Culture, Anthropology and the Enlightenment." *History of the Human Sciences* 18: 55–76.

Febvre, L., É. Tonnelat, M. Mauss, A. Niceforo, L. Weber, L. E. Mot, and E. T. L. Idée. 1930. *Civilisation—le mot et l'idée*. Paris: La Renaissance du livre.

Fisch, J. 1992. "Zivilisation, Kultur." In *Geschichtliche Grundbegriffe*, ed. O. Brunner, W. Conze, and R. Koselleck, 7:679–774. Stuttgart: Klett-Cotta.

Gong, P. 2012. "Cultural History Holds Back Chinese Research." *Nature* 481 (26 January): 411.

Gupta, A., and J. Ferguson. 1992. "Beyond 'Culture': Space, Identity, and the Politics of Difference." *Cultural Anthropology* 7(1): 6–23.

Hamesse, J. 1988. ""Collatio" et "reportatio": deux vocables spécifiques de la vie intellectuelle au moyen âge." In *Terminologie de la vie intellectuelle au Moyen Âge*, ed. O. Weijers, 78–87. Turnhout: Brepols.

Harris, S. J. 1989. "Transposing the Merton Thesis: Apostolic Spirituality and the Establishment of the Jesuit Scientific Tradition." *Science in Context*: 29–65.

Hart, R. 1999. "On the Problem of Chinese Science." In *The Science Studies Reader*, ed. M. Biagioli, 189–201. New York: Routledge.

Höpfl, H. 2004. *Jesuit Political Thought: The Society of Jesus and the State, c. 1540–1630*. Cambridge: Cambridge University Press.

Huarte de San Juan, J. 1575. *Examen de ingenios para las ciencias*. Baeza: Juan Bautista Montova.

Huarte de San Juan, J. 1593. *Examen de ingenios para las ciencias*. Antwerp: Plantin.

Hughes, G. 2010. *Political Correctness: A History of Semantics and Culture*. Oxford: Wiley-Blackwell.

Ji, F. 2004. *Linguistic Engineering: Language and Politics in Mao's China*. Honolulu: University of Hawai'i Press.

Judt, T. 2005. *Postwar: A History of Europe since 1954*. New York: Penguin.

Keller, E. F. 2002. *Making Sense of Life: Explaining Biological Development with Models, Metaphors, and Machines*. Cambridge, MA: Harvard University Press.

Keesing, R. M. 1994. "Theories of Culture Revisited." In *Assessing Cultural Anthropology*, ed. R. Borofsky, 301–310. New York: McGraw-Hill.

Klemperer, V. 2000. *Language of the Third Reich*. London: Athlone Press.

Knobloch, J., H. Moser, W. Schmidt-Hidding, M. Wandruszka, L. Weisgerber, and M. Woltner, eds. 1967. *Kultur Und Zivilisation*. Vol. 3 of *Europäische Schlüsselwörter*. München: Hueber.

Knorr Cetina, K. 1999. *Epistemic Cultures: How the Sciences Make Knowledge. Measurement*. Cambridge, MA: Harvard University Press.

Koselleck, R. 1989a. "Linguistic Change and the History of Events." *Journal of Modern History* 61(4): 649–666.

Koselleck, R. 1989b. "Social History and Conceptual History." *International Journal of Politics, Culture and Society* 2(3): 308–325.

Koselleck, R. 2002. *The Practice of Conceptual History: Timing History, Spacing Concepts*. Stanford, CA: Stanford University Press.

Koselleck, R. 2006. *Begriffsgeschichten*. Frankfurt am Main: Suhrkamp.

Kuper, A. 1999. *Culture: The Anthropologists' Account*. Cambridge, MA: Harvard University Press.

Lea, J. 2009. *Political Correctness and Higher Education: British and American Perspectives*. London: Routledge.

Lloyd, G. 1990. *Demystifying Mentalities*. Cambridge: Cambridge University Press.

Niedermann, J. 1941. *Kultur: Werden und Wandlungen des Begriffs und seiner Ersatzbegriffe von Cicero Bis Herder*. Florence: Bibliopolis.

O'Malley, J. W. 2000. "How Humanistic Is the Jesuit Tradition? From the 1599 Ratio Studiorum to Now." In *Jesuit Education 21: Conference Proceedings on the Future of Jesuit Higher Education*, ed. M. R. Tripole, 189–201. Philadelphia: St. Joseph's University Press.

Palonen, K. 2002. "The History of Concepts as a Style of Political Theorizing: Quentin Skinner's and Reinhart Koselleck's Subversion of Normative Political Theory." *European Journal of Political Theory* 1(1): 91–106.

Persio, A. 1576. *Trattato dell'ingegno dell'uomo*. Venice: Aldus Manutius.

Pico della Mirandola, G. 1557. "Oratio de hominis dignitate." In *Opera omnia*, 313–331. Basel: Heinrich Petri.

Possevino, A. 1593. *Bibliotheca selecta qua agitur de ratione studiorum in historia, in disciplinis, in salute omnium procuranda*. 2 vols. Rome: Typographia Apostólica Vaticana.

Possevino, A. 1598. *Coltura de gl'ingegni: Nella quale con molta dottrina et giuditio si mostrano li doni che ni gl'ingegni dell'huomo ha posto iddio, la varietà et inclinatione loro, e di dove nasce et come si conosca, li modi, e mezi d'essercitarli per le discipline, ...* Vicenza: Giorgio Greco.

Possevino, A. 1603. *Bibliotheca selecta de ratione studiorum, ad disciplinas, et ad salutem omnium gentium procurandam: recognita novissime ab eodem, et aucta, et in duos tomos distributa*. Venetiis: Altobellus Salicatius.

Possevino, A. 1604a. *Cultura ingeniorum e biblioteca selecta: Examen ingeniorum ioannis huartis expenditur*. Venetiis: Ioannes Baptista Ciottus.

Possevino, A. 1604b. *Il soldato christiano con nuove aggiunte et la forma di un vero principe et principessa*. Venezia: Domenico Imberti.

Possevino, A. 1605. *Cultura ingeniorum, e bibliotheca selecta Antonii Possevini: Examen ingeniorum io; huartis expenditur*. Paris: Claudius Chappelet.

Possevino, A. 1610. *Cultura ingeniorum: Examen ingeniorum ioannis huartis expenditur*. Coloniae Agrippinae: Ioannes Gymnicus.

Rheinberger, H.-J. 1995. "From Experimental Systems to Cultures of Experimentation." In *Concepts, Theories, and Rationality in the Biological Sciences*, ed. G. Wolters and J. G. Lennox, 123–136. Pittsburgh: University of Pittsburgh Press.

Rheinberger, H.-J. 1997. *Toward a History of Epistemic Things: Synthesizing Proteins in the Test Tube*. Stanford, CA: Stanford University Press.

Rheinberger, H.-J. 2011. "Consistency from the Perspective of an Experimental Systems Approach to the Sciences and Their Epistemic Objects." *Manuscrito* 34(1): 307–321.

Richter, M. 1990. "Reconstructing the History of Political Languages: Pocock, Skinner, and the Geschichtliche Grundbegriffe." *History and Theory* 29(1): 38–70.

Rocher, D. 1969. "Les mots culture et civilisation en français et en allemand." *Etudes germaniques* 24(2): 171–193.

Said, E. W. 1978. *Orientalism*. New York: Vintage Books.

Said, E. W. 1993. *Culture and Imperialism*. New York: Vintage Books.

Sheehan, J. 1978. "Begriffsgeschichte: Theory and Practice." *Journal of Modern History* 50(2): 312–319.

Skinner, Q. 1969. "Meaning and Understanding in the History of Ideas." *History and Theory* 8(1): 3–53.

Skinner, Q. 1980. "Language and Social Change." In *The State of the Language*, ed. L. Michaels and C. Ricks, 562–578. Berkeley: University of California Press.

Skinner, Q. 2006. "How Many Concepts of Liberty?" Paper presented at the Vice Chancellor's Distinguished Lecture, University of Sydney, July 24, 2006. Accessed August 31, 2013. http://sydney.edu.au/podcasts/2006/Skinner.mp3.

Skinner, Q. 2008. *Hobbes and Republican Liberty*. Cambridge: Cambridge University Press.

Stocking, G. W., Jr. 1968. *Race, Culture and Evolution: Essay in the History of Anthropology*. Chicago: University of Chicago Press.

Vermeir, K. 2013. "Philosophy and Genealogy: Ways of Writing the History of Philosophy." In *Philosophy and Its History: New Essays on the Methods and Aims of Research in the History of Philosophy*, ed. M. Lærke, E. Schliesser, and J. E. H. Smith, 50–70. Oxford: Oxford University Press.

Vierhaus, R. 1972. "Bildung." In *Geschichtliche Grundbegriffe*, ed. O. Brunner, W. Conze, and R. Koselleck, 1:508–551. Stuttgart: Ernst Klett Verlag.

Virues-Ortega, J., G. Buela-Casal, M. T. Carrasco-Lazareno, P. D. Rivero-Davila, and R. Quevedo-Blasco. 2011. "A Systematic Archival Inquiry on Juan Huarte de San Juan (1529–88)." *History of the Human Sciences* 24(5): 21–47.

Williams, R. 1960. *Culture and Society, 1780–1950*. New York: Doubleday.

Williams, R. 1981. *Politics and Letters: Interviews with New Left Review*. London: Verso.

Williams, R. 1983. *Keywords: A Vocabulary of Culture and Society*. Rev. ed. New York: Oxford University Press.

Williams, R. 1989. *Resources of Hope*. London: Verso.

Williams, R. 1995. *The Sociology of Culture*. Chicago: University of Chicago Press.

Wilson, J. 1995. *The Myth of Political Correctness: The Conservative Attack on High Education*. Durham, NC: Duke University Press.

Zara, A. 1615. *Anatomía ingeniorum & scientiarum*. Venice: Ambrosius Dei.

Zedelmaier, H. 1992. *Bibliotheca universalis und Bibliotheca selecta: Das Problem der Ordnung des gelehrten Wissens in der frühen Neuzeit*. Köln: Böhlau.

10
—————

From Quarry to Paper
Cuvier's Three Epistemological Cultures

The palaeotherium is much less famous than the mammoth or *T. rex*, yet its discovery by Georges Cuvier in 1798 was quite an important scientific event. In fact, it marks the true birth of paleontology. My purpose in this chapter is not to explain why the reconstruction of the Montmartre animal played a major role in the creation of this new discipline or, more precisely, in the elaboration of the history of the earth that is still associated with the name of Cuvier. Martin Rudwick (2005, 2008) has given a brilliant account of this role in his work on the reconstruction of geohistory between 1770 and 1840. My interests focus on a more limited question: I want to examine what kind of epistemological cultures Cuvier mobilized to transform a technical problem of fossil osteology into a key point in his research strategy. The notion of culture I use in this context is also somewhat limited: it refers to some shared systems of practices and values passed down by training and tradition. In contrast to the Kuhnian notion of paradigm, such a notion does not entail either a consensus about what good science should be or the imposition of any specific conceptual framework.

In fact, as Cuvier's case shows, the cultural resources used for shaping a research strategy were quite heterogeneous and even at times incompatible. It would be risky to assume that all contemporary scientific practitioners shared the same views about science and to reconstruct a sort of systematic epistemology of the age from occasional epistemological stances adopted by some scientists in specific contexts. This does not mean that there were no ideas about what science is and what it should be that were used polemically or for justification. It is indeed an important task for historians of science to describe these ideas, how they diffused, and their diffusion and usage inside and outside the scientific world. However, most of the actual practitioners were committed to gathering data and solving problems in very narrow research fields and did not claim to develop consistent epistemological views. In other words, their sense of what mattered in this context was very often opportunistic.

I want to emphasize that there is nothing pejorative in this statement. Cuvier has often been considered inconsistent, or even cynical, because in his research work he defended a neutral science of facts against makers of systems like Jean-Baptiste Lamarck and at the same time promoted preconceptions about the history of the earth and life in programmatic summaries. This kind of judgment misses the point. As I hope to show, Cuvier's attitude was coherent both in supporting a positivist approach in natural history and in developing ideological views in line with biblical statements. In each case, referring to general ideas about scientific method or about relations between science and religion contributed to his main purpose, which was to transform natural history into a science as powerful and legitimate as the physical sciences.

Therefore, the notion of epistemological culture can be useful for departing from a normative conception of epistemology and for developing a pragmatic approach focused on the ways real actors used epistemological statements and values in their research works. From this viewpoint it can be helpful to associate epistemological cultures with the working spaces in which they become effective. The notion of "working space" has been often used by historians of science. What I mean is nevertheless different from what has been generally considered as a scientific working space. The paradigmatic historiographic example of such a scientific working space is the laboratory, which is a type of location characterized by some specific traits, such as being a closed artificial site devoted to experiment. Historians of science have also considered many other types of location, such as the observatory, the garden, the cabinet of curiosities, and even the open field, all of which significantly differ from the laboratory. But these working spaces share the common feature of being defined mainly by their location (Galison and Thompson 1999). To avoid this limitation, I adopt a more abstract point of view and consider a working space as a situated *dispositif*.

The notion of dispositif was elaborated by Michel Foucault, especially in his 1977 book *Discipline and Punish*. Foucault characterizes the dispositif of the prison as the combination of heterogeneous elements mobilized to create a technological system for controlling bodies. More generally, for Foucault, a dispositif is a tool, an apparatus, that may relate to discourses but can also refer to institutions, practices, epistemological values, social roles, and objects. Heterogeneity and effectiveness are its two main features. Foucault uses the notion of dispositif to describe big institutions like the army or prisons, but in my view it is even more relevant for small and transitory arrangements shaped and handled by specific actors in specific contexts.

Thus, a working space is a kind of dispositif that implies a sense of the situation. Unlike a location such as a laboratory, which is defined by a univocal position, a situation is structured by a set of relations that can be explored on different spatiotemporal scales. Therefore, a working space can be also described as a network of interactions in specific geohistorical settings. It may be achieved by a single person, as in Cuvier's case. However, it is more generally the product of collective actions, for instance, when a group of scientists reshape their common environment. From this point of view, the notion of working space can be related to the Latourian notion of actor-network. The role of cultures should be considered by historians of science in relation with such actor-centered arrangements in order to escape woolly aprioristic cultural generalizations. Epistemological cultures, in this sense, refer to the fact that constructing a scientific working space needs shared practices and values deriving from tradition. For instance, Cuvier's working space mobilized three different epistemological cultures. Each of them played a specific role in its gradual elaboration.

The first, which I call museum culture, is specific to collectors, curators, and taxonomers. Eighteenth-century comparative anatomy developed in close connection with this kind of epistemological culture. Conveyed in a tradition that had dominated the world of natural history from the sixteenth century onward, museum culture implied complex practices for gathering, exploiting, and keeping information, as well as basic epistemic values such as precision, completeness, and tidiness. As a true cabinet naturalist, Cuvier shaped his working space along these lines. His priorities were collecting, selecting, organizing, and displaying.

However, Cuvier did not restrict himself to the traditions of museum culture in making his own working space. In his normative and polemical texts, he also relied heavily on the epistemological culture dominating the physical sciences in France in his time, which for convenience I call late Newtonianism. For most of the eighteenth century Newtonianism had been irrelevant to natural history. It concerned physics and its mathematization and put laws of nature at the very center of scientific research, whereas natural history dealt with describing forms and organizations of natural beings. Astronomy had been the paradigmatic example of a Newtonian science, in which all the facts can be deduced from laws expressed in mathematical terms. During the last decades of the eighteenth century, Parisian scientists embarked on the program of late Newtonianism. They took astronomy as a model for reforming all the physical sciences, including chemistry. Even natural history was affected: René-Just Haüy formulated laws in crystallography, and Félix Vicq d'Azyr tried to develop the same kind of approach in comparative anatomy. When Cuvier entered the field of natural history, he followed in their footsteps and adopted Pierre-Simon Laplace, the leading proponent of late Newtonianism after the French Revolution, as his spiritual father, dedicating his 1812 *Research on Fossil Bones* to him. Late Newtonianism can be defined as an epistemological culture. Its adoption implied importing new values and methods in disciplines such as natural history that were imbued with the Baconian spirit. Conflict with the system of norms and values traditionally attached to these disciplines was obviously a risk. As I describe in this chapter, such conflict was certainly the case for Cuvier's comparative anatomy.

Cuvier did not just combine museum culture with Newtonianism. He also made use of quite different epistemological resources drawn from historical and antiquarian scholarship. From the Renaissance onward, specific values and methods had developed in these fields of research. They concerned ways of dealing with vestiges of the past that time had spoiled and altered. How could they be unearthed and reconstructed in their authentic condition, and how could they be dated? Beyond technical issues, this kind of scholarship raised fundamental questions about the history of mankind and its relation to the natural environment, and was compelled to deal with problems abutting religious concerns. It gave birth to a specific epistemological culture that I call antiquarianism.

Antiquarianism definitely shared values and methods with museum culture. However, it also conveyed religious and political implications that were absolutely alien to the epistemological culture of collectors. Moreover, it did not match easily with the kind of Newtonianism that Cuvier was keen to refer to when he spoke of scientific methods. Discrepancies among these three epistemological cultures, museum culture, Newtonianism, and anti-quarianism, played a major role in the development and transformation of Cuvier's working space. In fact, the aim of this chapter is to explore precisely how Cuvier mobilized these cultural resources to invent the new science of paleontology, and what effects their synthesis had on its development.

Making a Working Space of His Own

Biographers of Cuvier have already noted that his genius partly depended on his ability to shape an efficient, distinctive working space (Outram 1984; Rudwick 2005, 405). This achievement was the product of intense activity, both social and intellectual. Born at Montbéliard in 1769, Cuvier studied cameralism (a German science of administration) at the Academy of Stuttgart and returned to France just before the revolution. Hired in Normandy as a private tutor by a rich Protestant family, he started some research work there in natural history, botany, and zoology. Although this work won him recognition from a few Parisian savants, Cuvier was almost unknown when he arrived in Paris in January 1795. This situation changed dramatically within a couple of years. Taking advantage of the fluid situation following Robespierre's fall, he quickly climbed the ladder in the small world of Parisian natural history. One year after his arrival, at the age of twenty-five, he held the chair of natural history at the École Centrale du Panthéon and became both member of the First Class of the National Institute of Sciences and Arts established in 1795 to replace the former Royal Academy of Sciences and Jean-Claude Mertrud's deputy for teaching comparative anatomy at the museum. "A true mushroom, but a good one," joked Louis Jean-Marie Daubenton.

From the outset, the National Museum of Natural History was the very center of Cuvier's working space. This new institution, founded in 1793 to replace the former King's Garden, offered exceptional opportunities for an ambitious young naturalist like Cuvier, not only because of its numerous naturalist collections but also through its stimulating environment. Cuvier lived within its walls and was in constant contact with his colleagues, especially with his friend Étienne Geoffroy Saint-Hilaire, holder of the chair of zoology of vertebrates; with Lamarck, who held the chair of zoology of invertebrates; and with Barthélemy Faujas de Saint-Fond, who held the chair of geology. There was a mix of collaboration and competition between these savants, who shared some common institutional interests but differed both on scientific questions and on questions of leadership. In 1802, Cuvier convinced his colleagues to launch a specialized journal on natural history, the *Annales du Muséum National d'Histoire Naturelle*, which could publish lengthy scholarly papers with numerous illustrations. In fact, he himself made extensive use of this outlet for publishing his memoirs on gypsum fossils.

There is no doubt that the chair of comparative anatomy at the museum was a strategic position for Cuvier, while also being the centerpiece of his working space. Teaching duties and research facilities were attached to this chair. Cuvier gave an annual lecture course in comparative anatomy in the lecture room of the museum. He created a gallery of comparative anatomy close to his house for his research and opened it later to the public. He also used the menagerie established in the museum in 1795. He observed the animals and dissected them after their death. Cuvier had faithful assistants to help him: his laboratory assistants Emmanuel Rousseau and Charles-Léopold Laurillard; the anatomist Constant Duméril, who wrote up his lecture courses; and his brother Frédéric Cuvier, who was given the position of keeper of the menagerie in 1803. In his chair, Cuvier did not confine himself to continuing the program outlined by Vicq d'Azyr, the founder of modern comparative anatomy. He was much more ambitious. However opposed he was to Buffon's style, he embarked on research of comparable scope, which included the history of life and the history of the earth. He started a systematic study of the different species of mammals with this project in mind as soon as he settled in the museum. This was the context of his entry into the field of fossil bones.

But however important the museum was for him, Cuvier's working space stretched far beyond its walls. Cuvier was certainly not a field naturalist, but neither was he locked in an ivory tower. He was involved in the Parisian society of the Directorate and the Consulate, where intellectuals, officials, politicians, and bankers met regularly, and he was skilled in finding support for his career in this milieu (Outram 1984, 49–68). He also had a foot in several learned societies, such as the Société d'Histoire Naturelle, the Société Philomathique, and the Société des Observateurs de l'Homme, and he attended several Parisian salons. Thanks to all these positions, he had close connections with Parisian savants, naturalists, mathematicians, physicists, and chemists, as well as with antiquarians, historians, philosophers (*idéologues*), and politicians. Among the scientists of the old generation, Cuvier admired Laplace, whom he had met shortly after arriving in Paris. But he had no patron in natural history. In fact, from the moment he got his position in 1796, he considered himself a master in the field. Later, in the Consulate, Cuvier was elected permanent secretary for physical sciences of the First Class of the Institute. This election was the start of a series of nominations that made him one of the notables of the regime and a *cumulard* in the scientific world. Moreover, his positions at the institute and in the administration of public instruction gave him many opportunities to extend his network of scientific correspondents and protégés and establish his working space all over Napoleonic Europe. For Cuvier, it was a decisive means to secure his scientific power and to promote his scientific conceptions.

Cuvier did not neglect the public, either, which gave him an audience and could provide support in scientific debates with his colleagues. He was a great lecturer, and his lecture courses in the museum were very popular. He also lectured in the Collège de France, where he was appointed professor in 1800 (and where he let his assistant Jean-Claude Delamétherie lecture until 1812) and in the Paris Athénée, a private educational institution similar to the Royal Institution in London, where he gave a famous lecture about the history of the earth in 1805. His rhetorical skill was instrumental in popularizing his ideas in his writing as in his speaking. The *Discourse on the Revolutions of the Surfaces of the Earth*, first published in 1812 and frequently republished and translated, was among the most popular books written by scientists in the early nineteenth century. Its success can only be compared with that of the *System of the World*, the book Laplace wrote in 1796, which Cuvier considered a model for his own work.

Reconstructing Montmartre Animals

At the time Cuvier embarked upon his paleontological inquiry on the Parisian fossil mammals, his career was still in its early stages. The story started a few days after Bonaparte's departure to Egypt, in the spring of 1798. The expedition aimed to occupy that country and thus to ruin the English trade with India, but it also had a scientific aspect: no fewer than 150 savants, artists, and engineers took part in the expedition. Among them were several naturalists, including Geoffroy Saint-Hilaire. Cuvier had also been invited to join the expedition, but he had declined. As he explained later, he preferred to stay in Paris among the collections of the museum, which were considered the richest in the world. By then Cuvier had been in Paris for only three years, but he had already emerged as a leading figure in the small company of Parisian naturalists. From the first months of his stay, he had worked hard on comparative anatomy and zoological taxonomy. This research program on classification soon led him to study some mysterious cases of putative extinct species of mammals: the mammoth from Siberia; the Ohio animal (which he renamed the "mastodon" in 1812); the Paraguay animal, which he immediately called the "megatherium"; and some other strange animals. As early as 1796 he raised the possibility of extinct animals belonging to a vanished world. However, this first foray into paleontology remained limited. Because the museum possessed only very few fragmentary specimens of these fossils in its collection, all his conclusions were based on secondary documentation, mainly on drawings of fossils kept abroad.

The Montmartre fossils changed the situation dramatically. Remains of extinct animals were found not only in remote places but nearby and, so to speak, under the feet of the Parisians. Moreover, unlike the mammoth, the mastodon, and the megatherium, these fossils did not lie in superficial deposits; they were embedded in deep, hard rocks. In fact, people had quarried for plaster around Paris for centuries, and quarriers were accustomed to finding these amazing bones. Jean-Étienne Guettard (1768, 1–18) had already described some fossils extracted from Montmartre's gypsum, and the naturalist Robert de Lamanon had published an important paper in 1782 on the gypsum formation, its fossils, and its origin. Yet, surprising though it may seem, Cuvier, as he explained later, had absolutely no notion of the existence of these fossils before his meeting with a dealer in objects of natural history, François Vuarin. One day in June or July 1798, Vuarin showed him some fossil bones that had been found in gypsum quarries around Paris. Cuvier was very interested and asked him for other specimens. Vuarin gave him the names of some private collectors to whom he had already sold the same kind of fossils.

Cuvier promptly made his own investigations. Although the hill of Montmartre was just outside Paris, the field on which he started his inquiry was inside Paris: the world of the collectors. In search of gypsum fossils, Cuvier explored both public collections, especially in the cabinet of the School of Mines, and the private collections of his colleagues, such as Alexandre Brongniart and Faujas. Moreover, he used his connections with society to gain access to the cabinets of rich amateurs, such as the cabinet of the Marquis de Drée. Cuvier very soon had about a hundred bones at his disposal. He supplemented these pieces with new fossils, but he himself did very little fieldwork. He obtained some proxy specimens in the form of drawings from naturalists in the province and abroad, such as Adrian Camper in the Netherlands. Moreover, he paid quarriers to search for fossil bones and bring them to him (Rudwick 2004, 2005, 381–384).

The first piece Cuvier studied closely, in July or August 1798, was a lower jaw kept in the mineralogical cabinet of the School of Mines. The same piece seems to have been already described by Guettard in 1774. When Cuvier examined the fossil for the first time, he noticed a big canine tooth and inferred that it was the jaw of a dog (Cuvier 1798a, 138). But one month later he completely changed his mind: from then on, he believed the jaw belonged to an unknown animal different from any extant species. This decisive statement was the result of a work process that can be reconstructed, much as Cuvier reconstructed the animal.

In fact, the details of the dentition were initially invisible because the other teeth were still largely embedded in the stone. Fortunately, Cuvier was allowed to scratch out the stone to extricate these teeth. Thus, he discovered that the jaw had big molars quite different from the molars of a ruminant and that, similar to the jaw of a rhinoceros, an empty space separated the first molar from the canine tooth. It was, however, almost impossible to discern the incisors. By coincidence, another lower jaw kept in the private collection of Auguste Nicolas de Saint-Genis gave precise indications on these front teeth. Finally, Cuvier could reconstruct the whole dentition of the Montmartre animal, which was somewhere between the dentition of a rhinoceros and that of a tapir (figure 10.1).

Another impressive fossil played a major role in the initial reconstruction of the Montmartre animal: a head kept in the de Drée collection (figure 10.2), which had already been studied and roughly drawn by the naturalist Lamanon in 1780. By accurate examination, Cuvier deduced that it belonged to the same species as the lower jaw. He also showed that the animal had a small trunk, like a tapir. Other scant fossil bones found in various collections confirmed this initial deduction.

In his studies of the head, Cuvier used comparative anatomy. For instance, he systematically compared the teeth of the Montmartre animal with the teeth of some living species of herbivores, using his thorough knowledge of the dentition of mammals (see Cuvier 1805a, 3:139–168). With this method, it was not difficult for him to reconstruct the dentition. However, the reconstruction of the feet was far more difficult because it involved many different bones, some small and short and others long and fragile. Cuvier obtained his first results with the rear feet, which were much easier to reconstruct than the front feet. For this work he systematically applied what he later called the principle of the correlation of forms. Thus, Cuvier tried to put together different bones from the de Drée collection, which he could identify as bones of the tarsus: a calcaneum, an astragalus, a cuboid, and a scaphoid. He could reassemble these different bones by considering the positions and shapes of their articular facets (the parts that link them together). He also found a nice piece comprising a metatarsus and a digit in the same de Drée collec-

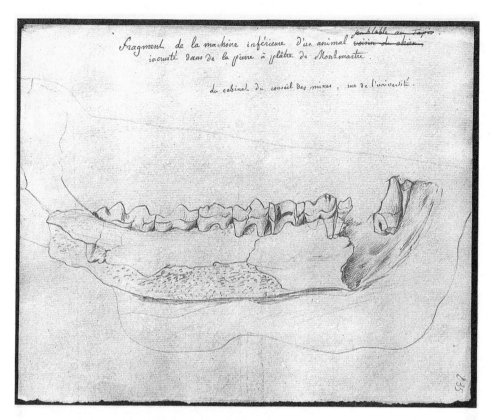

FIGURE 10.1. Jaw of a palaeotherium. Cuvier's drawing.

tion. With all these elements, he proposed a hypothetical reconstruction of a complete rear foot that showed two digits. Later on he discovered a wonderful complete piece in the stockroom of the museum, which he was able to scratch out of the matrix to extricate the bones. This find confirmed the existence of the two digits and the two metatarsi. Cuvier noted that no extant species had this kind of foot.

Cuvier reached this conclusion within a few weeks (Rudwick 1977, 285–291). Yet, the puzzle proved to be more complicated than he had at first thought. Cuvier soon discovered that some rear feet were smaller and had not two but three digits. He deduced this last fact from the close examination of the facets of some bones belonging to the tarsus, especially cuboids and scaphoids. Finally, he found the essential pieces of this kind of foot put together in the same block of gypsum belonging to the de Drée collection. After having made a drawing of this block, he extracted the bones and

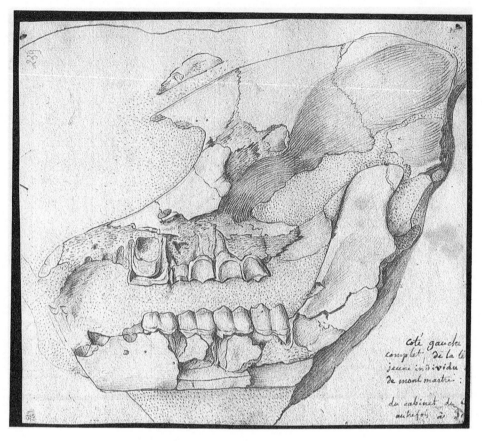

FIGURE 10.2. Head of a palaeotherium. Cuvier's drawing.

pasted them together with wax. He achieved the same kind of assemblage with other pieces. These reconstructions actually confirmed the existence of some three-toed rear feet. Finally, in a paper published in the fall of 1798, Cuvier announced the discovery of three different extinct species of mammals in the Parisian gypsum: an animal with two-toed rear feet, whose size was comparable to that of a horse, and two smaller animals with three-toed rear feet, one the size of a pig and the other the size of a hare (Cuvier 1798b). Two years later, he mentioned six extinct species, distinguished by size and the number of their digits (Cuvier 1800a).

At this stage, the relation between the rear feet and the head was still unclear. Cuvier had initially conjectured that the two-toed rear feet belonged to the same Montmartre animal as the jaw and the skull. In fact, the only reason for this attribution was that their respective sizes fitted together. On the other hand, as he explained later, Cuvier was concerned about several fossil teeth that he discovered in some collections: although they were the same size as the molar teeth of the Montmartre animal, their shapes were slightly different. Finally, he obtained a new piece, a jaw with molars, which clearly revealed the existence of a species without canine teeth. In reconsidering many pieces kept in different collections, Cuvier realized that this kind of species without canine teeth was far from uncommon among the fossils of the Paris gypsum. This confronted him with a difficult problem. He had finally discovered that there were two kinds of jaws, some with canine teeth and others without, and that there were also two types of foot, some with two digits and others with three. But which jaw could be associated with each type of foot? Cuvier proposed three means to solve this problem: comparing the respective sizes of the jaws and the feet, pairing the jaws and the feet according to their frequency among the fossil bones, and using the zoological affinities. First, he noticed that all the bones that were as small as the bones of a hare were jaws without canines or from two-toed feet. It was an initial argument for systematically associating the two kinds of fossil. The respective frequency of the different fossil bones led to the same conclusion. Finally, the analogies between the jaw and the rear feet of the Montmartre animal and the jaw and the rear feet of a tapir provided the most conclusive argument for the existence of two kinds of species, the first without canines and with two-toed rear feet, and the second with canines and three-toed rear feet.

Cuvier could not have obtained such proof without mobilizing significant social, material, and intellectual resources. Much of his task, in his research on the fossil bones, had therefore consisted of building an adequate working space. Collecting the fossil bones required having a network of informants, composed not only of professional scientists and amateurs but also of officials, traders, and practical men. Preparing and studying the pieces mobilized specific skills, some peculiar to Cuvier and others shared by his team of assistants. Moreover, exploiting and publicizing the results implied scientific expertise and a specific scientific status. In the dispositif that Cuvier created for his research on fossil bones, most elements belonged to museum culture. This was the case for most of his working practices, such as dissecting, measuring, cleaning, scratching, pasting, assembling, drawing, corresponding, and classifying. As a sedentary naturalist, Cuvier was an outstanding representative of this old tradition, which he raised to its highest level.

Scratching was especially delicate, because the fossil bones required extraction from their gypsum gangue and were extremely crumbly. Drawing appeared to be still more decisive. Cuvier used it to keep records of specimens that he did not own, but also as a mean to reconstruct the structure of the animal (Rudwick 2000, 2005, 381–384). As a matter of fact, scratching was such a destructive operation that it was often necessary to draw the fossil before removing certain parts of it. Moreover, the drawings made comparisons between bones much easier. This technical use of drawings required a style that favors distinctness and faithfulness. As Rudwick has already stressed, Cuvier was an excellent artist. His drawings were remarkably accurate and suggestive, especially compared with the pictures made in the eighteenth century. For instance, Cuvier represented the skull of the de Drée collection in a much more exact and precise manner than Lamanon had done twenty years before. His drawing also showed that the skull had deteriorated in the course of a few decades, which reminds us how fragile the gypsum fossils were.

Publishing the Discoveries

Dorinda Outram (1984, 67) has remarked that Cuvier, after refusing the opportunity to visit Egypt, became less and less a public savant, reorienting himself toward the production of specialist science for an audience of his peers. His election as a permanent secretary of the First Class of the Institute in 1803 seems to mark the peak of this trend. His retreat behind the walls of scientific institutions was exactly contemporaneous with his careful reconstitution of the extinct animals. Cuvier, who concentrated on this time-consuming research work, did not seem eager to publish his results. Thus, after having announced the publication of a book on fossils of quadrupeds in a call for an international collaboration in paleontology (Cuvier 1800a), he decided to postpone the project because he had to confirm his initial result by gathering new materials. It would have taken several years for his book to be ready for publication if Faujas, his colleague at the museum, had not published a book on geology in 1803 (Faujas 1803a) and several memoirs on fossil bones in the *Annales du Muséum National d'Histoire Naturelle*. As the holder of the chair of geology and a great collector of fossils, Faujas, who had been a protégé of Buffon, was in direct competition with Cuvier in the emerging domain of paleontology.

Faujas expressed skepticism about the very notion of extinct species and claimed that the supposedly extinct animals like the mammoths and the Montmartre animals might be still living in remote parts of the world (Faujas 1803a, 5–7, 205, 308–309; 1803b, 194–195). This attack triggered a quick reaction from Cuvier, which developed into what a contemporary called a "pen war" between the two men.[1] Immediately after the publication of Faujas's book, Cuvier read a series of four lectures on the Parisian fossil bones at the academy. The lectures were published some months later as three articles in the *Annales* (Cuvier 1804a, 1804b, 1804c). In these memoirs, Cuvier exposed his research work on the heads and feet of the different species of Montmartre animals. For the first time, he gave scientific names to the genera he had identified: *Palaeotherium* for the animal with canine teeth and three-toed rear feet, and *Anoplotherium* for the animal with no canine teeth and two-toed rear feet.

Yet, publishing the rest of his paleontological research proved more chaotic than Cuvier had anticipated. An unexpected discovery in the fall of 1803 inspired an important change in the order of his presentations. Workers had excavated the first quasi-complete skeleton of a midsize palaeotherium in a quarry of Pantin, a small village close to Paris and Montmartre. Despite the fact that it had no skull and no feet, this skeleton confirmed the reconstruction Cuvier had proposed in his first memoirs. It also gave a number of new indications about the long bones and the trunk of the palaeotherium. Cuvier published its description as soon as he could in the *Annales* (Cuvier 1804d). Yet, he was much less successful with the front feet, for which fossil bones were much scarcer. He remained deadlocked until he was finally able to extricate an almost complete tree-toed front foot from a piece in the de Drée collection, probably with the help of Laurillard. This section, which cost him great effort, was finally published in summer 1805 (Cuvier 1805b).

Meanwhile, Cuvier had made a particularly significant discovery. Workers had found a gypsum block containing the skeleton of a small animal, which Cuvier immediately identified by its teeth as an opossum (genus *Didelphis*), resembling animals that still lived in South America. Faujas had claimed in his book that the two faunas of the Old World and the New World were entirely independent, and he thus considered it impossible that a fossil animal found in Europe should belong to a genus specific to the fauna of the New World. It was a direct blow to Cuvier, who had already identified some fossil bones found in the south of France as belonging to a kind of tapir considered at that time to be living only in South America (Cuvier 1803). Cuvier used this discovery as an opportunity to carry out a spectacular paleontological experiment to prove the correctness of his hypotheses and the soundness of his method of reconstruction (Cuvier 1804f) (figure 10.3). He summoned several colleagues of the institute to witness the procedure that would dramatically confirm his identification. An opossum is a marsupial,

FIGURE 10.3. The opossum of Montmartre, Cuvier (1804f), plate 19 (no pagination).

and like all marsupials, it has two marsupial bones that sustain its pouch. These bones were not visible in the specimen. However, Cuvier conjectured that they were embedded in the block. He scratched out the stone in front of the witnesses—and the two bones actually appeared! With this experiment, Cuvier refuted Faujas's assertion. But he also had other objectives in mind. He wanted to show how technically delicate the study of the Parisian fossils was: "The lineaments are so lightly printed that we must examine them very closely in order to grasp them" (Cuvier 1804f, 1–2). Moreover, the performance aimed to illustrate the usefulness of comparative anatomy for paleontological research.

In his first lecture on comparative anatomy, published in 1800, Cuvier had developed general ideas about functions of organs, and the different systems of organs defined by these functions. Laws governing relations between these systems, he explained, are based on their mutual functional dependencies ("animal economy"). They have "the same necessity as laws of metaphysics or mathematics, for it is evident that the harmony between organs acting on one another is a necessary condition of the existence of the being to which they belong and that, if one of these functions were modified in a manner incompatible with the modifications of the others, this being could not exist" (Cuvier 1800b, 1:47). Moreover, such "laws of coexistence" exert control on the different parts of the same system of organs. Examining, for instance, the organs of movement, he noticed that "almost no bone varies in its facets, its curvatures, its protrusions without the other bones being subject to proportionate variations; and from the view of one of them, one can also deduce, until a certain point, the view of the whole skeleton" (1:56–57). In fact, Cuvier applied this rule to deduce the existence of the marsupial bones from his examination of the dentition of the fossil. Only because its teeth were identical with those of a modern opossum could he predict with certainty that some marsupial bones were hidden under the surface of the gypsum block.

But at the end of his memoir on the opossum, Cuvier went further, making the strong claim "that there is no science that could not become almost geometrical; chemists have shown it recently for their science, and I hope that the time is not far ahead when the same will be said for the anatomists" (Cuvier 1804f, 16). In writing these words, Cuvier associated comparative anatomy with the Newtonian spirit of his fellow physicists even more than previously. But insofar as these laws did not play any role in his prediction, his point at that time was less to underline the importance of natural laws in comparative anatomy than to oppose his comparative anatomy— supposedly founded on a rigorous analysis of facts—to what he scornfully called the "fantastic edifices" of the systems on the theory of earth.[2] In the same vein, Cuvier launched a final attack in 1806 on the makers of geological systems and their "ethereal castles," which seems to have reignited controversy (Cuvier 1807a).[3] This quarrel was soon followed by the discovery of two wonderful skeletons of *Anoplotherium* that entirely confirmed the hypothetical reconstruction that Cuvier had made from scattered and incomplete fragments. Its publication in February 1807 (Cuvier 1807f), after four other memoirs on phalanges—long bones of the legs and scapulas (Cuvier 1807b, 1807c, 1807d, 1807e)—completed the series of memoirs on the fossil bones from the Parisian gypsum quarries.

In total, between January 1804 and February 1807 Cuvier had written five lengthy memoirs specifically devoted to *Palaeotherium* and *Anoplotherium*. Preparing these papers was an essential part of his research activities at that time. This raises the question of the links between his work at the bench and his work at the writing table. At first glance, lack of sources seems to make it difficult to draw a clear distinction between the two. Yet, even if working manuscripts and drafts are missing, except for occasional rare pages, I think it is possible to discern how Cuvier was able to move from the world of things to the world of words. I define his way of doing this as a projection of his working space onto a paper space. The challenge was to convey the complex chain of reasoning that led him to reconstruct the extinct species from broken scattered pieces of bone. Cuvier himself noted that the plates of his memoirs, drawn and etched as soon as he received the objects, were in a mess, and that this had the advantage of giving a sense of the kind of chaos from which he had to extract his animals (Cuvier 1812, 3: sixth memoir, n.p.).[4] Although he admitted his memoirs would have been clearer if he had waited for complete pieces before writing, Cuvier thought that, as they were, they offered better proof to the reader of the effectiveness of his method of reconstruction.

Cuvier's memoirs had many illustrations, and much of the text amounted to a commentary of these drawings. We can thus consider the drawings as the base of his writing work. But, drawing was also essential in Cuvier's practice for studying the fossils. In fact, the drawings of his printed memoirs directly derived from the drawings he made during his research work. The case of the rear feet shows clearly that there was no major difference between these two kinds of drawings. For instance, of the two drawings representing the tarsus and the metatarsus of the *Palaeotherium minus* ("petit animal de Montmartre") (figures 10.4 and 10.5), one shows the bones embedded in a block of gypsum from the de Drée collection, and the second shows the same bones extricated from the gypsum, rearranged in their original positions, and pasted together with wax. These two drawings directly contributed to the reconstruction of the tarsus. Yet, the same drawings, etched and printed in the memoir on the rear feet, were simply used as visual support for a commentary about the three-toed feet of the *Palaeotherium minus*. More generally, the drawings were used in the printed memoirs to represent the material reality of the fossil bones. They were first and foremost proxy specimens, to be seen and scrutinized. In very many cases the commentary itself was written only to help the reader to notice some interesting features of the bones.

Yet however important the drawings were, they would have been useless without the text. Put in a rather arbitrary order at the end of each memoir, by themselves the drawings were passive, meaningless objects of examination. Only the text gave them meaning and life, because it conveyed the momentum of the research and the argument of the proof. Cuvier wrote his memoirs during his investigations. Thus, the writing table was a true continuation of the experimental bench. Cuvier described his approach using both narratives and arguments. In fact, he constantly hovered between systematic reasoning based on the principles of anatomy and creative tinkering spurred by the discovery of new bones. This tension can be seen as a very effective rhetorical tool, instilling a feeling of suspense in the mind of the reader. More prosaically, it might reflect the ambiguity of Cuvier's approach: on one hand, a strong adherence to deductive modes of thinking; on the other hand, an exceptional aptitude for observing facts and objects and deducing unforeseen conclusions from almost invisible clues. Therefore, reading the text was essential not only for interpreting the drawings but also for grasping a context of discovery that was both contingent and necessary. It was as if the working space itself was poured into a graphic-textual mold.

The way Cuvier used drawings points to a more general fact that is worth emphasizing: even if his entire work on the Parisian fossil bones since 1798 was wrapped in the sort of museum culture that had developed for a long time at the King's Garden cabinet of natural history, Cuvier developed a style that neatly departed from that of his predecessors. As he explained in the letter to Mertrud opening his *Leçons d'anatomie comparée* (1800b, 1:i–xxii), he dismissed bookish compilations of former observations and claimed to base all his descriptions on his own direct observations (1:xv–xvii). The reason for such a requirement was the precision and exactitude he thought necessary in comparative anatomy. Such an ambitious program required Cuvier to organize his dispositif systematically for this purpose. At an epistemological level, it pushed him to distinguish himself from the more traditionally inclined naturalists by referring to values attached to the sort of Newtonianism that had prevailed among Parisian physicists. Yet, Cuvier was much more cautious in this respect when he wrote his technical memoirs on the gypsum fossils than when he lectured on comparative anatomy. He did not hesitate to speak of natural laws comparable to mathematical laws in his lectures, but he merely stressed the decisive role of facts in his research work on extinct species. In short, museum culture still dominated Cuvier's epistemology, while Newtonianism remained more of a nuance of thought.

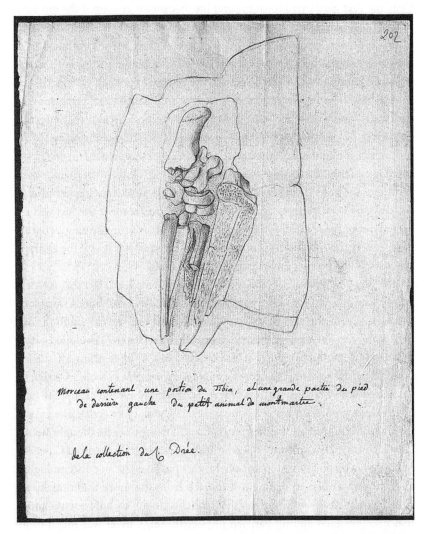

FIGURE 10.4 AND FIGURE 10.5. Rear feet of a *Palaeotherium minus*. Cuvier's drawings.

en dessus   en dessous.

Le tarse et une portion du métatarse du petit animal de Montmartre
cs n ont été tous tirés d'un morceau de plâtre de la collection du C. Drée,
représenté planche préc.d; et collés avec de la cire.

a a. l'astragale. bb. le calcaneum. cc. le cuboïde. DD. le scaphoïde
e e. le cunéïforme du doigt moyen. f f. celui du doigt interne
ɜ ɜ. le métatarse du doigt interne. h h. celui du doigt moyen. i i. celui
du doigt interne   ils sont rompus tous les trois.

¼. plus grand que nature.

Inventing Paleontology

The publication of Cuvier's research on fossil bones in the *Annales du Muséum National d'Histoire Naturelle* was not the end of the story. From the beginning, Cuvier had located his paleontological inquiry in the wider context of the history of the earth. As early as 1796, after having shown that the mammoth was an extinct species, distinct from the elephants, he claimed that this fact proved "the existence of a world previous to ours, destroyed by some kind of catastrophe" (Cuvier 1796, 445). Again in 1800 he emphasized the importance of the fossils of mammals for the theory of the earth in his call for an international collaboration on paleontology (Cuvier 1800a). Yet, in his research work on gypsum fossils he made a point to confine himself to technicalities and avoid any allusion to general geological theories. He remarked only at the end of his memoir on the fossil opossum that its results refuted all systems relating to the theory of the earth. Denouncing daydreams was part of his rhetoric at that time.

However cautious he was vis-à-vis these kinds of system, there is no doubt that his motives in developing his research on the Parisian fossil bones went far beyond the reconstruction of some extinct species and, from the start, touched on fundamental questions about the history of the earth. Accordingly, he was eager to understand the formation of the gypsum layers in order to situate Montmartre's animals in a broader geological picture. With this in mind, in 1803 he engaged in collaboration with the geologist Brongniart about the geology of the Parisian Basin. Brongniart was probably the more active in the field of their joint research, but Cuvier provided some of the main ideas, especially those concerning the alternation of marine and freshwater formations. The result was the *Essai sur la géographie minéralogique des environs de Paris*, which appeared in 1811 (Cuvier and Brongniart 1811).

The following year Cuvier published his *Recherches sur les ossemens fossiles de quadrupeds* (*Research on Fossil Bones*; Cuvier 1812), a vast treatise in four volumes comprising almost all his papers on paleontology published in the *Annales* between 1803 and 1809. The first volume contained his geological monograph on the Parisian Basin. The second volume consisted of memoirs on fossil bones from the Quaternary period, such as the mammoth and the megatherium, discovered on the earth's surface or in superficial deposits. The third volume focused especially on the Montmartre's animals from the Tertiary period, and the fourth volume on some mammals and reptiles whose fossil bones had been found elsewhere, including the famous Maestricht animal, or mosasaur, from the Secondary period. The ordering of the volumes had a dramatic effect. In fact, Cuvier changed the meaning of his entire paleontological research by including the monograph on the geology of the Parisian Basin in the first volume of his treatise. This reversal favored the global and historical approach of geology at the expense of the local and technical approach of comparative anatomy.

The change was considerably reinforced by the "Preliminary Discourse," which opened the first volume, in which Cuvier developed at length ideas he had only sketched before. He explained in nontechnical terms the methods he had used to reconstruct the fossil animals and expounded his general view about the history of the earth. Fossil bones were presented as proof of the existence of some extinct species in the past that had no direct relation to extant species. He refuted Lamarck's transformist theory and attributed the extinction of species to some kind of sudden revolution or catastrophe. On the basis of the human records of antiquity, he claimed that the last revolution, which he assumed to be the same as the deluge, happened about ten thousand years ago.

The "Preliminary Discourse" had a tremendous impact all over Europe, far beyond the scientific world. It was translated into English and published as a book very quickly. In 1821, Cuvier produced a new version of his *Discourse*, published separately and translated into many languages. To this day, Cuvier's name and renown have been closely associated with this well-known work and its dogmatic, erroneous statements, whereas his *Research on Fossil Bones* has been forgotten and neglected, even by specialists and historians of paleontology. In fact, in writing his "Discourse" Cuvier put his previous research work on fossil bones in a new context of reception and interpretation. The targeted reader was no longer the expert in the field or the fellow scientist but the amateur interested in general ideas of the kind associated with the name of Buffon. Thus, Cuvier underwent a fundamental shift from his own working space toward public space. This led him to air views he had concealed in his previous technical works. Henceforth, he relied heavily on two epistemological cultures that were alien to the traditional culture of natural history: Newtonianism and antiquarianism. The first belonged to the sphere of mathematics and physics, and the second related to history and archaeology.

As I have pointed out, Cuvier's Newtonianism was closely related to his conception of comparative anatomy. For Cuvier, the rationale of comparative anatomy must be sought in physiology because the anatomical structure depends on the role each of its components plays in conserving the whole animal. Thus, the hierarchy and coordination of functions explain the coexistence of organs and provide the principle of the subordination of characters, which Cuvier (1795) already used in 1795. Cuvier borrowed this principle from the taxonomic method Antoine Laurent de Jussieu had invented for botany (Daudin 1926, 13–35). Cuvier thought its application would make it possible, by means of functional reasoning, to deduce the whole anatomical structure—the general plan, as he called it—from the study of some of its parts. However, comparative anatomy involving extinct species of mammals was restricted to osteology. Therefore, physiology could not serve as a direct guide for studying each bone and reconstructing the whole structure. Cuvier had to derive a specific method for vertebrate paleontology from the general principle of the subordination of characters. This principle, which he defined as "the principle of the correlation of forms" in his *Preliminary Discourse*, consisted in studying the necessary connections between organs, for instance, the bones, and deducing from these connections a way of reasoning, which he called a calculus, for reconstructing the whole skeleton from fragmentary bones.

As many commentators have remarked, rhetoric was omnipresent in Cuvier's work. Cuvier spoke like an orator in his public lectures and wrote in a literary style, using long periodic sentences and all the resources of rhetoric. There is still a debate among specialists about the interpretation of his methodology in comparative anatomy: were the principles of subordination of characters and of correlation of forms rhetorical developments intended to indicate a deductive science like mechanics? Or were these methods effective tools in his research work? It is beyond doubt that the *Preliminary Discourse* tended to exaggerate some traits of Cuvier's research work to make them more convincing and impressive. The famous statement made early on by Cuvier in the *Leçons d'anatomie comparée*, that it is possible to reconstruct the whole animal from the bones of its foot alone, is obviously an example of such rhetorical exaggeration. But as I have shown, in fact, Cuvier extensively used the articular facets to reconstruct the tarsus and the whole foot in his work on the palaeotherium. This case was indeed the paradigmatic example of the principle of correlation of forms.

This principle, as it was de facto used in his research work on fossil bones, was essentially empirical. It expressed the very method of traditional comparative anatomy, which consisted in making comparisons not only between different species but also between different organs belonging to the same species. It gave Cuvier a sort of guide to reassemble the different organs in order to reconstruct the whole animal. But the inferential reasoning was purely inductive. Cuvier profoundly changed his point of view in his *Research on Fossil Bones*. From then on, the starting point was the geological environment and the conditions of life. Cuvier claimed it was possible up to a certain point to deduce the anatomical structure of living beings from the consideration of the environment they inhabit, inasmuch as this structure is functionally determined by their conditions of life. Thus, Cuvier no longer considered the correlation of forms as an empirical principle—it was a rational one, which expressed the entire dependence of each part on the whole system. Many commentators have deduced from this that the principle of the correlation of forms was pure rhetoric and that Cuvier had not made use of it in his concrete research on fossil bones. In my view, this interpretation results from a confusion between the empirical and the rational principle. A close reading of his memoirs on the gypsum fossils shows clearly that Cuvier used a systematic approach combining the comparison of bones belonging to different species with the empirical principle of the correlation of forms in order to reconstruct the animals.

Even though Cuvier saw paleontology as a sort of derivative of comparative anatomy, it also involved geology and history. Therefore, the *Preliminary Discourse* emphasized its role in the stratigraphic study of secondary formations. The link between paleontology and geology was still more fundamental for interpreting fossils as monuments of the past, that is, for using fossils as a means to reconstruct the whole history of the earth. Antiquarianism played its role at this point. From the beginning, Cuvier's aim had been to reconstruct not only extinct species but also, through this fauna, the whole past of the earth. Such an objective entailed the history of mankind because there was nothing certain at that time concerning chronology in the remote past. Rudwick has notably explained how decisive the antiquarian spirit was for the development of Cuvier's geology and, more generally, for the emergence of what Rudwick has called "geohistory" at the beginning of the nineteenth century. Thus, Cuvier was very interested in orientalism, even if he chose not to go to Egypt himself. He studied mummified animals from Egypt, especially the ibis, and through this discovered an argument against transformism (Cuvier 1800c, 1804e). It is striking that he chose this issue to open his treatise of 1812, just after the *Preliminary Discourse* and before the geological description of the environs of Paris. Moreover, Cuvier was convinced that historical textual documents could give invaluable information about the history of the earth. Fossil fauna of Montmartre, he explained, proved that revolutions had occurred in the earth's history, and several facts suggested that these revolutions were generally sudden and violent. He asserted, on the basis of the biblical account, which he considered a reliable source, that the last revolution had been the deluge.

To us, this latter statement looks like a pure hypothesis, if not an ideological prejudice. For Cuvier, however, catastrophism had been at the very center of his research program about extinct species from the beginning and the Bible was a powerful argument for this theory. What changed with the *Preliminary Discourse* is that, from then on, he placed this idea at the forefront. It was not a minor change, because such a position implied a rejection of all forms of transformism and eternalism, which had strong support among French naturalists. If Cuvier had already expressed his catastrophism in the popular lectures at the Athénée, he had avoided making explicit statements on this point in his scientific papers, except for some brief remarks at the end of the lengthy papers he published in 1806 on mammoths and mastodons. Rather, up until then, his favorite stance at the institute had been to criticize any geological speculations that had no scientific basis. It is all the more notable that he gave up this cautious posture in publishing the *Research on Fossil Bones* and its "Preliminary Discourse" by addressing both his fellow scientists and the general educated readership.

Reference to new epistemological cultures helped Cuvier to take up the challenge. On one hand, by referring to the atheist Laplace and comparing laws of comparative anatomy with laws of astronomy, he made it clear both that he did not indulge in religious divagations and that he adhered to the same methods and values as his fellow physicists. This was the role of Newtonianism. On the other hand, he provided a strong argument for believers who claimed that science could support the mosaic revelation by equating textual evidence of the Bible with paleontological evidence of fossils. Antiquarianism supplied the justification. Despite his rhetorical virtuosity, however, Cuvier could not entirely eliminate the genuine tension between these two epistemological cultures. He wavered between a deterministic conception, which inspired his whole work in comparative anatomy, and a contingent conception, which inspired his work on geology. Paleontology was exactly in the middle. Such a tension was fruitful in a sense, but it led him to elaborate a theory associating fixity with catastrophism, which turned out to be an impasse. However, such a combination of universalism and historical contingency also contributed to the success of Cuvier's work in public opinion, because it expressed, at the level of scientific research, a tension that existed in the political situation of the time. Cuvier was ironically dubbed the "Napoleon of natural history" with good reason. The Napoleonic regime tried to combine change with stability and to reconcile revolution with the ancien régime, and the Enlightenment with religion. Cuvier went in the same direction, with the same ultimate result: failure in both cases.

Notes

All translations are my own.

1. Letter of the naturalist Félix de Roissy to Louis-Benjamin Fleuriau de Bellevue, 17 Pluviôse XII (February 7, 1804), recently put up for auction.

2. Modern commentators have repeatedly used this memoir on the opossum to illustrate the discrepancies between Cuvier's great theoretical claim about the application of the principle of correlation of forms in comparative anatomy and his actual practice in paleontology. In fact, Cuvier himself did not appeal to any zoological law for predicting the existence of the marsupial bones. Thomas Huxley seems to have been the first to make this erroneous supposition in his essay *On the Method of Zadig* (1883). He was followed later by Henri Daudin (see Daudin 1926 and William Coleman 1964).

3. On the controversy that might have followed this report, see Duvernoy 1833, 100–101.

4. "Sixième mémoire, servant de supplément aux cinq premiers, dans lequel on décrit des morceaux nouvellement tirés des carrières, et propres à compléter les précédens."

References

Coleman W. 1964. *G. Cuvier, Zoologist: A Study in the History of Evolution Theory*. Cambridge, MA: Harvard University Press.

Cuvier, G. 1795. "Mémoire sur une nouvelle division des mammifères, et sur les principes qui doivent servir de base dans cette sorte de travail, lu à la Société d'histoire naturelle le premier floréal de l'an troisième." *Magasin Encyclopédique* 2: 162–190.

Cuvier, G. 1796. "Mémoire sur les espèces d'éléphans tant vivantes que fossiles, lu à la séance publique de l'Institut national le 15 germinal an IV." *Magasin Encyclopédique* 3: 440–445.

Cuvier, G. 1798a. "Extrait d'un Mémoire sur les ossemens fossiles de quadrupèdes." *Bulletin des sciences par la Société Philomathique* 18: 137–139.

Cuvier, G. 1798b. "Sur les ossemens qui se trouvent dans le gypse de Montmartre." *Bulletin des sciences par la Société Philomathique* 20: 154–155.

Cuvier, G. 1800a. "Extrait d'un ouvrage sur les espèces de quadrupèdes dont on a trouvé les ossemens à l'intérieur de la terre, adressé aux savans et aux amateurs des sciences." *Journal de physique* 52: 253–267.

Cuvier, G. 1800b. *Leçons d'anatomie comparée*. Vols. 1 and 2. Paris: Baudoin.

Cuvier, G. 1800c. "Mémoire sur l'ibis des anciens Égyptiens." *Journal de physique, de chimie et d'histoire naturelle* 51: 184–92.

Cuvier, G. 1803. "Sur quelques dents et os trouvés en France, qui paraissent avoir appartenu à des animaux du genre tapir." *Annales du Muséum National d'Histoire Naturelle* 3 (13th cahier: Vendémiaire an XII): 132–143.

Cuvier, G. 1804a. "Sur les espèces d'animaux dont proviennent les os fossiles répandus dans la pierre à plâtre des environs de Paris. Premier mémoire: Restitution de la tête." *Annales du Muséum National d'Histoire Naturelle* 3 (16th cahier: Nivôse an XII): 275–303.

Cuvier, G. 1804b. "Sur les espèces d'animaux dont proviennent les os fossiles répandus dans la pierre à plâtre des environs de Paris. Deuxième mémoire: Examen des dents et des portions de têtes éparses dans nos carrières à plâtres, qui diffèrent du Palaeotherium medium, soit par l'espèce, soit même par le genre." *Annales du Muséum National d'Histoire Naturelle* 3 (17th cahier: Pluviôse an XII): 364–387.

Cuvier, G. 1804c. "Sur les espèces d'animaux dont proviennent les os fossiles répandus dans la pierre à plâtre des environs de Paris. Troisième mémoire: Restitution des pieds." *Annales du Muséum National d'Histoire Naturelle* 3 (18th cahier: Ventôse an XII): 442–472.

Cuvier, G. 1804d. "Cinquième mémoire: Sur les os du tronc." *Annales du Muséum National d'Histoire Naturelle* 4 (19th cahier: Germinal an XII): 65–75.

Cuvier, G. 1804e. "Mémoire sur l'ibis des anciens Égyptiens." *Annales du Muséum National d'Histoire Naturelle* 4 (20th cahier: Floréal an XII): 116–135.

Cuvier, G. 1804f. "Sur le squelette presque entier d'un petit quadrupède du genre des sarigues trouvé dans la pierre à plâtre des environs de Paris." *Annales du Muséum National d'Histoire Naturelle* 5 (28th cahier: Nivôse an XIII): 277–292.

Cuvier, G. 1805a. *Leçons d'anatomie comparée.* Vols. 3–5. Paris: Baudoin.

Cuvier, G. 1805b. "Troisième mémoire, deuxième section: Restitution des pieds de devant." *Annales du Muséum National d'Histoire Naturelle* 6 (34th cahier: Messidor an XIII): 253–283.

Cuvier, G. 1807a. "Rapport sur un ouvrage manuscrit de M. André, ci-devant connu sous le nom de P. Chrysologue de Gy, lequel ouvrage est intitulé Théorie de la surface actuelle de la Terre (lu à l'Institut le 11 août 1806) (with Hauÿ and Lelièvre)." *Mémoires de la classe de mathématiques et de physiques de l'Institut de France* 8: 128–145.

Cuvier, G. 1807b. "Suite des recherches sur les fossiles des environs de Paris: Troisième mémoire. 3e section: Les phalanges." *Annales du Muséum National d'Histoire Naturelle* 9 (49th cahier: January 1807): 10–15.

Cuvier, G. 1807c. "Suite des recherches sur les fossiles des environs de Paris: Quatrième mémoire. 1re section: Sur les os des extrémités." *Annales du Muséum National d'Histoire Naturelle* 9 (49th cahier: January): 16–44.

Cuvier, G. 1807d. "Suite des recherches sur les fossiles des environs de Paris: Quatrième mémoire. 2e section: Sur les os longs des extrémités antérieures." *Annales du Muséum National d'Histoire Naturelle* 9 (49th cahier: January): 89–102.

Cuvier, G. 1807e. "Suite des recherches sur les fossiles des environs de Paris: Quatrième mémoire. 3e section: Sur les omoplates et les bassins." *Annales du Muséum National d'Histoire Naturelle* 9 (50th cahier: February): 205–215.

Cuvier, G. 1807f. "Suite des recherches sur les fossiles des environs de Paris: Cinquième mémoire. 2e section: Description de deux squelettes presque entiers d'anoplothérium commune." *Annales du Muséum National d'Histoire Naturelle* 9 (50th cahier: February): 272–282.

Cuvier, G. 1812. *Recherches sur les ossemens fossiles de quadrupèdes*. 4 vols. Paris: Deterville.

Cuvier, G., and Brongniart, A. 1811. *Essai sur la géographie minéralogique des environs de Paris avec une carte géognostique, et des coupes de terrain*. Paris: Baudoin.

Daudin, H. 1926. *Cuvier et Lamarck: Les classes zoologiques et l'idée de série animale (1790–1830)*. Paris: Alcan.

Duvernoy, G. 1833. *Notice historique sur les ouvrages et la vie de M. Le Baron Cuvier*. Paris: F. G. Levrault.

Faujas de Saint-Fond, B. 1803a. *Essai de géologie ou Mémoires pour servir à l'histoire naturelle du globe*. Paris: Patris.

Faujas de Saint-Fond, B. 1803b. "Mémoire sur deux espèces de bœufs dont on trouve les crânes fossiles en Allemagne, en France, en Angleterre, dans le nord de l'Amérique et dans d'autres contrées." *Annales du Muséum National d'Histoire Naturelle* 2 (9th cahier : Prairial an 11): 188–200.

Foucault, M. 1977a. *Discipline and Punish: The Birth of Prison*. New York: Pantheon Books.

Foucault, M. 1977b. "Entrevue: Le jeu de Michel Foucault." *Ornicar: Bulletin périodique du champ freudien*, 10 (July): 63. *Dits et Ecrits* 3, text 206, 298–329.

Galison, P., and E. Thompson, eds. 1999. *The Architecture of Science*. Cambridge, MA: MIT Press.

Guettard, J.-E. 1768. *Mémoires sur différentes parties de la physique, de l'histoire naturelle, des sciences et des arts, etc.* Vol. 1. Paris: Prault.

Huxley, T. 1883. "On the Method of Zadig." In *Collected Essays: Science and Hebrew Tradition*, 4: 1–23. London: Macmillan.

Lamanon, R. de. 1782. "Description de divers fossiles trouvés dans les carrières de Montmartre près Paris: Et vues générales sur la formation des pierres gypseuses." *Observations et mémoires sur la physique, . . . ou Journal de physique* 19: 173–194.

Outram, D. 1984. *Georges Cuvier: Vocation, Science and Authority in Post-Revolutionary France*. Manchester: Manchester University Press.

Rudwick, M. 1977. *Georges Cuvier, Fossil Bones, and Geological Catastrophes*. Chicago: University of Chicago Press.

Rudwick, M. 2000. "Georges Cuvier's Paper Museum of Fossil Bones." *Archives of Natural History* 27: 51–68.

Rudwick, M. 2004. "Researches on Fossil Bones: Georges Cuvier and the Collecting of International Allies." In *The New Science of Geology: Studies in the Earth Sciences in the Age of Revolution*, 1–12. Burlington, VT: Ashgate.

Rudwick, M. J. S. 2005. *Bursting the Limits of Time: The Reconstruction of Geohistory in the Age of Revolution*. Chicago: University of Chicago Press.

Rudwick, M. J. S. 2008. *Worlds before Adam: The Reconstruction of Geohistory in the Age of Reform*. Chicago: University of Chicago Press.

HANS-JÖRG RHEINBERGER

11
—

Cultures of Experimentation

This paper is to be seen as a sequel to and expansion of my earlier work on experimental systems. The move from the micro-level to the meso-level of experimentation, from a local to a more regional perspective, is part of what I would like to conceive of as a particular form of a historical epistemology of experimentation working its way from below. In short, experimental cultures can be characterized as ensembles or bundles of experimental systems sharing one or more constituents—technical, material, social, or biological (e.g., model organisms in the case of the life sciences). The notion of sharing is important here. It is what justifies talking of cultures in this context. Aspects of this meso-level of experimentation have been discussed under the notion of "styles" of scientific practice (Hacking 1992) or "ways of knowing" and "ways of working" (Pickstone 2000, 2011).[1] The particular feature I would like to capture by the term "culture," however, is the aspect of material interaction between systems of experimentation—an interaction that engenders meaning while unfolding, rather than the other way around, by instantiating or embodying ideals. Counter to the generally accepted notion of culture that is tightly bound to the realm of the symbolic (Geertz [1973] 2000), my use of the concept here ties it to the materialities of the scientific working process.

In the first part of this chapter, I take a closer look, albeit still in the form of a rough sketch, at a particular experimental culture in the life sciences: in vitro or test tube experimentation. In vitro experimentation was of crucial importance to the life sciences for the better part of the twentieth century. I characterize and follow some of the different forms taken by the analysis of biological functions in the test tube. I also trace the opposition that these experimental moves encountered over time.

An early form of in vitro experimentation was based on homogenates of cells in parallel with efforts to develop tissue and cell cultures in the test tube. From the 1930s on, fractionation of cellular contents by high speed centrifugation became predominant, in conjunction with the purification of particular cellular substances. In the wake of World War II, radioactive tracing added yet another dimension to the test tube assessment of metabolic processes. In the second part of this chapter I set the concept of experimental culture against a broader discussion about the notion of culture in modernity, and sound out the more narrowly conceived historiographical potential of the concept of experimental culture. In contrast to disciplinary history, with its institutional concerns, a cultural history in this sense stresses the epistemic dimension involved in dealing with scientific communities.

Experimental Cultures: In Vitro

In vitro experimentation had a deep and lasting impact on the life sciences of the twentieth century. Indeed, it is the in vitro or test tube culture of biological experimentation that made the conjunctures of biology, physics, and chemistry possible, which prepared the ground for the molecularization of the life sciences. The linguistic distinction between "in vitro" and "in vivo" is itself a product of the development of biological experimentation toward the end of the nineteenth century. The *Oxford English Dictionary* quotes George Gould's 1894 *Illustrated Dictionary of Medicine, Biology, and Allied Sciences* as the earliest source for a definition of "in vitro": "In the glass; applied to phenomena that are observed in experiments carried out in the laboratory with microorganisms, digestive ferments, and other agents, but that may not necessarily occur within the living body" (623b). The definition is as ad hoc and unspecific as it is telling; besides pointing to the origin of in vitro work in microbiology, it directly addresses the specter behind all test tube manipulation of biological things. Put in the form of a question, is what we observe outside the body, or cell, the same as what occurs inside the intact organism? And with that question, a boundary is addressed that is no longer concerned, as it was around 1800, with the ontological threshold between the organic and the inorganic that helped biology to acquire the status of a nineteenth-century science sui generis. Around 1900, what was at stake was the fixation of the epistemic conditions under which it is (still) possible for processes occurring within the organism to manifest outside the organism and thus to become accessible to analytic investigation. The *enjeu* is the creation of test tube environments in which—in an always precarious manner—biological entities are exposed to measurements, entities that are otherwise hidden from the scientific gaze, buried deep in the cell or the organism as a whole. In what follows I very briefly sketch a few stages and exemplary configurations in the historical development of test tube culture (an encompassing history of this culture is still missing). In vitro experimentation had a decisive impact on the reconfigurations of the life sciences during the twentieth century—not only would the hybrid discipline of biochemistry have been impossible without this way of carrying out biological experiments (see, as the locus classicus, Kohler 1982), but molecular biology as a whole in its now classical phase would come to rest on its premises.

Homogenates

In its canonized history, the birth of modern in vitro biochemistry is usually marked by Eduard Buchner's report on the "Alcoholic Fermentation without Yeast Cells," a paper published in 1897.[2] Certainly, the recovery of extracts from animals, as well as from plants, for medical, technical, and alimentary purposes was a procedure going back to antiquity. It was thus not simply the manipulation of organic substances *extra corpore* that defined the new test tube culture. Following Herbert Friedmann (1997, 108), what was really new was the claim and demonstration that tissue extracts could provide and represent not just what have been called natural products—compounds synthesized in tissues and organs—but natural processes as well. If one takes Friedmann's remark seriously that, "from now on, extract repeats or mirrors process" (108), what was at stake in the transition from a living system to a test tube system was not just the transition from biology to chemistry or, to put it differently, from physiological processes to organic substances, but rather a reduplication of life—a life, so to speak, under other conditions. This process of repetition or of mirroring life, however, implies a permanent threat of distortion that confronts its practitioners with the equally permanent question of how far, in the mirror, we are still able to see nature and how to decide whether we still do or do not.

Stabilizing the liquid external milieu revealed itself as the decisive task around which the in vitro culture, mainly of enzyme activity analysis, was formed at the beginning of the twentieth century. A generation before, the French physiologist Claude Bernard ([1878–1879] 1966, 202–203) had seen what he called the "fluid internal milieu," in which the elements of the tissues were bathing, to be decisive for the occurrence of the processes essential to life. He was convinced that the "organic syntheses" he held to be characteristic of living beings could occur only in this milieu. "I believe to have been the first," he stressed, "to insist on the idea that for the animal there are actually two milieus: an *external milieu* in which the organism is placed, and an *internal milieu* in which the elements of the tissues live" (113). Consequently, Bernard assumed that organic syntheses could be assessed only through experiments on—and in—the living animal. Bernard became known and famed for the development of an in vivo physiology based on sophisticated surgical techniques, along with tricks to follow the transformation of substances within the organs of the animal body (Holmes 1974). Of course, he made invasive interventions from outside, but these were used and meant to produce and record physiological changes that happened inside the living organism. For Bernard, the test tube was counterindicated at this point—it was good only for quantifying substances ex post, after they had been processed in the body itself.

A new generation of biologists and chemists around the turn of the century started to search for an equivalent to Bernard's internal milieu that would allow them to turn the inside out. Gradually, an international community formed centered on the analysis of enzyme function. It included, to name just a few prominent figures, Auguste Fernbach and Louis Hubert at the Pasteur Institute in Paris, Søren Sørensen at the Carlsberg Laboratories in Copenhagen, and Leonor Michaelis and Maud Menten at the Charité in Berlin. Conditions were now sought under which isolated enzymes, especially enzymes that were not secreted by glands into body cavities but were contained within cells themselves, would retain or even enhance their metabolic function.[3] To build up such environments—not by mimicking the cellular contents proper but by functionally replacing the cytoplasm of the living cell—became one of the biggest technical challenges of test tube biology of the early twentieth century. To quote Friedmann (1997, 110), what was at stake was the realization of something completely different from organic chemistry with its drastic syntheses under high pressure and temperature, namely, "the duplication of a cellular chemical process outside an intact biological system." Such duplication was not imperative to rebuild the complete "internal milieu" in the test tube, true to nature; rather, what mattered was to create an environment in which a biological function could be represented in isolation and, ideally, be followed in its intermediate steps. A completely new culture of dealing with biological processes was emerging. It remained controversial over the decades, but it began more and more to dominate the form and the ambience of the biological laboratories around the world.

Most of the processes that became tractable in the context of the new test tube biochemistry of the early twentieth century—prominent among them anaerobic alcoholic fermentation—were revealed to be rather complex metabolic cascades that extended over many intermediate steps, and in which several enzymes or enzyme complexes and coenzymes intervened. Often these enzymes also turned out to be bound to subcellular structures, such as membranes, from which they could not be separated without some or even a complete loss of function. The question of whether the course and pattern of action of a process in the test tube corresponded to the biological process within the cell could, as a rule, not be answered unequivocally. The experimental answers had to be negotiated again and again, and this permanent questioning kept the research process going. But it also meant that simply replacing physiological chemistry in the older sense with the new test tube biochemistry remained problematic for many biologically oriented chemists. As late as 1940, just to take one prominent example, Richard Willstätter, who had already received the Nobel Prize for his research on chlorophyll in 1915, made this clear in quite drastic words, in one of his last papers, written together with Margarete Rohdewald, on the enzymatic systems of the conversion of sugars: "The postmortem changes occur in an uneven fashion, irregularly, without meaning. The meat machine is the cliff of physiological chemistry. Observations on the muscle homogenate or with extracts do not allow straightforward conclusions on anything about the living processes in the muscle" (Willstätter and Rohdewald 1940, 20). And with reference to Buchner's basic experiments of almost half a century before, Rohdewald and Willstätter stated on the basis of their own experiments: "The opinion that macerated yeast sap would react with carbohydrate in the same manner as living yeast was quite arbitrary. The same holds for Buchner's press sap" (51–52).

Tissue and Cell Cultures

Given these remarks, it is not at all astonishing to see that a second in vitro culture, running in parallel to the test tube enzymology on the basis of cellular extracts, gradually took shape: the extracorporeal culture of tissues and even of single cells derived in intact form from multicellular beings. Here, the notion of culture takes on the meaning of a specific form of cultivation. The term "culture" was used in precisely that sense by those who applied in vitro techniques, from the beginning of the twentieth century onward, to study isolated cells of higher organisms. They followed the medical and biotechnological microbiologists' lead with their methods of breeding "pure" bacterial cultures in fluid and on solid media, on the one hand (Summers 1991), and the protozoologists' culturing efforts of unicellular higher organisms such as yeasts, amoebas, and protozoa, on the other (see, e.g., Rheinberger 2010, esp. chap. 5). What was at stake here was the creation of an external milieu in which intact cells that had been isolated from their tissue connections showed phenomena of growth and were induced to divide. Alternatively, tissue slices that had been isolated from organs, such as the liver, were assayed for aspects of their metabolic turnover that could be quantitatively measured with micro-methods in a controlled artificial environment.

A few examples, in historical order, illustrate the case. Between 1906 and 1910 the embryologist Ross Harrison, first at Johns Hopkins and then at Yale, succeeded in observing the outgrowth of axons in explants of neural tissue from frog embryos in an in vitro culture. What proved to be decisive for his success was the choice of substrate: the cells began to grow when Harrison embedded the tissue in clotted lymph (Harrison 1910; Keshishian 2004). Alexis Carrel at the Rockefeller Institute in New York, to whom the concept of "tissue culture" can be traced, continued along these lines and created the basis for culturing human tissue and cells derived from such tissues, in particular heart muscle cells. He even spoke in this context of building "a new type of body in which to grow a cell" (Landecker 2004). In this medical environment, as Hannah Landecker has observed and described in detail under the suggestive title of *Culturing Life* (Landecker 2007), isolated cells soon advanced to as many tools for diagnostic and even therapeutic purposes.

In Germany, Otto Warburg worked with intact cells in the test tube to elucidate the process of biological oxidation. Together with Otto Meyerhof at the University of Heidelberg, he first endeavored to analyze respiration in extracts of bacteria. However, he soon realized that he could get results with broken cells only when he left them unfractionated, and even then only in an irregular fashion, whereas extracts free of cell debris yielded no results at all. Consequently, as Petra Werner (1988, 131) put it in her biography of the cell physiologist, Warburg saw himself challenged to "use the whole cell as a 'test tube.'" Warburg (1928, 1) summarized his experimental experiences as follows: "Since experience teaches us that one cannot separate the catalysts of living substance—the ferments—from their accompanying inactive compounds, it suggests abandoning the methods of preparative chemistry and analyzing the ferments under their most natural action conditions, that is, in the living cell itself." He thus constructed the famous Warburg apparatus, consisting of a water bath, manometer, reaction vessel, and shaking gear. Between the wars, it became one of the most important—and emblematic—pieces of laboratory equipment for in vitro physiology based on tissue slices.

Warburg's contemporary, Alfred Kühn, himself a pioneer in experimental gene physiology, was another example of those who remained skeptical toward destroying cells to study their reactions, as we learn, for example, from a letter to his former coworker Ernst Caspari, who had been forced to leave Nazi Germany and immigrated to the United States. Caspari wrote to him after World War II about experiments in which he tried to follow eye pigment formation in fruit fly (*Drosophila*) homogenates instead of whole animals. In a letter to Caspari in response dated December 1946, Kühn tellingly exclaimed, "I cannot believe that the whole synthesis . . . can occur in a tissue homogenate."[4] In agreement with Willstätter and Warburg, among many others at this time, Kühn (1938, 107) considered the intact organism, or at the very least the intact cell, "a laboratory of a peculiar chemical regime."

Cell Fractionation

For about four decades into the twentieth century, the alternatives for analyzing the biological functions performed by cells were either to prepare raw tissue extracts or to keep intact cells or conglomerates of cells—tissue slices—as "black boxes" in the test tube and to microchemically measure the various gases and other small chemical compounds resulting from their turnover. It was only with the advent of high-speed centrifugation in the 1930s that a few cell biologists, mostly working in medical contexts and seeking to open the black box of the cell, began not only to extract the content of whole cells but also to centrifuge homogenates to separate the cell sap into more or less well-defined structural or functional fractions. "Differential fractionation" became their catchword.

Various physical methods and instruments, such as mixers, mills, mortars, vibrators, and ultrasound devices, were explored to break up cells as gently as possible and to keep the cell contents as intact as one could. Solubilizers and detergents, even lytic enzymes, were alternately tried to see whether they would do the job. These efforts stood in the tradition of homogenate preparation. The tissue of choice was, first and foremost, the liver of clinical test animals such as guinea pigs, rats, and mice (Gaudillière and Löwy 1998; Rader 2004), because these animals were readily available in large amounts and liver tissue was particularly active in regeneration and therefore thought to exhibit particularly robust activity patterns.

In addition, one needed to explore the osmotic conditions under which the isolated cellular contents such as organelles—nuclei and mitochondria—were able to keep their structure and remain functionally active. To meet this double condition—gentle homogenization and native fractionation with respect to cellular organelles—buffers and salts of a widely different nature were tested. Finally, one needed to determine the centrifugation conditions—length of run, rounds per minute, form and inclination angle of the centrifuge tubes, and so forth—under which the cell sap could be separated into fractions neatly isolated from one another.[5]

The Belgian physician and cytologist Albert Claude, working at the Rockefeller Institute in New York, has to be considered as *the* pioneer of cell fractionation. During the late 1930s and early 1940s he used high-speed centrifugation to isolate the tumor agent of Rous sarcoma from chicken tissue. But to his surprise the controls with healthy tissue that he ran in parallel to his infectious preparations contained a pellet that, in terms of chemical composition, was not distinguishable from the fraction containing the tumor agent. The latter appeared to be embedded in a sedimentable cellular component of unknown constitution and function, and it seemed hopeless to separate them. The fact that the normal cellular component was the measurable and manipulable part of the mixture led Claude out of cancer research altogether and into what came to be established and addressed as in vitro cytology (see Löwy 1990; Rheinberger 1993). Starting from a complete homogenate—and proceeding according to the principle that during the whole preparation procedure no fraction was to be discarded—Claude established four different cellular fractions: a quickly sedimenting fraction containing cell wall fragments and nuclei, a mitochondrial fraction, a microsomal fraction containing particles that were no longer visible under the light microscope, and a nonsedimentable supernatant containing proteins and the rest of the cytoplasm. Claude (1975, 433) referred to his in vitro cell research as a "quantitative method of analysis according to a balance sheet principle."

Claude was of course not alone in setting the stage for this kind of in vitro cytology. He could build on the work of Robert Bensley and Normand Hoerr at the University of Chicago, who had pioneered the characterization of a particular cellular organelle, the mitochondrion. At the same time in Europe, the group around Jean Brachet at the Free University of Brussels was studying the fractionation and isolation of microsomes, starting from yeast cell extracts, in parallel (and in competition) with Claude (see Rheinberger 1997a). Yet, despite the fact that the Rockefeller group around Rollin Hotchkiss, George Hogeboom, and Walter Schneider had developed a test regime based on enzyme activity assays into a sophisticated procedure they called "biochemical mapping" (see Palade 1951, 144), the functional characterization of these isolates proved a tedious and protracted task.

Radioactive Tracing

A more versatile in vitro functional analysis of these fractions became available only with the introduction of yet another new technology—a technology that revolutionized all of biological in vitro research around the middle of the twentieth century: the radioactive labeling of individual molecules and molecular components that play a role in metabolic and genetic turnover. After measuring devices had been developed that were sensitive enough to efficiently register the decay of tritium (^{3}H) and radioactive carbon (^{14}C), and after isotopes had become available en masse as a by-product of reactor technology (Creager 2013), the method of radioactive labeling functioned like some sort of biochemical electron microscope. However, the introduction of radioactivity into the laboratory cultures of biology not only had an impact on the experimental systems in which it was used, that is, on the experimental culture of in vitro cytology that was gradually reshaping itself as molecular biology; it also had an impact on the laboratory architecture and on laboratory life as a whole. The possibility of measuring radioactive traces of minimal intensity in biological samples crucially depended, not least, on keeping the environment of the experiment uncontaminated. This requirement of a strictly clean working environment not only resulted in a completely new laboratory regime but had also a massive influence on the very design of the experiments.

These few remarks may suffice to indicate that the technology of radioactive tracing cannot be reduced to an instrument, such as the liquid scintillation counter, or to a substance, such as radioactive building blocks of macromolecules. It represented a structure that with its various facets came to impregnate and permeate a whole experimental culture. The essentials of this culture can be briefly characterized as follows. First, it established an indicator principle in the analysis of metabolic processes and thus elevated the in vitro analysis of cellular functions to a new level. Radioactive tracing in the test tube meant the possibility of going beyond the boundaries of chemical measurements that had previously relied on the availability of microgram mounts of substances in purified form—radioactive measurement could proceed in an essentially unpurified background. On top of that, the sensitivity of measurement was now higher by almost six orders of magnitude than it had been for standard microchemical quantification. Second, the tracer technology was itself a driving force for the development of new radioactive measurement technologies, such as the liquid scintillation counter, where their integration into the experimental systems of biological chemistry affected not only their size but also their design (see Rheinberger 2010, chap. 9). Third, the technology resulted in something like a point of material mediation of the know-how among biologists, chemists, physicists, and engineers. It was a technology that in its very design—physics of isotopes, chemistry of liquid scintillation, electronics, and biological specimen preparation— required bringing together different areas of expertise and therefore also transcended, and tended even toward the abolition of, traditional disciplinary boundaries.

Taken together, the combination of differential cell fractionation and radioactive labeling led the culture of in vitro biochemistry, with its structurally or functionally oriented experimental systems, into the domain of the cellular synthesis of macromolecules and, with that, into the center of molecular biology as it emerged after World War II (Rheinberger 1997b). This experimental background to molecular biology has only belatedly received due consideration in its historiography.[6] What we are confronted with is a technical enhancement that reversed (at least to some extent) the massive damping down of biological functions and performance resulting from going in vitro. Put succinctly, what the cell sap lost in terms of biological activity it regained in radioactivity. The creation of an extracellular environment for the intracellular processing of genetic information transfer will probably remain the prime example for the kind of test tube biology that became characteristic for the decades after World War II. As a result, during the 1970s the membrane-free microsomes of bacteria—ribosomes—became the first cellular organelle dissected into its components in the test tube and reconstituted under appropriate buffer and incubation conditions into fully active particles. The in vitro assembly of functionally active ribosomes, as well as the determination of the partial reactions of the complex cycle of protein synthesis they performed, was thus something like the apotheosis of test tube molecular biology (see Rheinberger 2004). Since then, gene technology, itself a result of the development of molecular biology, is finding its way back into the cell and, with its contemporary molecular tools, choosing anew—but in a different way—the intact cell and even the organism as the space of its experimental intervention.

In assuming an experimental cultures perspective, we can derive two historiographical lessons from these examples. First, experimental cultures are themselves historical entities. Their epistemically productive life span is limited. They can coexist for a time, become marginalized, or become replaced altogether: fractionation replaced simple homogenization, and cell culturing coexisted with homogenization and fractionation. Today, cell culturing proliferates,[7] and other forms of in vitro experimentation have become marginalized. Second, these cultures form bridges between and among themselves and, in the process, tend to cross institutional academic barriers. They do not do away with these structures, but they can reconfigure their boundaries. In the long run, the cultures of in vitro experimentation had an enormous impact on the traffic between biology and medicine in the twentieth century. Today's biomedicine is largely a result of this traffic.

Culture: Some General Epistemological Reflections

The term "culture" has a wide scope of meanings. In its broadest sense its modern use amounts to distinguishing things made by humans from those that are not, that is, from those occurring in nature. The Hungarian-born sociologist Karl Mannheim argued that the distinction itself had evolved historically and in its full exposition came to form the core of the cultural self-understanding of modernity. For us moderns, Mannheim (1980, 48–49) argued, "being and meaning, reality and value have been broken up and become disintegrated for our experience," but it is only through this disintegration that a "determination of culture as nonnature has really become concrete und internally consequent." Henceforth, culture is what is understood and conceived as being engaged in a "mental-historical trajectory" (*geistig-historischer Werdegang*). In contrast, "nature, forming the opposite to modern 'culture,' thus being its correlate, is something that is completely free of meaning, of value, a mere substrate of possible sense. Actually it comprises the totality of all those determinations that do not pertain to the cultural. Nature is thus that which is impenetrable by the spiritual, indifferent to value, not subjected to mental-historical development."

If one speaks, as I do, about cultures of the natural sciences, of cultures of experimentation aiming at an understanding of just that nature supposed to be devoid of meaning, the seeming paradox points to a view of the sciences that situates itself beyond that modern dichotomy. It transports and reflects the effort and the attempt to locate the sciences—that is, the knowledge of nature—not simply on the side of nature itself, not simply on the side of the objects of study. In the discourse of the sciences about themselves, and the public image they project about themselves, this happens again and again. In contrast, the arguments made for *Science as Practice and Culture* (Pickering 1992) and for *Epistemic Cultures* (Knorr Cetina 2000) are aimed at perceiving not only the scientific institutions but also scientific knowledge itself as a cultural phenomenon in all its historicity. This goes a decisive step beyond Mannheim, who, in his critical reflection upon the modern dichotomy between nature and culture, did not question that it was just the knowledge of what he called the "historical-cultural reality" that had to be seen as a product of and related to its respective cultural-historical standpoint and perspective. Natural scientific knowledge for Mannheim (1980, 110–111) was still "only inasmuch bound to its own history, as the later knowledge presupposes all those results that preceded it, in a way as its necessary premises." Mannheim thus conceded historicity to natural scientific knowledge, but a purely internal one, a history free from culture, if you will.

Pierre Bourdieu characterized the dilemma by describing this distinction as the inescapable "double face" of scientific knowledge. In his *Méditations pascaliennes* (1997, 130–131) he put it as follows:

A realistic vision of history forbids a fictive transgression of the uncrossable limits of that history and prompts us to examine how, and under what historical conditions, truths can be squeezed out of history that are not reducible to that history. One has to admit that reason did not fall from heaven like a mysterious gift destined to remain inexplicable and that it is therefore historical through and through. But one is not at all forced to conclude, as is usually done, that it is reducible to history. It is in history, and in history alone, that one must look for the principle of the relative independence of reason from history whose product it nevertheless is—or more precisely, one has to look at the genuine, but very specific, historical logic according to which those universes of exception have been instituted in which the singular history of reason accomplishes itself.

Experimental Cultures

Whether we hold with the cautions of Bourdieu or we prefer a more radical variant of historical epistemology, we have to enter into a closer discussion about the use of the concept of culture for characterizing the sciences and their development. Generally speaking, it presents a challenge to the modern nature–culture divide. I prefer a descriptive assessment to start with: I have spoken of experimental cultures in the precise sense of ensembles of experimental systems. If experimental systems are to be seen as the smallest functional units of modern experimental research, units that can be localized at the level of individual laboratories, then experimental cultures can be seen as consisting of conglomerates of related experimental systems able to communicate with one another across their boundaries. Such ensembles have to meet at least three conditions, which are certainly necessary; whether they are also sufficient is a matter open to discussion. First, there has to be a certain overlap in the technologies that these clusters of experimental systems use and on which they rest. Experimental cultures share research technologies. Second, there has to be something like a flow of matter among systems forming an experimental culture; that is, they have to share either concrete material objects or experimental environments in which these objects are embedded. Third, an experimental culture is characterized by a circulation of scientists moving among adjacent experimental systems to combine their specific, already existing know-how or to bring their specific know-how to bear on slightly different experimental contexts. In this sense, experimental cultures are seen as more or less coherent fields of scientific research consisting of networks of experimental systems. What holds them together are material forms of sharing things, patterns of circulation. Thus, they stand for and characterize a certain kind of epistemic cohesion that is distinctly different from what is usually described with the concept of scientific discipline— whose focus is generally on forms of institutionalization. The concept of experimental culture, in contrast, focuses on the process of research. In addition, disciplines are frequently—albeit not always—characterized ontologically by the realm of their objects. Experimental cultures, in contrast, are epistemologically defined by opening up a characteristic approach to a realm of objects. The in vitro cultures of twentieth-century biological experimentation are exemplary in this respect.

Gaston Bachelard had something similar in mind when he demanded that a historical epistemology deserving its name should take seriously the fact that the dynamics of contemporary scientific knowledge production rest on a very mobile regionalization. According to Bachelard (1949, 132), the modern sciences create what he called "kernels of apodixis." They are, however, in all their unconditionality and at times esoteric exclusiveness, temporally finite as well as spatially restricted. At the limit, each of these kernels requires its particular epistemological attention. In this respect, Bachelard (1940, 12, 14) spoke of a "dispersed philosophy"—a distributed or "differential" epistemology. To express it in his own words: "We must attempt a rationalism that is concrete and in line with the precision of particular experiments. It is also necessary that our rationalism is sufficiently *open* to receive new determinations from the experiment" (Bachelard 1949, 4). An epistemology that aspires to grasp the dynamics of scientific work at the frontier to the unknown is bound to be as mobile and risky as the work itself that it tries to understand.

Bachelard (1949, 132–133) also spoke of "cantons," regions, or domains in a "city of knowledge." For him, such cantons or quarters were like islands of scientific culture with their own codes, semantics, and forms of emergence. For them, he himself used the term "culture," and he gave it a very specific meaning: an "access to an emergence" (*une accession à une émergence*; 133). Bachelardian scientific cultures are thus specific epistemic milieus in which new knowledge can be created, in which unprecedented things can happen. They are thus cultures of innovation. If, as in the view of Bachelard, the artistic act of creating new forms of art is more tainted by the individual, then this is different for the scientific act of creating new knowledge. Scientific "emergences," to quote again from Bachelard's book *Le rationalisme appliqué*, "have a decidedly social constitution" (133). This means that they take the form of cultures also in the sense of scientific communalities—communities that deal with the phenomena of their epistemic interest and desire in specific, shared ways. These cultures engage in the game of supplementation of phenomena, of approaches, and of concepts that are characteristic for their particular area of research, and so keep it going. The more narrowly such areas are defined, the easier conventions, forms of measurement, ways of description, and grids of classification can be modified or altered, and eventually also transmitted to other realms of research. Regionalization creates epistemic flexibility. Bachelard saw no deplorable loss of the synthetic gaze in the fragmentation into different cultures of contemporary research—he rather viewed it as a prerequisite for the tremendous fertility of the sciences of his day.

In conclusion, cultures of experimentation can be described as ways of dealing with the respective objects of research that are historically determined and exert their power in historically circumscribed conjunctures. Often enough, it is only through them that phenomena can be brought to manifestation and are thus transformed into tractable objects of research. Cultures of experimentation live from and nourish those epistemic core events Bachelard called emergence. They are the working environments in which the production of new knowledge is enacted. Like experimental systems, they are, as ensembles of such systems, structures that one has to describe in their concreteness and historical contingency—in which epistemic, technical, and social moments are inextricably intertwined. They are concretions, not abstractions. Particular experimental cultures can dominate whole epochs in the development of a science. In this sense they could be addressed as "leading cultures." The in vitro cultures of experimentation described briefly in this chapter have played exactly this role in bringing molecular biology to the fore.

Notes

All translations are my own.

1. See also the particular twist to the notion of style in chapter 13 in this volume.

2. The priority dispute around this event is not surveyed here; see, e.g., Kohler 1971 and Wainwright 2003.

3. See Fernbach and Hubert 1900, Sørensen 1909, Michaelis and Davidsohn 1911, and Michaelis and Menten 1913. For a disciplinary history of biochemistry, see Fruton 1972 and Kohler 1982.

4. Alfred Kühn to Ernst Caspari, December 27, 1946, Kühn Papers, Archive of the Institute for the History of Medicine, University of Heidelberg.

5. For examples and a methodological discussion, see Bechtel 1996, esp. sec. 4.

6. But see Morange 2000; see also Cohen 1984, Fruton 1999, and the other chapters in this volume.

7. See chapter 5 in this volume, in which cell culturing is central to the case studies on biological engineering. It would be interesting to use this perspective to compare the model systems described there.

References

Bachelard, G. 1940. *La philosophie du non*. Paris: Presses Universitaires de France.

Bachelard, G. 1949. *Le rationalisme appliqué*. Paris: Presses Universitaires de France.

Bechtel, W. 1966. "What Should a Connectionist Philosophy of Science Look Like?" In *The Churchlands and Their Critics*, ed. Robert N. McCauley, 121–144. Oxford: Blackwell.

Bernard, C. [1878–1879] 1966. *Leçons sur les phénomènes de la vie communs aux animaux et aux végétaux: Librairie Philosophique*. Paris: J. Vrin.

Bourdieu, P. 1997. *Méditations pascaliennes*. Paris: Seuil.

Buchner, E. 1897. "Alkoholische Gährung ohne Hefezellen." *Berichte der Deutschen Chemischen Gesellschaft* 30: 117–124.

Claude, A. 1975. "The Coming Age of the Cell." *Science* 189: 433–435.

Cohen, S. 1984. "The Biochemical Origins of Molecular Biology." *Trends in Biochemical Sciences* 9: 334–336.

Creager, A. N. H. 2013. *Life Atomic: A History of Radioisotopes in Science and Medicine*. Chicago: University of Chicago Press.

Fernbach, A., and L. Hubert. 1900. "De l'influence des phosphates et de quelques autres matières minérales sur la diastase protéolytique du malt." *Comptes Rendus des Séances de l'Académie des Sciences* 131: 293–295.

Friedmann, H. C. 1997. "From Friedrich Wöhler's Urine to Eduard Buchner's Alcohol." In *New Beer in an Old Bottle: Eduard Buchner and the Growth of Biochemical Knowledge*, ed. A. Cornish-Bowden, 67–122. Valencia: University of Valencia.

Fruton, J. S. 1972. *Molecules and Life: Historical Essays on the Interplay of Chemistry and Biology*. New York: Wiley.

Fruton, J. S. 1999. *Proteins, Enzymes, Genes: The Interplay of Chemistry and Biology*. New Haven, CT: Yale University Press.

Gaudillière, J.-P., and I. Löwy, eds. 1998. *The Invisible Industrialist*. London: Palgrave Macmillan.

Geertz, C. [1973] 2000. *The Interpretation of Cultures*. New York: Basic Books.

Gould, G. M. 1894. *Illustrated Dictionary of Medicine, Biology, and Allied Sciences*. Philadelphia: Blakiston.

Hacking, I. 1992. "'Style' for Historians and Philosophers." *Studies in History and Philosophy of Science* 23: 1–20.

Harrison, R. G. 1910. "The Outgrowth of the Nerve Fiber as a Mode of Protoplasmic Movement." *Journal of Experimental Zoology* 9: 787–846.

Holmes, F. L. 1974. *Claude Bernard and Animal Chemistry*. Cambridge, MA: Harvard University Press.

Keshishian, H. 2004. "Ross Harrison's 'The Outgrowth of the Nerve Fiber as a Mode of Protoplasmic Movement.'" *Journal of Experimental Zoology* 301A: 201–203.

Kohler, R. E. 1971. "The Background to Eduard Buchner's Discovery of Cell-Free Fermentation." *Journal of the History of Biology* 4: 35–61

Kohler, R. E. 1982. *From Medical Chemistry to Biochemistry: The Making of a Biomedical Discipline*. Cambridge: Cambridge University Press.

Knorr Cetina, K. 2000. *Epistemic Cultures: How the Sciences Make Knowledge*. Cambridge, MA: Harvard University Press.

Kühn, A. 1938. "Grenzprobleme zwischen Vererbungsforschung und Chemie." *Berichte der Deutschen Chemischen Gesellschaft* 71: 107–114.

Landecker, H. 2004. "Building 'A New Type of Body in Which to Grow a Cell': Tissue Culture at the Rockefeller Institute." In *Creating a Tradition of Biomedical Research: Contributions to the History of the Rockefeller University*, ed. Darwin H. Stapleton, 151–174. New York: Rockefeller University Press.

Landecker, H. 2007. *Culturing Life: How Cells Became Technologies*. Cambridge, MA: Harvard University Press.

Löwy, I. 1990. "Variances in Meaning in Discovery Accounts: The Case of Contemporary Biology." *Historical Studies in the Physical and Biological Sciences* 21: 87–121.

Mannheim, K. 1980. *Strukturen des Denkens*. Frankfurt am Main: Suhrkamp.

Michaelis, L, and H. Davidsohn. 1911. "Die Abhängigkeit der Trypsinwirkung von der Wasserstoffionenkonzentration." *Biochemische Zeitschrift* 36: 280–290.

Michaelis, L., and M. Menten. 1913. "Die Kinetik der Invertinwirkung." *Biochemische Zeitschrift* 49: 333–369.

Morange, M. 2000. *A History of Molecular Biology*. Cambridge, MA: Harvard University Press.

Palade, G. E. 1951. "Interacellular Distribution of Acid Phosphatase in Rat Liver Cells." *Archives of Biochemistry* 30: 144–158.

Pickering, A., ed. 1992. *Science as Practice and Culture*. Chicago: University of Chicago Press.

Pickstone, J. V. 2000. *Ways of Knowing: A New History of Science, Technology and Medicine*. Manchester: Manchester University Press.

Pickstone, J. V. 2011. "A Brief Introduction to Ways of Knowing and Ways of Working." *History of Science* 49: 235–245.

Rader, K. 2004. *Making Mice: Standardizing Animals for American Biomedical Research, 1900–1955*. Princeton, NJ: Princeton University Press.

Rheinberger, H.-J. 1993. "Vom Mikrosom zum Ribosom: 'Strategien' der 'Repräsentation' 1935–1955." In *Die Experimentalisierung des Lebens: Experimentalsysteme in den biologischen Wissenschaften 1850/1950*, ed. Hans-Jörg Rheinberger and Michael Hagner, 162–187. Berlin: Akademie.

Rheinberger, H.-J. 1997a. "Cytoplasmic Particles in Brussels (Jean Brachet, Hubert Chantrenne, Raymond Jeener) and at Rockefeller (Albert Claude), 1935–1955." *History and Philosophy of the Life Sciences* 19: 47–67.

Rheinberger, H.-J. 1997b. *Toward a History of Epistemic Things: Synthesizing Proteins in the Test Tube*. Stanford, CA: Stanford University Press.

Rheinberger, H.-J. 2004. "A History of Protein Synthesis and Ribosome Research." In *Protein Synthesis and Ribosome Structure: Translating the Genome*, ed. Knud H. Nierhaus and Daniel N. Wilson, 1–51. Weinheim: Wiley-VCH.

Rheinberger, H.-J. 2010. *An Epistemology of the Concrete: Twentieth-Century Histories of Life*. Durham, NC: Duke University Press.

Sørensen, S. P. L. 1909. "Enzymstudien. II. Mitteilung: Über die Messung und die Bedeutung der Wasserstoffionenkonzentration bei enzymatischen Prozessen." *Biochemische Zeitschrift* 21: 131–304.

Summers, W. C. 1991. "From Culture as Organism to Organism as Cell: Historical Origin of Bacterial Genetics." *Journal of the History of Biology* 24: 171–190.

Wainwright, M. 2003. "Early History of Microbiology." *Advances in Applied Microbiology* 52: 333–355.

Warburg, O. 1928. *Über die katalytischen Wirkungen der lebendigen Substanz: Arbeiten aus dem Kaiser Wilhelm-Institut für Biologie Berlin-Dahlem*. Berlin: Julius Springer.

Werner, P. 1988. *Otto Warburg: Von der Zellphysiologie zur Krebsforschung*. Berlin: Neues Leben.

Willstätter, R., and M. Rohdewald. 1940. "Über enzymatische Systeme der Zucker-Umwandlung im Muskel." *Enzymologia* 8: 1–63.

12
———

The People's War against Earthquakes
Cultures of Mass Science in Mao's China

By a strange coincidence, more than ten powerful earthquakes (roughly M7 or above) struck China during the tumultuous years of the Cultural Revolution, from 1966 to 1976. The series of massive earthquakes started with the Xingtai earthquake in March 1966 and culminated with the Tangshan earthquake in July 1976, which killed a quarter of a million people.[1] Because much of China is located between two highly active earthquake zones—one running roughly along the Pacific Rim and the other stretching from the Himalayas to Southeast Asia—earthquakes are not uncommon in many parts of China. Nevertheless, it is unusual that major earthquakes hit China proper with such frequency and intensity as in the 1960s and 1970s. This period of natural disasters overlapped with a tempestuous period of modern Chinese history. Fear and anxiety about earthquakes simmered in the environment of social and political uncertainties. On the one hand, the Chinese experienced violent social and political turmoil; on the other hand, they worried that earthquakes might strike at any time.

The Xingtai earthquake, which caused more than 8,000 deaths and 50,000 injuries, was the catalyst for a nationwide campaign to wage war against earthquakes. Zhou Enlai, the premier, visited the disaster site and summoned a group of scientists to discuss ways of fighting earthquakes. Until then, Chinese scientists had not done much in seismology, and many of them considered earthquake prediction beyond the current state of science and technology. A few prominent figures in Chinese science—notably the geologist Li Siguang and the geophysicist Weng Wenbo—strongly advocated earthquake prediction, however. Their voices proved decisive. Subsequently, Li would play a leading role in organizing the institutions and directing research in seismology.[2]

There were immediate and obvious reasons for the political leadership to be concerned about the threat of earthquakes. Natural disasters could severely damage society, state, and economy. At the time, there were also practical reasons that encouraged the Chinese to emphasize earthquake prediction. Most houses and buildings in China were not earthquake resistant. Major cities were so densely populated that it would be impractical to tear down and rebuild the houses according to the principles of seismological engineering. It would take too long and cost too much. In comparison, it seemed to make more sense to invest in prevention and defense: developing methods to predict earthquakes, educating people about earthquake prediction and defense, organizing the people for mass emergencies, and so on. This policy, of course, depended heavily on an assumed feasibility of earthquake prediction, yet once one accepted the premise, the choice became obvious.[3]

Such practical considerations were important, but there were other reasons. Traditionally, earthquakes and other natural calamities possessed political and cultural significance. A big natural disaster, especially something dramatic like a major earthquake, was considered an ominous sign of major political changes, such as the end of a reign. This anxiety could easily be amplified by the social and political commotion during the Cultural Revolution. Campaigns against "superstitions" had been a longtime policy of the Chinese Communist Party, yet certain traditional beliefs, such as those associated with earthquakes and other "abnormal disasters" (*zaiyi* 灾异), posed more serious threats to the party-state than did others. Not surprisingly, the party-state was eager to inculcate the masses that earthquakes were natural phenomena, that they could be predicted, and that people should not have superstitious ideas about them.[4] The party-state did not want its political legitimacy to be questioned. It might also be true that after the catastrophe of the Great Leap Forward in the late 1950s and early 1960s, the political leaders had become more cautious and responsive to natural calamities.[5] They were probably very concerned about the political ramifications of a major natural disaster. The fact that the nation's capital, Beijing, was located in an area that was seismically active only heightened the anxiety. If a powerful earthquake struck Beijing, the nation might be thrown into chaos and would become dangerously vulnerable to external aggression. Because at the time China considered itself under imminent military threat from the Soviet Union and the United States, this worry could not have been far from the minds of its leaders. Thus, the Chinese program of earthquake monitoring and defense has to be placed within this complex social, cultural, and political context.[6]

The purpose of this chapter is not to enter the debate over the feasibility of earthquake prediction but to examine the ideology, organization, and practice of the program of earthquake monitoring, prediction, and defense during the Cultural Revolution, with a focus on the culture of mass science. As this chapter shows, China's program of earthquake monitoring and prediction was inseparable from the notion of "mass science" (*qunzhong kexue*) or what was sometimes called "the people's science" (*renmin kexue*).[7] There was no official distinction between the two terms, though depending on the context they could have distinct connotations. *Qunzhong* was, first of all, a political category, as in the *qunzhong luxian* or the mass line, whereas *renmin* had more of a sense of the common people or even the *Volk*. It will be clear that the Chinese approach to earthquake prediction drew on the connotations of both terms, namely, both "mass" and "folk," and that this combination was not only political but also epistemological.

To understand the Chinese program of earthquake monitoring and prediction, this chapter considers the culture of mass science during the Cultural Revolution at three levels: the culture of mass science itself as scientific knowledge and practice, the political and cultural context in which mass science took place (especially that of Maoism, which laid the ideological foundation of the Cultural Revolution), and the general question of what this case study may tell us about the opportunities and pitfalls in studying science in a non-Western society. At the first level, I examine the main components of mass science and explain how the different components were supposed to work together to produce scientific results. The "culture of mass science," in this sense, refers to an ensemble of institutions, practices, actors, expert knowledge, instruments, and epistemologies in the programs of earthquake prediction and monitoring. In rough terms, it addresses the problem of science as culture similar to Hans-Jörg Rheinberger's culture of experimentation (see chapter 11 in this volume), albeit in a radically different historical context.

At the second level, this study situates mass science in its most relevant political and cultural context, the Cultural Revolution, with a focus on Maoist thought and related political currents, such as nationalism. In considering these issues, I pay particular attention to the ideological underpinnings and symbolic representations of mass science. The purpose is to show how mass science was embedded in the ideological and political culture of the Cultural Revolution. Some of my discussion on the topic may be fruitfully compared with Guillaume Lachenal's discussion of the cultural politics of an African AIDS vaccine (see chapter 3 in this volume).

Finally, at the third level, this study considers certain important issues of "cultures without culturalism" with regard to the study of science in non-Western societies (as similarly discussed in chapter 2 in this volume).[8] The tendency to rely on preconceived culturalism to understand science in non-Western societies has not disappeared in the post-Said era. It is still frequently present, both in the ways Western commentators explain non-Western societies and in the ways non-Western societies represent themselves vis-à-vis the supposed West. The case study of Maoist mass science will enable us to reflect on these issues in a new light.

These three interpretive levels are interrelated. One cannot understand mass science without examining Maoism. Similarly, one cannot properly study science in Mao's China without being mindful about the pitfalls of culturalism. They are, however, distinct enough that they may be discussed separately for analytical purposes, which I do in the conclusion of this chapter.

Earthquake Monitoring and Prediction

During the ten years of the Cultural Revolution, many areas of scientific activity ran into difficulties. Yet seismology expanded rapidly. The State Seismological Bureau was founded in 1971. The number of seismological stations increased from a handful in the early 1960s to around 250 by the mid-1970s, not including numerous observation stations associated with mass science projects (Raleigh et al. 1977, esp. 237).[9] Most of the expert seismologists had a background in geology or geophysics; the top ones often had received their education abroad, in Europe or the United States before the communist era or in the Soviet Union during the 1950s. Those in the rank and file were younger scientists who had recently graduated from universities or institutes of science and technology or who had been extracted from the mines, oil fields, surveying offices, and other places where useful talents were found. Not surprisingly, there were scientific differences and political factions within the community: geologists and geophysicists did not see eye to eye on scientific matters; proponents of different seismological methods vied for status and recognition; the dominance of Li Siguang created tensions between his disciples and those of his rivals; and so forth.[10] All this was further compounded by the political infighting so pervasive during the Cultural Revolution. Overall, however, the seismological community in China functioned relatively well. It was largely shielded from outside political struggles because its mission was considered a state priority. For this reason, the seismological community, despite its internal differences, operated with a shared goal: earthquake prediction.

China was not alone in trying to develop methods of earthquake prediction. At the time, Japan, the Soviet Union, and the United States all pursued earthquake prediction or related seismological research.[11] Many of their research directions, if not particular techniques, were also quite similar, such as seismic mapping, geodetic measurements, foreshock sequences, and electromagnetic fields. But certain features distinguished China's approaches to earthquake prediction from those of the other nations. For instance, the Chinese researched heavily and systematically into historical records. Although other seismologists also used historical records, the Chinese were fortunate to have inherited a uniquely rich body of historical documents, and they combed through them to piece together a picture of the spatiotemporal distribution of earthquakes in China. By means of historical research, Chinese scientists hoped to find the recurring patterns of earthquakes over thousands of years (see Dizhen wenda bianxiezu 1977, 122; Li 1981, 167–244). Presumably, the knowledge would be helpful to long-term earthquake prediction. In this area of research, other nations could only look at China with envy. Although, as expected, most of the historical records dated back only to the more recent time periods and concentrated on the more densely inhabited parts of China, they still constituted an unrivaled body of textual documentation. The Chinese also mined the historical records for accounts of precursory signs of earthquakes, the significance of which will become clear below.

Another distinctive feature of Chinese seismology was its emphasis on short-term and imminent prediction, within a time frame of weeks or days. In contrast, most seismologists in other countries were less bold, or more realistic, in assessing the feasibility of short-term earthquake prediction:

they opted instead for research on long-term forecasting or hazard assessment based on geophysical models. The distinction between prediction and hazard assessment is crucial. Prediction in the Chinese approach meant pinpointing the precise place, time, and magnitude of an earthquake. If the margin of error was too big to allow appropriate emergency decisions, the prediction was useless. A statement such as, "There is a 50 percent chance of this province experiencing an earthquake of magnitude 7 or larger within the next 30 years," would not have done much good. One could hardly take any actions based on that piece of information, other than pushing for better urban planning and stricter building codes. A prediction should also come early enough for effective action. A warning that left no time for people to respond would be as good as no warning at all. Ideally, there would have to be enough time for preventive evacuation of an entire city. Therefore, a prediction is also different from advance detection of a coming earthquake. A method or instrument that could detect seismic waves that are already approaching would be of little help. The difference of a second or two would not save many lives.[12]

In the mid-1960s, a major American proposal to develop earthquake prediction failed to garner enough funding and was aborted (Geschwind 2001, chap. 6; see also Wyss 1999). In the 1970s, there was another surge of interest in earthquake prediction. Japan and the Soviet Union, too, had their own attempts at earthquake prediction. None of them, however, could boast the same level of confidence and emphasis on short-term prediction as China. The Chinese attitude owed much to the following set of factors: pressure from the political leadership, the relative isolation of Chinese scientists from the international scientific community, the dominance and self-assurance of a few scientific strongmen (e.g., Li Siguang), a sense of urgency among the scientists toward combating earthquakes, and a "can-do" spirit fueled by patriotic zeal. In the process, the Chinese developed certain areas of research that were very much their own, notably that of macroscopic phenomena.

With increasingly sophisticated technology, modern seismology depended principally on microscopic measurements, using highly sensitive instruments to monitor seismic activities. Although Chinese seismologists did not have the most advanced equipment, they pursued this area of research as well. However, they also devoted much attention to research on macroscopic phenomena (viz., premonitory phenomena that could be observed without using instruments), as well as other phenomena that could be observed with very simple tools (see, e.g., Huang and Deng 1978; Zhu and Jiang 1980). Their approach originated from three sources. First, China did not have a lot of advanced equipment or many seismological stations at first. Since it was impossible to cover a large country like China with so few observatories, Chinese scientists welcomed research methods that would compensate for the shortage of advanced observatories. Second, they believed that macroscopic phenomena could lead to pretty accurate predictions of earthquakes, whereas in technologically more developed countries such as the United States, the scientific establishment tended to presuppose that advanced technology rendered observations of macroscopic phenomena superfluous. Third, there was the political content: mass participation, mass mobilization (Schmalzer 2008, chaps. 3, 4; Mullaney 2012). Since Chinese seismologists believed that certain macroscopic phenomena could be symptomatic of earthquakes, they naturally wanted to have many observers out there looking for them. And, politically, if science was indeed of the masses and for the masses, then people should learn about science and do science. Earthquake prediction, therefore, provided a perfect opportunity for mass science.

Mass Participation in Science

The party-state mobilized the masses in the national enterprise of earthquake monitoring and defense.[13] The shortage of seismological stations and a belief in the possibility of short-term earthquake prediction explained only part of the ambition. The enterprise was at the same time scientific research, disaster control, national defense, political campaign, and nation building. Maoist programs of mass science, such as earthquake prediction, barefoot doctors, and various attacks on "elite science," were based on the tenets of integrating experts and the masses and combining indigenous and Western science. The underlying political doctrine asserted the class character of science, exalted everyday knowledge, and projected a particular vision of scientific and political modernity.

China devised an earthquake monitoring system, called the Collective Monitoring, Collective Defense program, that aimed to integrate both the experts, who staffed the seismological bureaus and stations, and the masses, who also played a role in earthquake monitoring. According to one estimate, there were also "several tens of thousands" of lay participants in the program (Raleigh et al. 1977, 237). The observation activities of the masses were supervised by the local party committees, schools, communes, and factories, rather than by the seismological bureaus. Since there were thousands of these observation units in the people's war against earthquakes, it is not surprising that their scientific and technological sophistication varied. The most advanced among them were well equipped and did not look different from formal seismological observatories. But far more common were simple observation units attached to middle schools, factories, commune stores, and post offices. One unit or team usually consisted of several individuals.[14]

I have conducted interviews with more than a dozen people who, as school students, had been involved in the earthquake prediction program. In most cases, they recalled the excitement of participating in the activities. They learned about geology, seismology, and emergency response plans. Teachers took them to selected places where they set up observation points. They felt that they were doing something valuable, important, exciting. One of them said that, as a ninth grader, "I was excited to be on night duty; they were rare opportunities to sleep overnight outside of my home, despite suffering from mosquito bites." These amateur groups not only conducted observations but also often made their own instruments. They were encouraged to be inventive. One imagines that it was rather like going to a science summer camp or participating in high school science fairs (if one must look for an American analogy). The difference is that the Chinese students were not doing kids' science or learning to do science simply as part of science education. They were taking part in an official national science project. In other words, they were actually doing science and contributing to the national effort against natural disasters.

Some of the more common objects for observation were ground tilt, telluric currents, geomagnetism, and well-water variations. In the late 1960s and early 1970s, many indigenous instruments were invented for making these observations. A young farmer invented the method of measuring telluric currents by using a simple device that consisted of an ampere meter and two electrodes planted underground at a distance apart. The intensity and variation of underground electric currents were thought to indicate earthquake activities. A seismologist was sent to evaluate the instrument. After some examination, the scientist confirmed the effectiveness of the device. The method was reported in the *Dizhen zhanxian*, an influential journal on earthquake studies at the time, and became widely adopted across the nation.[15] When in 2007 I interviewed the very scientist who had been assigned to investigate the instrument, he still took pride in having introduced the instrument to a wide audience. In the mid-1970s, a young technician invented a simple apparatus for measuring geomagnetism that consisted mainly of a magnetic needle, a mirror, a plastic bucket, and carbon powder (see figure 12.1).[16] The device was widely copied. Ground tilt could be measured

大地有磁场　震前显异常

地球具有磁场，在正常情况下，地球的磁场总是依照一定的规律缓慢地变化。但是，在地震孕育、发展过程中，由于地应力的变化，地下岩石的磁化强度也会发生变化，从而引起地磁场局部异常。

实践证明，利用地磁的异常变化来预报地震，能得到较好的效果。在测量地磁异常时一般测量其中三个要素：水平强度（水平面上的磁力强度）、垂直强度（铅垂线上的磁力强度）、磁偏角（磁场水平方向即磁针所指的方向与地理北极方向之间的夹角）的变化。地震台使用的磁变仪可以测量上述三个要素的相对变化值。广大群众在地震测报的实践中，也创造了不少土地磁仪如土磁偏角仪等，这些仪器只能测量磁偏角的变化。

地磁变化不仅与地震有关，而且还受太阳活动等其它因素的影响。因此，在发现地磁异常的时候，要认真分析情况，以免混淆。

认真调试地磁仪

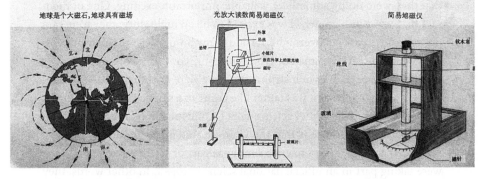

地球是个大磁石，地球具有磁场　　　　光放大读数简易地磁仪　　　　简易地磁仪

FIGURE 12.1. Page from an illustrated pamphlet on earthquake monitoring and defense, introducing the topic of geomagnetism and earthquakes.

with a plumb bob hanging from the ceiling and a piece of chalk (Anonymous 1975, n.p.; see also Liang 1970).

The text in figure 12.1 explains that underground seismic activities may cause changes in geostress, which in turn may cause anomalies in the local geomagnetic field. The text insists that practical experience has shown that the use of geomagnetism in earthquake prediction has produced very good results. It adds that the masses have created many kinds of indigenous instruments for geomagnetic observation and earthquake prediction. The four images show (counterclockwise) the Earth's magnetic field, two simple geomagnetic instruments for earthquake prediction, and two nonexpert participants testing geomagnetic instruments. Due to the state ideology of women's equality, it was common for a poster or pamphlet for science during the Cultural Revolution, to show women scientists or nonexpert women participating in scientific activities.

We may use a textbook for middle-school students as an example of these observation activities. The *Activities in Earthquake Monitoring and Predic-*

tion for Youngsters (*Shaonian dizhen cebao huodong*) was issued jointly by the Shanghai Seismological Office and several middle schools in 1978. The book consists of eight chapters, starting with an introductory chapter, "What Is an Earthquake?," and moving on to discuss, chapter by chapter, "The Macroscopic Phenomena of Earthquakes," "Underground Electric Currents," "The Electricity of Plants," "Geomagnetism," "Ground Deformation and Ground Tilt," "Geostress," and "Other" (which includes chemicals in water, radon levels, and ground temperature). The chapter on macroscopic phenomena covers animal behavior, plant behavior, underground water variations, ground deformation, sea level and tides, climate and weather changes, earthquake sounds, and earthquake light. In many of the chapters, homemade instruments and sample charts are introduced. Measuring the ground tilt with a bob pendulum, for example, requires the observer to measure and record the data three times a day (at 8 a.m., 1 p.m., and 6 p.m.). The data to be recorded include the tilt, room temperature, humidity, and weather. The observer is instructed to sign his or her name (Shaonian dizhen cebao huodong bianxie zu 1978).

This mass participation in scientific activity was not unique to earthquake studies, or to the Cultural Revolution era. During the Great Leap Forward, the people were also encouraged to invent new tools, machines, fertilizers, and methods for increasing agricultural and industrial production (Nongyebu nongju gaige bangongshi 1959). In fact, in some important ways, the agricultural meteorology campaign in the late 1950s foreshadowed what would happen in earthquake monitoring a decade later (see, e.g., Anonymous 1959; Jian 1959; Shangxisheng 1959). With the Great Leap Forward ending in misery, the frenzy of invention died down. Later, the new political and practical conditions revived the activity, this time with a concentration on earthquake prediction.

Most of the masses were not organized into active groups or teams manning observation points. But they were still participants in the earthquake prediction program. They received information about earthquakes through political and educational channels such as party offices, neighborhood organizations, production brigades, and commune units and through propaganda initiatives, including pocket-size pamphlets, mimeographed handouts, colorful posters, exhibitions, film screenings, school competitions, and propaganda tours to the countryside. They learned about the basic science of earthquakes, methods of earthquake prediction, emergency defense, and so on. Although most people did not use instruments or routinely keep a record of what they observed, they were urged to pay attention to any macroscopic seismic phenomena. If they noticed any unusual phenomenon that might indicate the coming of an earthquake, they were to report it immediately to the local offices—it was their duty to do so. Together the various levels of mass science would form a blanket warning system against earthquakes (Fan 2012a).

Arguably, the warning system was an extension of the surveillance society under communist rule. People were constantly being admonished to be vigilant, to watch out for enemies of the party-state, and to report any suspicious activities, only this time the enemy was a natural force. Of course, one should not assume that this system worked seamlessly. Political instabilities during the Cultural Revolution could be disruptive. Besides, people always had their own ideas, desires, aims, and lapses. Boredom, negligence, and indifference seeped in after the initial excitement. How many people could keep on checking the same unchanging meter and recording the readings day after day, thrice a day, without occasionally slacking off? Symbolic incentives—stationery or small sums of money—were provided. At the year's end, a celebratory meeting gathered people together; prizes, certificates, and the like were handed out. Even so, the outcome might be uncertain. Nevertheless, we should not underestimate the seriousness and dedication many people felt about earthquake monitoring. What they were expected to do was, after all, what they would want to do for themselves—protect themselves and their families. In an area of potential major earthquakes, this was a matter of life and death.

Approaches to Seismology

The main function of the warning systems was to watch out for precursory signs of earthquakes. The Chinese believed that some of these precursory signs were macroscopic phenomena that could be observed by common people. Therefore, the program of earthquake prediction incorporated folk wisdom and everyday observations that described "anomalous" natural phenomena that might indicate the coming of an earthquake: animal behavior, unseasonable weather, well water variations, underground temperature, and so on.[17] These anomalous signs indicated a forthcoming earthquake just as symptoms suggest a hidden but menacing disease. Some of the anomalies were obvious, such as foreshocks; others were obscure, such as weather conditions. Seen in this way, the Chinese way of earthquake prediction may be likened to a science of reading anomalous signs or symptoms of earthquakes, that is, a semiotics of earthquakes. However, observing and reading signs can be difficult. What an experienced or perceptive doctor can find out about the bodily disorder of a patient by checking the pulse or looking into the ears may escape another doctor completely. Similarly, the precursory signs of earthquakes could be obscure or elusive. The next section discusses what was involved in observing the precursory signs of earthquakes and how it worked in the Chinese project of mass participation in earthquake prediction. Here I comment on another methodological tendency of the Chinese approach to seismology, which might be characterized as phenomenological.

Although Chinese scientists also attempted to explain the causal relationships between these macroscopic phenomena and earthquakes, their approach prioritized correlating possible precursory phenomena and earthquakes rather than building geophysical models of seismicity. In this regard, their approach was primarily phenomenological. They were willing to cast their net wide, collecting and considering a broad range of possible symptoms without first systematically constructing a theoretical causal model. For example, there was a belief among the Chinese that droughts often preceded earthquakes, and one seismologist developed a method to correlate the relationship between droughts and earthquakes (Geng 1985).[18] His interest was in finding out how to use droughts or meteorological data to calculate the incidence of earthquakes. He labored through large quantities of data and derived certain formulas for predicting earthquakes based on the historical records of droughts. But what were the geological and physical models that might explain the relationship? The question seemed to him a little beside the point, and he did not have a lot to say about it. He simply responded that the flow of heat from inside the earth caused by seismic movements will have effects on the atmosphere, and that severe droughts thus indicate major seismic movements. There were many other prediction theories of a similar nature. The "doubling of geomagnetic variation" method, developed in the late 1960s, determined the corresponding relationship between geomagnetic variations related to the orbiting of the earth and the occurrences of earthquakes. According to the inventor of the theory, he first noticed a particular correspondence between the readings of geomagnetic meters in various locations and the instances of earthquakes. He then surmised that the geomagnetic variations were also related to the revolution of the earth around the sun, especially the resulting solid-earth tides. (Since the variation seemed to be seasonal, he attributed it to the influence of the sun rather than the moon.) Before long, he also noticed a correspondence between occurrences of earthquakes and of magnetic storms (caused by solar activity). Based on these discoveries and some calculation, he developed a formula for predicting earthquakes. In explaining the theory to me, he admitted that the causes of earthquakes were obscure, yet he also maintained that one of the factors lay in the relationship between the sun and the earth.[19]

These methods might seem crude and superficial to mainstream Western scientists. Yet, to many Chinese seismologists at the time, the principal issue was whether a method worked or not: if a method actually predicted earthquakes, it was valuable, regardless of how much or little it might reveal about deeper geophysical properties.

This is not to say that Western geology and seismology had no use for the phenomenological approach. On the contrary, phenomenology remained a fundamental component of the established methodology in geology. For instance, certain critical contributions to the theories of continental drift, and later plate tectonics, benefited from phenomenological reasoning (Oreskes 1999, 63–64, 81, 84, 95, and chap. 9). I recently asked a seismologist based in the United States for his general opinion about the Chinese theories of earthquake prediction. This seismologist had extensive contact with his Chinese colleagues in the late 1970s and early 1980s, and he once took some of the Chinese research seriously enough to investigate it. His reply was that many of the Chinese theories based on phenomenological reasoning were not testable. We do not have to accept the underlying implication of his comment—that is, there was a fundamental epistemic divide between what he considered as testable or untestable theories—to recognize that there was indeed a notable difference in interest, degree, emphasis, and style between Chinese and American seismological approaches. Compared with American seismology, Chinese research focused much more on the predictive rather than the explanatory power of a theory.

Mass Science and Everyday Knowledge

Overall, the phenomenological and semiotic approaches fit in well with Maoist mass science and the Collective Monitoring, Collective Defense program. Most of the lay participants had only very basic knowledge of modern geology and seismology or, in remote areas, hardly any at all; their main responsibility was to observe the macroscopic phenomena that might indicate the coming of an earthquake. To illustrate how this mass participation in scientific observation worked, I examine two particular areas of observation: underground water and animal behavior. In earthquake prediction, the Chinese paid much attention to underground water variations, such as unusual changes in the level, taste, appearance, and temperature of well water. The underlying rationale was that seismic activities might alter the chemical and physical qualities of underground water. There were numerous anecdotes about how well water looked and tasted different before an earthquake. One such account related that on the eve of an earthquake, a tofu factory was unable to make tofu using the water from their well (presumably because the chemical content of the water had changed) (Guo 1982, 53–54).[20] A rhyme frequently included in the material for science dissemination about earthquake prediction runs thus:

> Wells play a vital role,
> Earthquake signs they may tell;
> Their levels may strangely change,
> Their waters may bubble or be turbid
> With the taste becoming bitter or sweet;
> On finding such strange events,
> Report them quickly as you can. (adapted from Tang 1988, 154)[21]

Some wells were believed to be particularly sensitive and accurate in predicting earthquakes, and they acquired a reputation among the local population. These "treasure wells" were treated like windows into the deep bowels of the earth. Certain other symptoms received similar consideration. People were instructed to pay attention to ground temperature, unseasonable weather, and most significantly, animal behavior.

Many Chinese seismologists accepted that aberrant animal behavior could be seen as premonitory phenomena of an earthquake. Again, the reasoning was plausible: animals are often more sensitive than humans to changes in natural environments. The Chinese found that historical records and folk wisdom contained much evidence to support this notion, although they did little controlled research into the connections between earthquakes and animal behavior or physiology until the 1970s (Zhongguo Kexueyuan Shengwu Wuli suo Dizhen Zu 1977; Jiang and Chen 1993).

Animals thought to be useful for earthquake prediction included a wide array of pets, livestock, household pests, and wild animals; dogs, cats, cows, horses, pigs, chickens, pigeons, parakeets, ducks, mice, rabbits, fish, and snakes all appeared able to detect and respond to minute changes before an earthquake. One frequently cited example was the bizarre behavior of snakes in the weeks before the Haicheng earthquake in 1975. There were eyewitness accounts of snakes crawling out of their hibernation holes in the February cold of northern China and freezing to death (Zhongguo Kexueyuan Shengwu Wuli suo Dizhen Zu 1977, 66–67). Some people described vivid images of snakes struggling to come out even though their front portions were already numb with cold. Burrowing animals were believed to be particularly sensitive to changes underground. However, livestock and other animals that most people came in contact with every day could also serve as convenient instruments. A widely circulated rhyme on animal behavior and earthquake prediction runs:

> Before quakes, abnormal animal behavior may appear,
> The people should be prepared to fight the war
>   [against earthquakes];
> Cattle, sheep, horses, or mules may strangely refuse to
>   enter their sheds;
> Mice may start moving about during the day, showing great fear;
> Chickens may fly up into the trees, and pigs may
>   run around restlessly, even wrecking their sties;
> Ducks may refuse to go into the water, and dogs may bark madly.
>   (adapted from Tang 1988, 152; Jiang 1980)

The animals mentioned in the poem are all pets, livestock, or household pests. The content was gathered from folk knowledge, and the verse was intended mainly for people in the countryside; that is, people who worked in the field and were close to the natural environment. These people usually lived in areas without the best coverage of seismological networks, and they had to rely on themselves to predict earthquakes, including, in this instance, using animals as seismic detectors. In fact, they could contribute to the Collective Monitoring, Collective Defense program by utilizing their lay expertise. More than most people, farmers were able to notice occurrences of aberrant animal behavior. They were good observers because they were familiar with the routine or normal behavior of these animals. (In urban areas, zookeepers took on the role of expert observers of anomalous animal behavior.) People who handled livestock every day were particularly well equipped. That kind of expertise came from their daily labor, from their daily struggle with nature. It was based on an epistemology that derived from practice, work, and tradition.

Thus, monitoring macroscopic phenomena relied very much on everyday experience. Everyday experience was crucial, because it provided the background of normality against which the aberrant and the anomalous emerged. In this case, everyday experience was both a social and an individual acquisition. Most people did not experience significant earthquakes often enough to be able to compile premonitory symptoms; they had to start with the existing lore. Folk knowledge—traditional knowledge passed on to and circulated among a local population—was often community-based knowledge and skill. To-gether with other social and individual knowledge, folk knowledge, experi-ence, and skill constituted a kind of "lay expertise" (see Epstein 1995; Wynne, 1996). Although it did not come from formal, institutionalized education and training, it could be just as specialized as any other form of expertise. Being able to make a good observation required one to have accumulated suffi-cient experience, knowledge, and practice. Only when a person was familiar with the routine and the normal—either a particular weather pattern or the typical behavior of particular animals at a certain time and place and in a certain situation—could he or she immediately notice the more subtle kind of anomalies.

From the standpoint of Maoism, this expertise was not only social but also thoroughly political, because it was embedded in class and derived from labor. The expertise, however, was also often tied to a particular place, and the everyday knowledge was intensely local. The value of the knowledge owed much to its local character. This is reminiscent of the agricultural meteo-rology campaign in the late 1950s mentioned above, which also drew on the local experience and knowledge about natural forces. Since the crux of the Chinese approach to earthquake prediction was to pinpoint the arrival of an earthquake rather than to provide long-term hazard assessments, precision was key. Knowledge that corresponded precisely to a place would suppos-edly be useful for earthquake prediction in the particular place.

Conclusion: Culture, Politics, and Mass Science

If we accept that science, as cultural production and social practice, is not free of politics, we can say that, in this respect, science in communist China was hardly unique. We cannot treat science in any country as though it is devoid of politics: science is neither devoid of the micropolitics of the sci-entific community nor immune from the macropolitics of society and pol-ity. Maoist mass politics and mass science challenged the Weberian idea of technocracy and modernity, which the scientific establishment in Western societies holds up as the true characterization of itself—detached, objective, rational, impersonal, apolitical, governed by rules, and run by experts. Mao-ist thought and science attacked this notion of technocracy and insisted on the primacy of politics—that is, constant class struggle and revolutionary creativity from the masses (Whyte 1973). Thus, Maoism created a different vision of political and scientific modernity. Yet, Maoist mass science was also fiercely dogmatic. It was mostly top-down, despite its claim to represent the mass line. The party-state declared that earthquakes could be predicted. Nobody, then, could be skeptical. Anybody who doubted openly that earthquakes could be predicted would run the risk of being labeled a counterrevolutionary, an enemy of the party-state. Ultimately, the Collective Monitoring, Collective Defense program depended, in principle, on a highly regimented society. It was a program of national defense.

Yet, the Collective Monitoring, Collective Defense program was also a scientific program that can be fruitfully understood in cultural terms. Following the plan outlined in the introduction to this chapter, I discuss the Chinese approaches to earthquake monitoring and prediction at three levels: mass science as culture, mass science in culture, and mass science and cultural essentialism.

At the level of mass science as culture, we may regard mass science as an ensemble of institutions, actors, instruments, and epistemologies. The institutions of mass science consisted of the political, social, and scientific institutions that engaged in training, organizing, and deploying participants, including both experts and the masses. The Collective Monitoring, Collective Defense program, for example, comprised a wide variety of research institutes, scientific disciplines (e.g., geology and geophysics), schools, communes, factories, and government offices. Together, these components formed the institutional structure of the earthquake monitoring and defense program. The human actors who participated in the program varied widely, from schoolchildren to foreign-trained PhDs, from party secretaries to the sent-down youth, and from old farmers to zookeepers. The range and composition of participants signified scientific and political messages, as illustrated in the propaganda posters (e.g., figure 12.1). The spotlighting of women scientists— young, with bob haircuts, smiling, and wearing white coats—reflected the political tenor of "state feminism" of the time. The frequent appearance of ethnic minorities, often represented by women in traditional costumes, shored up the message of nation building.

The participants employed various kinds of expertise in monitoring and predicting earthquakes. Trained geologists did fieldwork and drew seismic maps. Farmers in a village applied their local, everyday knowledge about the livestock and natural environment. The instrumentation of mass science in this case involved a wide variety of instruments and associated knowledge and skill. The instruments included seismographs, seismic charts, and other apparatus commonly employed by seismologists across the world. Yet, there was also an array of simple, homemade instruments that purported to monitor, predict, or warn of coming earthquakes. Indeed, even animals and the human body were instruments for earthquake monitoring and prediction. These instruments functioned within a matrix of institutions, expertise, different sites, and epistemologies. Seismological stations and the areas they surveyed were important sites of research, monitoring, and analysis, but there were numerous other sites of scientific work. Stables, zoos, wells, school yards, mine shafts, commune stores, and even one's own home were all places of knowledge production in the program of earthquake monitoring and prediction. Regarding the preferred epistemologies of earthquake monitoring and prediction, the program relied heavily on mass observation of macroseismic phenomena. The network of earthquake monitoring cast a web of distributed cognition in the mode of collective monitoring. Correspondingly, what might be called phenomenology and semiotics were important to these approaches to earthquake monitoring and prediction.

Second, by taking apart the assemblages of the Collective Monitoring, Collective Defense program, we get a better understanding of mass science as culture and practice. Yet it is also necessary to place mass science within the ideological and cultural context of the Cultural Revolution. Although the scale, violence, and social turmoil of the Cultural Revolution were unprecedented, the belief in the power of a radical cultural revolution to revive a decaying China can be traced to the early twentieth century (Schwarcz 1986). Despairing over the plight of China in a modern world, young Chinese intellectuals of the 1910s and 1920s came to the conclusion that the survival of the Chinese nation depended on a radical break from its cultural tradition. In important ways, the Cultural Revolution drew upon this intellectual tradition. The most revealing example may have been the Smash the Old Fours campaign that aimed to destroy old customs, old culture, old habits, and old ideas (MacFarquhar and Schoenhals 2008, chap. 4). Nevertheless, despite such continuities, Maoism was very different in its forms, ideologies, and practices from that earlier movement.

Maoism served as the guiding ideology of the Cultural Revolution. Political propaganda was a fact of life in communist China, but during the Cultural Revolution the apotheosis of Mao and his thought reached a new, feverish pitch. Under these circumstances, "ideologically correct" science had to be grounded in official Marxism and Mao's thought. Science had to be true to dialectical materialism because this philosophy correctly described the fundamental principles of how nature works. The goal of science was to conquer nature, to free oneself from nature. Mao declared, "Natural science is one of man's weapons in his fight for freedom. For the purpose of attaining freedom in society, man must use social science to understand and change society and carry out social revolution. For the purpose of attaining freedom in the world of nature, man must use natural science to understand, conquer and change nature and thus attain freedom from nature" (Mao 1940).

Maoist thought also insisted that science must serve the people. To achieve that, science had to be practical, empirical, and utilitarian. Abstract theoretical science came too close to idealism, which was not only philosophically wrong but also scientifically unsound. Sciences that did not have any obvious utilitarian functions were, at best, intellectual toys of the elite. Maoism had a strong strand of anti-elitism. Mao himself said that "the lowliest are the smartest; the highest the most stupid" (Mao [1958] 1987, 236–238). The people learned from their experience and labor, from their long struggle with nature. Such hard-earned knowledge—concrete, reliable, and often ingenious despite its humble origins—was truly useful and valuable. Thus, science was inherently political. It was objective, but it was neither neutral nor value-free. Good science required mass participation. Maoist thought insisted that science "walks on two legs":[22] one leg was the masses, and the other, the experts or specialists—experts and the people ought to work together, and elite scientists must learn from the people. During the Cultural Revolution, many intellectuals and scientists were sent to the countryside to "learn from the people." Meanwhile, farmers and factory workers were called on to participate in scientific work at research institutes.

Furthermore, Maoist science also emphasized the need to combine both indigenous and Western science. Mao himself believed that there was a great deal of unique knowledge in the Chinese lore on medicine. This idea was reflected clearly in the barefoot doctors program (Fang 2012). The doctors— often villagers or village youths who had received a brief training—used traditional medicine and healing practices together with Western-style drugs. The program provided health care for rural areas where medical resources had been scarce. To some extent, the program aimed to serve practical needs. Nevertheless, it was also emphatically ideological and embodied such political ideas as communist society, national traditions, and mass science. Similarly, in earthquake studies in the 1960s and 1970s, the state championed the integration of folk knowledge and more technical seismology. Not surprisingly, there were tensions between the drive for a cultural revolution and the doctrine to endorse and utilize specific bodies of traditional Chinese knowledge. Maoist mass science tried to maintain its ideological purity by criticizing both traditional elite knowledge and what was characterized as superstition among the common people.

Third, considering mass science and cultural essentialism, the unusual form and claims of Maoist mass science might induce one to look for explanations in the perceived differences between Chinese and Western culture. Perhaps Chinese culture, whatever it is, lent itself to certain ways of doing science. Perhaps traditional Chinese worldviews and ideas of nature, as described by Joseph Needham and others, still played tricks on contemporary seismologists (e.g., Needham and Wang 1956; Henderson 1984). Was that, perhaps, why Chinese seismologists relied so heavily on correlation and phenomenology? The problem with this perspective is that it tends to take for granted certain conventional ideas of "Chinese culture" and essentialize what it labels as "Chinese culture." It uncritically prefers an overgeneralized, vague characterization to specific historical explanations. There is a jejune strand of reductionism in such definitions of Chinese or Western culture or civilization. Cultures are seen as largely separate, categorically different, and internally homogeneous entities. I do not think that culturalism of this kind is a productive tool for historical analysis. In the case of Maoist mass science, there is the further risk of Othering—that is, reducing it to a form or product of totalitarianism or other similar political models (Josephson 2005).[23] Totalitarianism itself is a disputed category and has been much criticized as oversimplification. It is hardly productive to lump diverse historical cases under such a flawed concept. Some have even used the term "oriental despotism" to characterize "totalitarian" communist China (e.g., Wittfolgel 1963). In this formulation, a modern political system became nothing more than the trappings of an unchanging ancient civilization. Used in this way, the model is decidedly orientalist, reductionist, essentializing, and ahistorical.

But of course, Chinese scientists themselves were hardly immune to culturalism. Some of them adopted essentialist cultural categories to describe their own work. They cited Chinese culture when explaining their styles or approaches in doing science. They apparently tried to create a distinctive identity for Chinese science and scientists. Although, in this case, the historical actors themselves fell into the trap of culturalism, it is important to differentiate nationalism from culturalism. Their feeling about cultural identity was also a strong affirmation of the ability of the Chinese—and the Chinese nation—to do science.[24] The main concern was scientific success, national self-confidence, and the political creed of self-reliance that was a matter of both choice and necessity at a time of considerable international isolation. To be sure, some Chinese approaches to seismology were quite distinctive.

One might even go so far as to say that there was a national style of Chinese seismology. (My point here is not to endorse the notion of national style, which may be debatable, but to maintain the analytical differences between national style and culturalism.)[25] Explanations of national styles of science should not boil down to culturalism or presume that nations have their own characters that may shape their respective ways of doing science. Instead of using an amorphous definition of national culture as an explanans here, it might be more fruitful to treat it as an explanandum and scrutinize the particular historical circumstances that contributed to the distinctive style of doing science.

With so many roadblocks and stop signs confronting us, is there a way out? How are we to understand the Chinese program of earthquake monitoring and studies? Instead of invoking outworn essentialist categories (e.g., an ahistorical definition of "Chinese culture"), one must first examine historical particulars. As I have demonstrated in this chapter, the Chinese program of earthquake monitoring and prediction was a case of a particular state and society (communist China during the Cultural Revolution) responding to a particular kind of natural disaster (earthquakes—devastating, elusive, and warlike in their effects and images) in a particular political situation (perceived internal and external threats in the 1960s and 1970s). We will not gain much insight by submerging historical analysis under the heavy fog of culturalism. This does not mean, however, that culture is not a useful concept in understanding the Chinese program of earthquake monitoring and prediction. On the contrary, culture is a potent concept for historical interpretation—it is capacious, flexible, and evocative. I hope that my discussions of mass science *as* culture and mass science *in* culture have confirmed its value as an interpretive tool. In conclusion, we probably do not want a cultural revolution that banishes concepts of culture from scholarship—least of all in the study of the Cultural Revolution—yet it may do some good to smash a few idols of culturalism and sweep the pieces into the dustbin of history.

Notes

I am indebted to Karine Chemla, Evelyn Fox Keller, and the other participants in the "Culture without Culturalism" workshops at Les Treilles for their valuable comments on earlier drafts of this chapter. Part of this article is based on Fan 2012a.

1. For a discussion of some of the earthquakes, see Ma et al. 1982. A couple of other powerful earthquakes occurred shortly after the Tangshan earthquake in southwestern China, but they caused much less damage.

2. My description of the meetings described in this paragraph is based on my interviews. See also Fang 1995, 133–134, 260–265. Li Siguang's opinions on earthquake prediction in the aftermath of the Xingtai earthquake can be found in his posthumous collection, Li 1977, 15–43, 94–167. Weng had made contributions to petroleum exploration in China and devoted much of his time to developing theories for forecasting natural disasters. Intellectually impressive and politically powerful, Li (MSc 1918; DSc 1931, University of Birmingham) dominated Chinese geology in the 1950s to 1970s. Weng (PhD 1939, Imperial College of London) wrote his bachelor's thesis on earthquake prediction and never lost his interest in the topic.

3. This paragraph is based on my interviews and assessments. This is not to say that the Chinese did not care about seismological engineering—they did. However, in this area, their main focus was on large industrial or civil engineering structures. At any rate, compared with earthquake prediction, their effort in seismological engineering was rather limited. Critical and historical studies of earthquakes in China along these lines are very underdeveloped. There are no works in China studies comparable to, e.g., Geschwind 2001 and Clancey 2006.

4. See, e.g., Gungdong Sheng Geming Weiyuanhui Dizhen Bangongshi and Guojia Dizhenju Guangzhou Dizhen Dadui 1977 (n.p.), esp. the page that declares in large font and bold type, "Earthquakes Can Be Predicted." The text on the page then begins with the following: "Earthquakes are knowable; they can be detected and predicted." Note, too, that there is a portrait of Zhang Heng (AD 78–139) on the page. In the Chinese literature on earthquakes and seismology at the time, Zhang was credited as a pioneer in the world history of seismology or was sometimes simply referred to as the first seismologist in the world. Zhang was said to have invented a seismoscope. The pictures of both him and his invention frequently appeared in Chinese publications in the 1960s and 1970s. The phenomenon had to do both with concerns about earthquakes and with a desire to construct a lineage of heroes of Chinese science and a heroic narrative of national science (see Fan 2008b). There is little scholarship on the politics of earthquakes in China that can be compared with, e.g., Walker 2008 or Healey 2011. See also Coen 2013.

5. The communist regime insisted that the widespread famine during the Great Leap Forward had been caused by natural disasters. In fact, it called the catastrophe the "Three Years of Natural Disasters." On the Great Leap Forward, see, e.g., Dikötter 2010, which is very problematic in some of its claims, and Manning and Wernheuer 2012, which provides a collection of case studies.

6. On the Sino-Soviet split, see Chen 2001.

7. The basic ideas were already sketched out in Mao Zedong's speech on the occasion of the founding of the Society for Research on Natural Science of the Shan-Gan-Ning border region in 1940. The Chinese Communist Party also paid much attention to the popularization of science. An important movement of mass science was the Mass Scientific Research in Agricultural Villages (Nongcun qunzhong kexue shiyan) movement during the Great Leap Forward, which encouraged farmers to do scientific experiments (Quanguo kexie 1964). All translations that appear in this chapter are my own.

8. I have offered some reflections on this issue in Fan 2008a, 2012c, and 2012b. See also Anderson 2012 and Bray 2012.

9. An interviewee told me that there must have been 100,000 people involved. It is impossible to give an exact number because it varied and depends on who should be counted. The exact number is less important than the fact that the program was very large and extensive and involved many people from all areas and levels.

10. There are no historical and critical studies of Chinese seismology in this era. I am compelled to rely on my own interviews and conversations with Chinese scientists who were active during the 1960s and 1970s.

11. See Ohtake 2004, Rikitake 1976, and Guowai dizhen bianji bu 1979, which is a collection of articles on earthquake prediction translated from foreign languages. See also Hough 2009.

12. For an interesting discussion of earthquake prediction, see the debate, "Is the Reliable Prediction of Individual Earthquakes a Realistic Scientific Goal?" sponsored by the journal *Nature*, February 25–April 8, 1999; see Main 1999. See also the debate on earthquake prediction in *Science* (Geller et al. 1997; Wyss 1997), as well as Hough 2009.

13. See, e.g., Jia 1968, Ningjin xian geming weiyuanhui 1968, Anonymous 1970, and *Beijing Shi fangzhen bangongshi* 1970.

14. Oike 1978, 163–217, provides an account of the structure of the Chinese enterprise of earthquake studies and prediction in the 1970s.

15. Beijing Shi yanqing xian gewei hui 1969; Hebei sheng Laoting 1970; Ningxia huizu zizhiqu Longde xian geming weiyuanhui fangzhen lingdao xiaozu bangongshi 1972.

16. For an illustration of the instrument, see Shaanxi sheng geweihui dizhenju bian 1977, 50–51. There were exhibitions of new inventions for earthquake prediction. See, e.g., Beijing Shi zhaokai jiangxue huoyong 1970, the report on such an event in Beijing. The inventions included devices to measure ground tilt, ground sound waves, and the well water level.

17. See, e.g., Guangdong Sheng Geming Weiyuanhui Dizhen Bangongshi and Guojia Dizhenju Guangzhou Dizhen Dadui 1973, 33–72, 1977; Tangshan dizhen qianzhao bianxie xiaozu 1977; Zhu and Jiang 1980.

18. Based also on my interviews with Geng Qingguo in September 2007. Geng developed his theory in the early 1970s.

19. Based on my interview with Zhang Tiezheng on October 10, 2007. Zhang said that later, in the mid-1970s, he began to think that magnetic storms might actually be caused by seismic activity. In any case, the way he developed his theory was finding regular series of empirical numbers informed by simple and intuitive thinking about astronomy, geology, and seismicity. In fact, some theories of earthquake prediction developed in the 1980s bordered on numerology (e.g., those developed by Weng Wenbo and Xu Daoyi), but by then the Chinese seismological community had changed so much that it showed little interest in this kind of approach.

20. Guo 1982 recounts a couple of similar incidents.

21. During the Mao era, it was common practice to use rhymes to disseminate information among the rural population and the less educated.

22. Originally, Mao Zedong raised the banner of "walking on two legs" for economic development in the Great Leap Forward, but the slogan was widely adopted

later for any kind of policy that emphasized a two-pronged or a balanced, combined approach.

23. For a critique of this literature, see Fan 2012d, which draws on the same material as Fan 2012a but addresses specifically the issues of science, citizens, and the state.

24. For science and national identity in modern China, see, e.g., Fan 2008b.

25. The concept of "national style" in the history of science stirred much discussion in the 1980s and 1990s. See, e.g., Harwood 1987. See also the special issue of *Science in Context* (Daston and Otte 1991), which includes case studies and critical assessments of national styles in the history of science. I have cautioned against viewing science in modern China in isolation; see Fan 2007.

References

Anderson, W. 2012. "Asia as Method in Science and Technology Studies." *East Asian Science, Technology and Society* 6: 445–451.

Anonymous. 1959. "Nongye qixiang shao, zu de gongzuo changsuo shi zai tianjian han qunzhong zhong" ("The Work Sites for Sentinels and Vanguard Groups in Agricultural Meteorology Are in the Agricultural Fields and among the Masses"). *Nongye qixiang* 8: 2–3.

Anonymous. 1970. "Shixian dizhen yubao bixu dagao qunzhong yundong" ("To Achieve Earthquake Prediction, We Must Vigorously Pursue the Mass Movement"). *Dizhen zhanxian*, no. 5–6: 6–7.

Anonymous. 1975. *Dizhen changshi* (*Basic Knowledge about Earthquakes*). Beijing: Renmin chubanshe.

Beijing Shi fangzhen bangongshi. 1970. "Guanyu kaizhan qunzhongxing dizhen gongzuo de yixie tihui" ("Some Thoughts about the Development of Mass Participation in Projects of Earthquake [Defense]"). No. 5–6: 8–11.

Beijing Shi yanqing xian gewei hui, Yanqing xian Zhangshan ying diqu dizhen lianhe diaocha zu. 1969. "Tu didian fangfa jieshao." *Dizhen zhanxian*, no. 6: 29–31.

Beijing Shi zhaokai jiangxue huoyong Mao Zedong sixiang chuangzhi cebao dizhen tu yiqi jiangyong hui. 1970. "Qunzhong shi zhenzheng yingxiong" ("The Mases Are Real Heroes"). *Dizhen zhanxian*, no. 5–6: 21–25.

Bray, F. 2012. "Only Connect: Comparative, National, and Global History as Frameworks for the History of Science and Technology in Asia." *East Asian Science, Technology and Society* 6: 233–241.

Chen, J. 2001. *Mao's China and the Cold War*. Chapel Hill: University of North Carolina Press.

Clancey, G. 2006. *Earthquake Nation: The Cultural Politics of Japanese Seismicity, 1868–1930*. Berkeley: University of California Press.

Coen, D. 2013. *The Earthquake Observers: Disaster Science from Lisbon to Richter*. Chicago: University of Chicago Press.

Daston, L., and M. Otte, eds. 1991. "Style in Science." Special issue of *Science in Context*, 4(2).

Dikötter, F. 2010. *Mao's Great Famine: The History of China's Most Devastating Catastrophe, 1958–62*. New York: Walker.

Dizhen wenda bianxiezu. 1977. *Dizhen wenda (Earthquakes Q&A)*. Beijing: Dizhi chubanshe.

Epstein, S. 1995. "The Construction of Lay Expertise: AIDS Activism and the Forging of Credibility in the Reform of Clinical Trials." *Science, Technology, and Human Values* 20: 408–437.

Fan, F. 2007. "Redrawing the Map: Science in Twentieth-Century China." *Isis* 98: 524–538.

Fan, F. 2008a. "East Asian STS: Fox or Hedgehog?" *East Asian Science, Technology and Society* 1: 243–247.

Fan, F. 2008b. "National Narrative and the Historiography of Chinese Science." *Gujin lunheng* 18: 199–210.

Fan, F. 2012a. "'Collective Monitoring, Collective Defense': Science, Earthquakes and Politics in Communist China." *Science in Context* 25: 127–154.

Fan, F. 2012b. "Doing East Asian STS Is Like Feeling an Elephant, and That Is a Good Thing." *East Asian Science, Technology and Society* 6: 487–491.

Fan, F. 2012c. "The Global Turn in the History of Science." *East Asian Science, Technology and Society* 6: 249–258.

Fan, F. 2012d. "Science, State, and Citizens: Notes from Another Shore." *Osiris* 27: 227–249.

Fang, X. 2012. *Barefoot Doctors and Western Medicine in China*. New York: University of Rochester Press.

Fang, Z., ed. 1995. *Zhou Enlai yu fangzhen jianzai (Zhou Enlai and the Defense against Earthquakes and the Reduction of Disaster Damages)*. Beijing: Zhongyang wenxian chubanshe.

Geller, R. J., D. D. Jackson, Y. Y. Kagan, and F. Mulargia. 1997. "Earthquakes Cannot Be Predicted." *Science* 275(5306): 1616.

Geng, Q. 1985. *Zhongguo hanzhen guanxi yanjiu (Studies of the Relationship between Droughts and Earthquakes)*. Beijing: Haiyang chubanshe.

Geschwind, C. H. 2001. *California Earthquakes: Science, Risk and the Politics of Hazard Mitigation*. Baltimore, MD: Johns Hopkins University Press.

Guangdong Sheng Geming Weiyuanhui Dizhen Bangongshi and Guojia Dizhenju Guangzhou Dizhen Dadui. 1973. *Dizhen ji qi yufang (Earthquakes and Earthquake Defense)*. Guangzhou: Guangdong renmin chubanshe.

Guangdong Sheng Geming Weiyuanhui Dizhen Bangongshi and Guojia Dizhenju Guangzhou Dizhen Dadui. 1977. *Dizhen zhishi huace (Illustrated Pamphlet on Earthquakes)*. Beijing: Dizhen chubanshe.

Guo, Q. 1982. *Zhenqian qiguan (Strange Phenomena of Earthquakes)*. Beijing: Dizhen chubanshe.

Guowai dizhen bianji bu. 1979. *Dizhen yubao han dizhen xue (Earthquake Prediction and Seismology)*. Beijing: Kexue jishu wenxian chubanshe.

Harwood, J. 1987. "National Styles in Science: Genetics in Germany and the United States between the World Wars." *Isis* 78: 390–414.

Healey, M. A. 2011. *The Ruins of the New Argentina: Peronism and the Remaking of San Juan after the 1944 Earthquake*. Durham, NC: Duke University Press.

Hebei sheng Laoting xian Hongwei zhongxue keyan xiaozu (Science Research Group, Red Guards Middle School, Laoting County, Hebei Province). 1970. "Tu didian

zidong jiluyi" ("An Automatic Recorder of the Telluric Current"). *Dizhen zhanxian,*
no. 5–6: 46–47.

Henderson, J. B. 1984. *The Development and Decline of Chinese Cosmology.* New York:
Columbia University Press.

Hough, S. 2009. *Predicting the Unpredictable: The Tumultuous Science of Earthquake
Prediction.* Princeton, NJ: Princeton University Press.

Huang, L., and H. Deng. 1978. *Di guang* (*Earthquake Light*). Beijing: Dizhen chubanshe.

Jia, E. 1968. "Gongnongbing bixu chengwei dizhen kexue jishu de zhuren" ("Workers,
Peasants, and Soldiers Must Become the Masters of Seismological Science and Technology"). *Dizhen zhanxian,* no. 8: 14–18.

Jian, B. 1959. "Zenyang liyong nongmin jingyan ji suan nongye qixiang zhibiao" ("How
to Use the Experience of the Peasants to Calculate the Indices of Agricultural Meteo-
rology"). *Nongye qixiang* 9: 4–6.

Jiang, J. 1980. *Qiyi de benling: qiantan dongwu yuzhi dizhen* (*Strange Talents: Introduction
to Animals' Abilities to Predict Earthquakes*). Beijing: Dizhen chubanshe.

Jiang, J., and D. Chen. 1993. *Dizhen shengwuxue gailun* (*A General Study of Seismological
Biology*). Beijing: Dizhen chubanshe.

Josephson, P. 2005. *Totalitarian Science and Technology.* Amherst, NY: Humanity Books.

Li, Shanbang. 1981. *Zhonguo dizhen* (*Earthquakes of China*). Beijing: Dizhen chubanshe.

Li, Siguang. 1977. *Lun dizhen* (*Discourse on Earthquakes*). Beijing: Dizhen chubanshe.

Liang, Z. 1970. "Jieshao liang zhong tuyiqi" ("Introducing Two Indigenous Instruments"). *Dizhen zhanxian,* no. 5–6: 45–46.

Ma, Z., et al. 1982. *Zhongguo jiu da dizhen* (*China's Nine Major Earthquakes*). Beijing:
Dizhen chubanshe.

MacFarquhar, R., and M. Schoenhals. 2008. *China's Last Revolution.* Cambridge, MA:
Belknap Press of Harvard University Press.

Main, I. 1999. Introduction to *Nature* debate: "Is the Reliable Prediction of Individual
Earthquakes a Realistic Scientific Goal?" *Nature,* February 25–April 8, www.nature
.com/nature/debates/earthquake/equake_frameset.html.

Manning, K. E., and F. Wernheuer, eds. 2012. *Eating Bitterness: New Perspectives on
China's Great Leap Forward and Famine.* Seattle: University of Washington Press.

Mao, Z. 1940. "Speech at the Inaugural Meeting of the Natural Science Research Society
of the Border Region (February 5, 1940)." Accessed March 10, 2012. www.marxists
.org/reference/archive/mao/works/redbook/ch22.htm.

Mao, Z. [1958] 1987. *Jianguo yilai Mao Zedong wengao* (*Mao Zedong's Writings since the
Founding of the Nation*). Vol. 7. Beijing: Zhongyang wenxian chubanshe.

Mullaney, T. S. 2012. "The Moveable Typewriter: How Chinese Typists Developed
Predictive Text during the Height of Maoism." *Technology and Culture* 53: 777–814.

Needham, J., and L. Wang. 1956. *History of Scientific Thought.* Vol. 2 of *Science and Civilization in China.* Cambridge: Cambridge University Press.

Ningjin xian geming weiyuanhui. 1968. "Kao Mao Zedong sixiang, kao guangda
qunzhong zai dizhen zhanxian shang dada yichang renmin zhanzhen" ("Relying on
Mao Zedong Thought, Relying on the Masses, Let Us Fight a People's War on the
Frontline against Earthquakes"). *Dizhen zhanxian,* no. 8: 26–29.

Ningxia huizu zizhiqu Longde xian geming weiyuanhui fangzhen lingdao xiaozu bangongshi. 1972. "Tu didian, dixia shui yubao dizhen xiaojie." ("Little Explanations of Earthquake Prediction Based on Tulluric Currents and Underground Water"). *Dizhen zhanxian*, no. 10: 35–39.

Nongyebu nongju gaige bangongshi, ed. 1959. *Tuyang jiehe dagao gongju gaige yundong de jingyan (Experiences from the Vigorous Movement to Improve [Agricultural] Tools Based on the Combination of Foreign and Indigenous Methods)*. Beijing: Nongye chubanshe.

Ohtake, M. 2004. "The Late Professor Takahiro Hagiwara: His Career with Earthquake Prediction." *Earth Planets Space* 56: ix–xiii.

Oike, K. 1978. *Chūgoku no jishin yochi (Earthquake Prediction in China)*. Tokyo: Nihon Hōsō Shuppan Kyōkai.

Oreskes, N. 1999. *The Rejection of Continental Drift: Theory and Method in American Earth Science*. New York: Oxford University Press.

Quanguo kexie. 1964. *Nongcun qunzhong kexue shiyan huodong jingyan jiaoliuhui fayan cailiao huibian (Promotional Materials Compiled from the Meeting to Exchange Experiences from Village Mass Scientific Experimental Activities)*. Beijing: Quanguo kexie.

Raleigh, B., and the Haicheng Earthquake Study Delegation. 1977. "Prediction of the Haicheng Earthquake." *EOS, Transactions, American Geophysical Union*, 58: 236–272.

Rikitake, T. 1976. *Earthquake Prediction*. New York: Elsevier.

Shaanxi sheng geweihui dizhenju bian. 1977. *Dizhen yuce yufang zhishi (Knowledge about Earthquake Monitoring and Defense)*. Xian: Shaanxi renmin chubanshe.

Shangxisheng, Q. J. 1959. "'Tuyang shengzhan' si jie he, yubao fuwu da yuejin" ("The Four Integrations at Provincial Stations Based on Indigenous and Foreign Methods, the Great Leap Forward of Forecast Services"). *Nongye qixiang* 1: 5–7.

Shaonian dizhen cebao huodong bianxie zu. 1978. *Shaonian dizhen cebao huodong (Earthquake Monitoring Activities for Youngsters)*. Shanghai: Xinhua shudian.

Schmalzer, S. 2008. *The People's Peking Man: Popular Science and Human Identity in Twentieth Century China*. Chicago: University of Chicago Press.

Schwarcz, V. 1986. *The Chinese Enlightenment: Intellectuals and the Legacy of the May Fourth Movement of 1919*. Berkeley: University of California Press.

Tang, X. 1988. *A General History of Earthquake Studies in China*. Beijing: Science Press.

Tangshan dizhen qianzhao bianxie xiaozu, ed. 1977. *Tangshan dizhen qianzhao (Precursory Phenomena of the Tangshan Earthquakes)*. Beijing: Dizhen chubanshe.

Walker, C. 2008. *Shaky Colonialism: The 1746 Earthquake-Tsunami in Lima, Peru, and Its Long Aftermath*. Durham, NC: Duke University Press.

Whyte, M. K. 1973. "Bureaucracy and Modernization in China: The Maoist Critique." *American Sociological Review* 38(2): 149–163.

Wittfolgel, K. 1963. *Oriental Despotism: A Comparative Study of Total Power*. New Haven, CT: Yale University Press.

Wynne, B. 1996. "May the Sheep Safely Graze? A Reflexive View of the Expert-Lay Knowledge Divide." In *Risk, Environment and Modernity: Towards a New Ecology*, ed. S. Lash, B. Szerszynski, and B. Wynne, 44–83. London: Sage.

Wyss, M. 1997. "Can Earthquakes Be Predicted?" *Science* 278(5337): 487–490.

Wyss, M. 1999. "Without Funding No Progress." *Nature*, March 11, www.nature.com /nature/debates/earthquake/equake_frameset.html.

Zhongguo Kexueyuan Shengqu Wuli suo Dizhen Zu. 1977. *Dongwu yu dizhen* (*Animals and Earthquakes*). Beijing: Dizhen chubanshe.

Zhu, J., and Jiang, Z. 1980. *Yufang dizhen de hongguan zhishi* (*Knowledge about Macroseismic Phenomena and Earthquake Defense*). Chengdu: Sichuan minzu chubanshe.

IV. WHAT IS AT STAKE?

CAROLINE EHRHARDT

13

E Uno Plures?

Unity and Diversity in Galois Theory, 1832–1900

French mathematician Évariste Galois died in 1832, leaving only sixty pages of mathematical manuscripts, including a memoir on the algebraic solution of equations that the French Académie des Sciences had rejected a few months earlier and several short articles published in scientific journals (see Ehrhardt 2010b, 2011). Galois's work then practically vanished from the mathematical scene until 1846, when portions of it appeared in a journal run by the French academician Joseph Liouville (Galois 1846; see also Ehrhardt 2010a). Today, however, Galois theory is a well-established field of algebra taught in nearly all departments of mathematics, and the name Galois continues to be associated with the forefront of research.

The strange career of Galois's work, particularly the striking contrast between its initial rejection and its posthumous fortune, raises questions about the social and cultural aspects of the making and the reading of a mathematical text: What is a mathematical proof, and what is a correct definition? What does it mean to understand a mathematical work?[1] Considered from a historical viewpoint, these questions do not simply bear on an immanent conceptual body of mathematical knowledge. On the contrary, addressing them requires that we shift the emphasis to social and cultural factors, such as the kinds of production or subject that mathematicians find relevant; the kinds of knowledge, methods, and know-how individual readers seek in a text; the ways one does mathematics; and the specific practices they involve. All these factors are correlated to the time and place of the making and the reading of the mathematical text. Hence the meaning of a mathematical paper is not something given intrinsically in the text and is not universal; rather, it takes shape in a particular cultural environment, where knowledge, practices, representations, epistemological values, institutions, and social configurations are brought together in a subjective form.[2]

In this chapter I examine the tensions between the local creation and the "universality" of Galois theory, looking at the way it evolved during the second half of the nineteenth century. This process was far from linear—it bore the imprint of the many European mathematicians who, from the publication of Galois's paper onward, worked on the theory of equations and developed their own interpretations of Galois. I compare Galois's memoir on the algebraic solution of equations and some of the various ways that it was read and used, and show that Galois's successors did far more than simply fill in the gaps in Galois's paper—each of them added several things and reorganized it. In fact, Galois's readers constructed not one but several theories out of his work.[3]

The first part of this chapter provides an overview of Galois's paper, focusing on its aims and on the mathematical notions and practices it involved. Then I offer an analysis of three interpretations of Galois's work, by Arthur Cayley, Richard Dedekind, and Joseph-Alfred Serret, highlighting significant differences among local ways of practicing mathematics. Finally, I delve into some of the textbooks published at the end of the nineteenth century to show how some local aspects of mathematical knowledge and practices persisted through time and place, whereas others disappeared. While these textbooks allowed international discussions about what mathematicians generally called Galois theory, many traces of the competing interpretations of Galois's works remained visible in them. Clearly, Galois theory was not yet a fully recognized subdiscipline.

A Glance at Galois's Paper

In his "Mémoire sur les conditions de résolubilité des équations par radicaux" ("Memoir on the Conditions That an Equation Be Solvable by Radicals"; Galois 1997, 43–71), Galois tried to solve a problem that had dogged mathematicians for centuries: finding a criterion to determine which equations could be solved by using an algebraic formula.[4] Methods already existed to solve all equations of degree 2, degree 3, and degree 4—in fact, they harked back to ancient Mesopotamia for equations of degree 2, and the first solutions for equations of degree 3 and 4 were found by Girolamo Cardan and Niccolo Tartaglia in the sixteenth century (Dahan-Dalmedico and Peiffer 1986, 72–118).[5] Yet the existence of such a method for equations of degree higher than 4 was still open to question at the end of the eighteenth century. Even the great mathematician Joseph-Louis Lagrange, the author of a widely acclaimed treatise on this topic, *Traité de la résolution des équations numériques de tous les degrés* (1797), had given up, admitting he had no solution to offer. Shortly before Galois tackled the question, Niels Henrik Abel demonstrated that, in the general case, equations of degree 5 and above could not be solved algebraically and even found a sufficient condition for an equation to be solvable by radicals. Galois went one step further, for he concluded his memoir with the statement that finding a necessary and sufficient condition was possible. Nevertheless, his work on this topic fitted into a long scholarly tradition.

Galois's memoir started with the explanation of the mathematical concepts to be used for his purposes (Galois 1997, 45–47). The first was the concept of substitution, which consists in permuting given numbers or letters—or, in the case of equations, roots. Since Lagrange had used the idea of permuting roots extensively in his renowned research on the theory of equations, this kind of practice was familiar to the French mathematicians of that period. Indeed, Lagrange's book was not only widely read by professional researchers but also studied in secondary schools (*collèges royaux*) (Belhoste 1995, 86, 89, 106). Another famous academician, Augustin-Louis Cauchy, had worked on the topic of substitutions in a more general way and had offered a first conceptualization that may have inspired Galois (Dahan 1980). The second concept was that of adjunction. This consists in adding a new quantity to the ones already given in order to solve an equation that has no solution among the numbers considered initially. In Galois's time this idea was used in the particular cases of negative numbers adjoined to positive numbers and the imaginary number i (the square root of -1) adjoined to real numbers. Even if it was not conceptualized, the idea was explained in textbooks used in secondary schools (Bézout 1812; Lacroix 1799, 1800). It also appeared, developed in a more theoretical way, in Abel's research, which began to gain renown in 1830. Hence, up to that point, Galois's memoir shared its fundamental basis with other mathematical texts of its time (Ehrhardt 2011, 55–86).

Having stated these principles, Galois developed his proof in three stages. First, he defined a new object—one based on the equation and made with some substitutions of the roots—which he called the "group of the equation." Second, he studied what happened to this group when one added a new quantity by adjunction and showed that the group then split into smaller groups. Finally, he determined the criterion he was looking for, first using the language of groups—an equation could be solved by radicals if and only if, after having repeated the second stage a sufficient number of times, only one element remained in the final group—and second using the classical language of equations.

Thus, the concept of "group" was the keystone of all Galois's mathematical thinking, and that was precisely the source of its originality. In fact, one can find the word "group" as a mere synonym of "set" in older mathematical texts dealing with equations, but for Galois it meant more than that, since it had a true mathematical meaning. However, this new mathematical notion was not defined precisely in Galois's work as one would do today. Instead, the groups were implicitly determined through a set of mathematical practices used in the proof of the theorems and from which the mathematical meaning actually emerged.

On the one hand, Galois explained what he called the group of an equation using a tabular representation of all the elements:

> Whatever the given equation may be, one will always be able to find a rational function V of the roots such that all the roots will be rational functions of V. That said, let us consider the irreducible equation of which V is a root. . . . Let $V, V', V'', \ldots, V^{(n-1)}$ be the roots of this equation.
> Let $\varphi V, \varphi_1 V, \varphi_2 V, \ldots, \varphi_{m-1} V$ be the roots of the proposed equation. Let us write down the following n permutations of the roots:

(V)	$\varphi V,$	$\varphi_1 V,$	$\varphi_2 V, \ldots,$	$\varphi_{m-1} V,$
(V')	$\varphi V',$	$\varphi_1 V',$	$\varphi_2 V', \ldots,$	$\varphi_{m-1} V',$
(V'')	$\varphi V'',$	$\varphi_1 V'',$	$\varphi_2 V'', \ldots,$	$\varphi_{m-1} V'',$
$\cdots$				
$(V^{(n-1)})$	$\varphi V^{(n-1)},$	$\varphi_1 V^{(n-1)},$	$\varphi_2 V^{(n-1)}, \ldots$	$\varphi_{m-1} V^{(n-1)}:$

> I say that this group of permutations enjoys the specified property. (Galois 1997, 53; trans. Neumann 2011, 117)[6]

Galois used this tabular notation as a way to visualize a complex set and to arrange its elements to enable readers to understand how it is organized. It was already familiar from Lagrange's research (Lagrange 1797, sec. 107).

On the other hand, Galois sometimes described groups with a single capital letter. This notation led to different mathematical practices associated with groups. When using the table notation, the group is a set of elements arranged in lines or columns. With the single-letter notation, the "group" refers to a single mathematical object, not to a cluster; there is no need to describe the elements contained in the group exactly:

> When a group G contains another H, the group G can be partitioned into groups, each of which is obtained by operating on the permutations of H with one and the same substitution, so that $G = H + HS + HS' + \ldots$ (Galois 1997, 174–175; trans. Neumann 2011, 85)

In the version of Galois's paper published in 1846, the concepts, theorems, and argumentative structure remained the same as in the version of the work that the French academicians had found unclear in 1831. However, subsequent European mathematicians studied, completed, and developed Galois's work. Although no global change occurred in the field of algebra between the 1830s and the 1850s, each of them had his own reasons for examining Galois's ideas more closely. Their social and cultural environment, as well as their personal positions and mathematical trajectories, accounted for their interest in Galois's research and influenced their interpretations of it (Ehrhardt 2012).

Arthur Cayley: Groups as Generic Sets

In 1854, the English mathematician Arthur Cayley published a paper in the *Philosophical Magazine* with the title "On the Theory of Groups, As depending on the Symbolic Equation $\theta^n = 1$." He began this paper with a definition of the word "group" and indicated in a footnote, "The idea of group, applied to substitutions, [was] due to Galois":

> A set of symbols: 1, α, β, . . . all of them different, and such as the product of any two of them (no matter in what order), or the product of anyone of them into himself, belongs to the set is said to be a group. It follows that if . . . the symbols of the group are multiplied together so as to form a table, thus:

		1	α	β	..
	1	1	α	β	..
Nearer factors	α	α	α^2	$\beta\alpha$	
	β	β	$\alpha\beta$	β^2	
	$\vdots$				

that as well each line as each column of the square will contain all the symbols 1, α, β. (Cayley 1854, 41)

For Cayley, a group was a set of symbols that stood for operations. Each of them could be, for instance, one of the permutations that Galois represented with a column or a line of letters. The role of the table was to express the results of the different products one could make with these symbols.

Once this definition was stated, Cayley proceeded to seek all the groups with four or six elements, trying to determine their common pattern. In fact, as Cayley explained in the final part of his paper, he was interested in the "notion of a group"[7] because it was generic; that is, it could apply in very different mathematical situations (ranging from algebraic equations to elliptic functions, matrices, or quaternions).[8] His point was to show that "systems of this form were of frequent occurrence in analysis, and it was only on account of their extreme simplicity that they had not been expressively remarked" (Cayley 1854, 43). Hence, Cayley defined and used the group idea from a much more general and abstract standpoint than Galois did. Moreover, the way Cayley depicted the groups was different from Galois's presentation. Whereas Galois's tables focused on the internal organization of the group and on common points between elements, Cayley's table focused on operations, that is to say, on the way elements interacted.

Clearly, Cayley did not share a definition with Galois, or a visual representation of groups, or a way to apply the group concept. He did not import the general framework of the theory of equations from Galois's work, either. Therefore, Cayley's interpretation implied a complete conceptual change, as well as new representations associated with Galois's memoir.[9] If we take a closer look at Cayley's intellectual and social environment, Cambridge University in the Victorian age, we can gain a better understanding of this phenomenon.

In the nineteenth century, Cambridge was much more than a place of education: it was home to a specific mathematical tradition that structured a whole scientific community, both socially and culturally (Warwick 2003, 49–285). In the 1850s it was certainly the place in Europe where algebra was the least synonymous with the theory of equations. From the 1810s onward, mathematics in Cambridge was profoundly renewed by a group of mathematicians often called the English Algebraic School. For them, the relevant objects of mathematics were no longer concrete entities (e.g., numbers) but the symbolic operations one could define on these objects (Novy 1968; Koppelman 1971; Durand-Richard 1996). When Cayley used the "regular" chains of algebraic reasoning and made calculations with the symbols that composed his groups, in fact he applied ideas and methods described, for example, in George Peacock's *Treatise of Algebra* (1830) or in Charles Babbage's *Essay towards the Calculus of Functions* (1815). In addition, the tabular symbolism employed by Cayley to describe how the groups worked was very similar to that used by William R. Hamilton (1848) on quaternions or even to that found in Augustus de Morgan's *Formal Logic* (1847). In other words, this choice of notation was linked to a specific mathematical practice that was typical of the scholarly culture Cayley belonged to. Moreover, the mathematicians of the English Algebraic School had an epistemological way of thinking about the "essence" of algebra that they put into practice in mathematical works that developed new objects and new methods. Cayley's generation may have been less concerned with seeking the foundations of algebra, but the fact that Cayley saw the concept of the group as a kind of generic concept that would enable mathematicians to unify many particular situations reveals that he also tried to uncover the "true reasons" of these situations that were concealed by their specificity.

Cayley's understanding of Galois's text and of dealing with groups was therefore rooted in the culture and practices of the English algebraic school. It was so deeply rooted that Cayley "saw" in Galois's paper something that would have remained invisible if he had only looked at the results and theorems. However, this cultural environment did not shape Cayley's interest in groups, which originated in his research on caustics, that is, the patterns generated by rays of light incident on surfaces. In fact, Cayley's interest in groups remained slight, and he published very few papers on the topic. Nor did the familiarity between Cayley's research on groups and British symbolical algebra ensure the success of his research, which went almost unnoticed (Crilly 2006, 185).

Richard Dedekind: Abstract Groups

The German mathematician Richard Dedekind was the first to create a teaching course out of Galois's research (Dedekind 1981). The course, taught to a very small number of students, was held between 1856 and 1858 at Göttingen University, where Dedekind was a privatdocent, a senior outside lecturer. It remained unpublished until 1981.

It is quite obvious that Dedekind knew the paper published in Liouville's journal, but we can also assume that he knew Cayley's work. Indeed, in one of his previous papers, Dedekind referenced a work by Boole published in the same issue of the same journal as Cayley's paper (Dedekind 1855). This means that Dedekind had two different approaches to the concept of group at his disposal, and he actually used both of them. That Dedekind was a reader of Galois can be seen in the very title of the course (Galois Theory) and in the order of his lessons, which follows Galois's memoir. One consequence was that Dedekind only considered groups of substitutions, which was sufficient in the framework of the theory of equations. At the same time, Dedekind gave a definition of groups, which Galois had not done, and which was close to Cayley's.[10] He also retained from Cayley's approach the idea that the elements could be anything, even if his primary focus was on substitutions (Dedekind 1981, 63). However, Dedekind did not retain Cayley's notations, or his general aim of classifying groups containing a given number of elements. Instead, he borrowed from Galois the single-letter notation to represent the groups—a notation that precisely allowed erasure of any specificity that substitutions could provide to the groups.

With this general definition and this notation, Dedekind did not need a graphical representation to show that a given set was actually a group—he simply had to return to his first definition and prove it was verified (Dedekind 1981, 83–84). For that purpose, Dedekind did not do calculation *within* the group, but *with* the group, just as Galois did in the second example above. Hence this single-letter notation was associated in Dedekind's course with a proof practice different from Galois's. It consisted in characterizing the sets by the properties he had defined before, and not by the visual representation of the organization of its elements. More precisely, this notation allowed Dedekind to deal with groups just as if they were numbers and to use an analogy procedure not only with algebraic calculation, as Galois had done, but also with number theory. For instance, Dedekind defined the "divisors" of a group and the greatest common divisor of several groups (Dedekind 1981, 65, 67, 68, 78).

The result of Dedekind's reading was to incorporate Galois's theorems and objects in a framework that stretches from number theory to the idea that the right way to characterize an object is neither to write nor to represent it but to define it in advance with general properties that are supposed to show its "real nature." This conclusion tends to affirm the links between Dedekind's reading of Galois's work and the social and cultural environment he was working in.

The epistemological focus that led Dedekind to think about the very nature of the adjoined quantities of Galois's theory could be seen as a concern of German mathematicians at that time (Schubring 1996). Some German mathematicians linked this concern about the "essence" (*das Wesen*) of mathematical entities to a specific mathematical practice that consisted in avoiding calculation as much as possible and establishing a priori the general rules and principles that would lead to particular results (Ferreiros 1999). From this viewpoint, a mathematical theory should be defined not by a compilation of results and theorems but by an axiomatic construction of the objects it treated, a construction that allowed accessing of their "real nature." According to Johann Peter Gustav Lejeune Dirichlet, who was professor in Göttingen when Dedekind gave his seminar, this tendency to "substitute thinking for calculating" (Lejeune Dirichlet 1897, 245; Sinaceur 1990) was increasingly pronounced in modern mathematics. The German mathematician Gotthold Eisenstein even wrote that "the essential principle of the latest mathematical school" was to "enclose a problem as a whole by a brilliant method," which consisted in "presenting a formula, that [contained] the full scope of truths of a whole complex" right from the beginning (Rudio 1895, 158–159; trans. Schmitz 2004, 8). Similarly, Bernhard Riemann, who was also a privatdocent at Göttingen at that time, wrote that he was trying to underline the fundamental principles that allowed him to define new functions (Riemann [1851] 1898, 47–48).

This trend was not necessarily dominant in the mathematics of that time, and the mathematicians who promoted it continued to make many calculations in their research, but it influenced Dedekind, who was still a young mathematician in 1856–1858. Indeed, because he still considered himself a student at this time, Dedekind attended Dirichlet's lecture about number theory, as well as Riemann's seminar, and was very deeply influenced by them (Dugac 1976, 14–23).[11] In other words, the mathematical practice Dedekind used while teaching Galois's ideas reproduced a specific way to do mathematics that must have seemed familiar, natural, and legitimate to him. This was due not only to the specific social and cultural environment in which he worked but also to the research training he had received from his two mentors. As with Cayley's, Dedekind's interpretation of Galois's work bore the marks of the place and time of its making, as well as the traces of its author's own development.

Joseph-Alfred Serret: Galois Theory for French Students

Joseph-Alfred Serret's reading of Galois's texts was included in a textbook of advanced algebra (*Cours d'algèbre supérieure*) that went through numerous editions.[12] However, Galois played a rather minor role in the first two editions of the book.

Only one lemma (i.e., a subsidiary proposition) by Galois was expounded in the first edition of Serret's textbook in 1849. Serret presented it as a way to overcome the difficulty of calculation encountered when trying to apply a theorem by Lagrange. He reproduced Galois's statement but changed its proof: while Galois's proof had two stages and ended with the theoretical result announced, Serret added a third stage, where he described an effective method of calculation. In other words, Serret's reading profoundly changed the very meaning of Galois's lemma, for it gave both a practical dimension and an intrinsic mathematical interest to a result that was originally theoretical and intermediary. In the second edition, in 1854, Serret added a new proof of Galois's final theorem by the mathematician Charles Hermite, but this proof did not rely at all on the ideas and concepts of Galois's original memoir (Serret 1854, 569–575).

Although Serret acknowledged the significance of Galois's final result, he did not find it necessary to develop Galois's methods in the first two editions of the book, or to clarify its theoretical difficulties. The limited attention paid to Galois in these first two editions is certainly also because Serret had to give priority to Liouville, who had announced as early as 1843 that he would write about Galois's works (Liouville 1843). Twenty years later Liouville had not published anything, and Serret was freed from that obligation (Bertrand 1899). However, it is significant that, choosing to underline the final result and the bridges to Lagrange's, approach, Serret at first kept only the elements of Galois's memoir that linked it to the scholarly tradition of the theory of equations. In fact, it was only in the third edition of 1866 that the general framework of Galois's memoir and the notion of group appeared in Serret's textbook.

Although Galois's memoir was fully expounded on this occasion, Serret did not simply copy it. First, he did not make an extensive use of the concept of groups; instead, he preferred another notion, the *systèmes de substitutions conjuguées*. Developed by the French mathematician Cauchy in the 1840s, it is mathematically equivalent to that of the group but involves different practices (Dahan 1980). For instance, it is much more centered on combinatorial methods than the notion of the group as a set of elements. Second, when writing the proofs of the theorems, Serret provided all the details and intermediary results—even writing all the elements in each group—to help the reader see the real workings of Galois's approach. Even if Serret did not mention effective calculation, his interpretation of Galois's memoir amounted to a practical method for deciding whether an algebraic equation can be solved or not. In both cases, the way Serret interpreted Galois's memoir led him to use only the tabular representation of groups, namely, the one that was closer to the usual practices of the theory of equations and adapted to show the actual contents of the groups. In that, Serret's reading of Galois's memoir, in common with the other two readings analyzed above by Cayley and Dedekind, is strongly related to the mathematical culture in which it was rooted.

First, in France in Galois's time—and this was true twenty years later as well—mathematicians did not really differentiate between algebra and the theory of equations. In short, according to the synthesis made by Lagrange in the 1790s, doing algebra meant doing calculations on algebraic functions and counting permutations of roots with combinatorial methods (Ehrhardt 2011, 55–87; 2012, 54–59). The content of Serret's textbook illustrated this cultural environment. In particular, the book highlighted the issue of the number of values a function could take when its variables were permuted. Initiated by Lagrange and continued by Cauchy, this question became a fashionable field of research in Paris after 1845, when the new generation of French mathematicians, such as Joseph Bertrand (1845), Émile Mathieu (1860), and Serret (1850), started to work on it.[13] It finally resulted in a lasting link between the theory of equations and a specific use and theorization of the concept of substitution in French mathematical research, which left aside the notion of group defined by Galois.

Second, until the 1870s, the French mathematical milieu was mostly organized around the École Polytechnique. Training nearly all future mathematicians, this school influenced their practices and values. In particular, as an engineering school, it encouraged a permanent dialogue between theoretical knowledge and concrete applications (Belhoste 2001, 2003, 265–303). The French mathematicians shared a value system in which concrete results were important and in which calculation was favored above abstract reasoning in proof. Hence, Serret's reading of Galois pursued a way of thinking about equations that was totally in keeping with Serret's training. Its practical dimension also tallied with the knowledge and know-how that the students using the textbook could understand easily.

Galois's Theory at the End of the Nineteenth Century

Cayley, Dedekind, and Serret were far from the only ones interested in Galois's texts after 1850. Other mathematicians set out to fill the gaps in Galois's work, trying to reinforce the theoretical bases of that work or to make a whole new theory. Their readings of it corresponded to their directions and interests—they were also embedded in their own specific scholarly cultures. For instance, in 1870 the French mathematician Camille Jordan devoted a whole treatise to the theory of groups of substitutions, which gave Galois's works a much bigger role than Serret's textbook did (Jordan 1870). Jordan emphasized the generality and the theoretical framework of Galois's work, and the issue of solving equations became one of the applications of Galois's theory. Jordan's treatise and Serret's textbook quickly became works of international reference on the theory of equations and the issue of substitutions, so many mathematicians after 1870 knew Galois's theory from these two books.

Every mathematician who worked on Galois's paper modified it.[14] The change at the result level was more visible, as Galois's theorems were formulated in different ways. On closer inspection one can see that Galois's paper was also modified at the practice level: readers did not write the same kind of proofs and did not handle the objects in the same way. Moreover, each of them developed particular representations of Galois's work, explaining what its core should be or including it in another theory. Importantly, all these mathematicians destroyed the unity of Galois's original text, revealing three characteristics that are usually implicit in such a text: its results, theorems, and objects; the methods and know-how associated with them; and the representations, interpretations, and values linked with them. In all these facets, readers found ample room for creative interpretation of Galois's original work.

After 1870, the theory of groups developed independently of the theory of equations. It became a very active field of research, and links to Galois's research became increasingly weak. However, at the same time, mathematicians started talking about Galois theory as an independent field of research; they were aware of Galois's original memoir and could say what it was about. Yet each would probably have given a different description.

Many textbooks and papers based on lecture courses on Galois theory (or including Galois theory) were published between 1880 and 1910. It was in these books that the making of Galois theory actually occurred, for it was there that learned and specialized knowledge was transformed into knowledge that was taught and then widely shared. These texts provide a unique view of the tension between the making of Galois theory locally and its conversion into a theory claiming universality. First, all these books and articles relied on existing research, and most of them quoted the same reference works. Some of them were widely used across Europe and in the United States. Consequently, they played an important role in the circulation of Galois theory and the diffusion of shared principles and practices. Second, most of them stemmed from the author's teaching in his university. These books necessarily relied on the particular educational habits of the academic environment in which they were produced.

A sign of this universality in the making is that the names of the first readers disappeared under Galois's. Indeed, the authors of these textbooks presented all the theorems as Galois's theorems, to the extent that it seemed as if the mathematicians who succeeded him had only illuminated some of his ideas and concepts, without adding anything and without changing content or meaning. The name "Galois" now stood for a synthesis of the work done since the 1850s. It remains to define the exact nature of this synthesis. To this end I now analyze two widely used textbooks written in Germany: *Substitutionentheorie und ihre Anwendung auf die Algebra* (*Theory of Substitutions and Its Application to Algebra*) by Eugen Netto, published in 1882, and *Lehrbuch der Algebra* (*Algebra Textbook*) by Heinrich Weber, published in 1895. I then consider two other textbooks that used Netto's and Weber's synthesis, written somewhat later in France, to show how complex and multiform the theory bearing Galois's name could be at that time.

Two Textbooks in German

Both Netto and Weber were researchers in algebra. Netto defended his dissertation in Berlin and went on to teach in the city's gymnasium (high school) and at the University of Strasbourg before finally returning to Berlin University as a full professor in 1882. In the introduction to his book he presented himself as a disciple of the Berliner mathematician Leopold Kronecker (Netto [1892] 1964, iv). Weber belonged to the Königsberg school of mathematical physics and started collaborating with Dedekind around 1880. That was how he knew of the lecture notes for Dedekind's seminar of 1856–1858, which first stimulated his interest in Galois theory (Weber 1895, vii).

In the preface to his book Netto explained that he wanted to present a particular method of algebraic research; his purpose was to give the theory of substitutions a concrete foundation, to make the proofs simpler, and to present the contents more precisely than usual. Given the subject, Netto could hardly avoid using Jordan's treatise, which had become a reference in the field. However, whereas Jordan referred heavily to Galois's works, in particular to the notion of groups of substitutions, Netto avoided using Galois's name to define the results related to Galois theory, and he did not keep the structure of Galois's paper, either. In fact, the whole treatise is written around the concept of "rational domains" developed by Kronecker (1882).[15]

A consequence of Netto's choice to use rational domains instead of groups was that the proofs were written very differently from the work of Galois and Jordan. In fact, Netto preferred using algebraic calculation on equations and making the relations between the roots and the coefficients explicit instead of looking for groups associated to equations (see, e.g., Netto [1892] 1964, 180). Actually, Netto's way of thinking about equations was wholly influenced by Kronecker, who defined algebra as the art of abstract calculation and dismissed Galois's method as uninteresting (Kronecker 1853). From that viewpoint, Netto's book reflected the Berliner cultural environment in which the author had been trained and was now teaching (see Begehr et al. 1998). Nevertheless, and despite an epistemological viewpoint that almost made nonsense of Galois's research, Netto's book still contained traces of something from Galois's paper: Galois's theorems had been cleansed of their proofs and global structure, but they were still present on the pages of the textbook. In fact, Netto had associated a system of practices, objects, and statements with the original *Mémoire sur les conditions de résolubilité des équations par radicaux*, which differed from Galois's point of view but was still coherent enough to be seen as theory.

Weber wanted his algebra course to lay the groundwork for what he defined as "modern algebra," that is, group theory and number theory. This aim fitted perfectly with his own mathematical works and cooperations, as he had just written a paper with Dedekind that was to become a milestone in the development of "modern" mathematical concepts (Dedekind and Weber 1882; Weber 1893). Nevertheless, the first volume of Weber's textbook, which included Galois theory, remained a course about the theory of equations, and its content was not very different from that of Serret's book. In fact, the *Lehrbuch der Algebra* was neither at odds with classical algebra nor a mere copy of it; one could call it a reinterpretation with the concepts and language of the late nineteenth century of an algebraic tradition inherited from the eighteenth century that had the theory of equations at its core. From that perspective, it was the pedagogical nature of the book, rather than Weber's own mathematical preferences, that had guided its structure and contents: it was written and organized to familiarize a reader who would master the classical algebra with the modern one, while rewriting things he already knew in another language.

Notably, what Weber did with Galois theory and with the solution of equations was very different from what one could read in the mathematical literature at that time. To begin with, it was the first textbook to develop the concept of field, which was theorized by Dedekind in 1871 and became one of the foundations of Galois theory (Dedekind 1871). By contrast, the group concept was not developed for itself in this first volume. Second, Weber redrew the boundaries of Galois theory. In fact, the chapter titled "Galois Theory" linked, on a theoretical plane, the concepts of field, groups of substitutions, and adjunction. But in that chapter Weber did not state Galois's final theorem, namely, the condition for an equation to be solvable by radicals.

In fact, Weber dealt with the issue of algebraic solution in a separate chapter, expounding it in a completely new manner. There, he claimed that "the theory of groups provides the sharpest illumination of this issue" and used the language of modern algebra to formulate the process of decomposition of the group by successive adjunctions explained by Galois (Weber 1898, 692–693).[16] However, his explanation referred to Jordan's treatise rather than to Galois's work, and in the end the "theorem of Galois," as Weber called it, was no longer related to the original one: once a result about equations, it had become a result about groups and equations that had specific properties—respectively, being "linear" and "metacyclic"—properties that had not been studied at all by Galois itself (Weber 1898, 714). Underlying these words was a whole conceptual technology that did not even exist in Galois's time. Using the new language of modern algebra, Weber practically erased the original contents of Galois's memoir and defined Galois theory as a conceptual and general theory, while the theory of equations became an application of it. In doing so, he was actually closer to Jordan than to Galois.

Finally, Netto's and Weber's books both rested on the mathematical environment and preferences of their author—Berlin on one side, Göttingen on the other—but cannot be reduced to them, as both made intensive use of Jordan's works. If we look at the Galois theory that each of them sought to transmit to the reader, we can see two different objects that are also different from the theory Galois had tried to build. Netto kept the theorems, whereas Weber retained the conceptual framework that Galois had used to think about equation solving. For the former, Galois theory was just an appendix to the theory of equations; for the latter, it was an independent and abstract theory that tallied with the modernization of algebra he was defending. These books played an important part in the making of Galois theory as a subdiscipline but initiated two different ways of practicing it, one relying on calculation and the other on abstract concepts.

Two French Textbooks

In France, Serret's *Cours d'algèbre supérieure* remained the main reference work on Galois theory until the end of the nineteenth century and was reedited several times. This means that Galois theory remained mostly a practical theory about the solution of equations, characterized by the theorems of Galois but not by his methods or his way of reasoning; the conceptual side of Galois's paper was put aside, and mathematicians could use this "French Galois theory" without thinking about the relation between a group and an equation.

In 1895 two new textbooks were published with the aim of writing an updated version of Galois theory: *Leçons sur la résolution algébrique des équations* by Henri Vogt, and *Introduction à l'étude de la théorie des nombres et de l'algèbre supérieure* by Émile Borel and Jules Drach, which included an algebra section written by Drach. These books had similar foundations. The authors were former students of the École Normale Supérieure. They had probably been encouraged to write their books by the same man, their former professor Jules Tannery, who wrote a preface for each of them. Tannery believed the books would be "occupying a space that was empty and crying out to be filled." According to him, they would update Galois theory, first, because Serret's approach was no longer sufficient, and second, because they would make the research made in Germany available to French students (Tannery, preface to Vogt 1895, viii).[17]

Vogt's and Drach's interest in Galois theory derived from their doctoral research, more precisely from the idea that this theory could be applied to analysis.[18] Moreover, Vogt and Drach used the same material—the books by Serret, Jordan, Weber, and Netto—and the same concepts: groups and rational domains; both reproduced Galois's results. One might expect to find a unified Galois theory in these books, but despite all their common features, Galois theory retained its multifaceted character in these two textbooks.

Vogt's book was presented not as a treatise about Galois theory but as a textbook on the theory of equations. Yet everything in it was related to Galois—the introduction, the table of contents, even the name of the theorems or definitions: Galois theory as presented in this book was not just a part of the classical theory of equations, it *was* a modern theory of equations, since the whole structure was organized around Galois's works, and all the concepts and ideas linked to them were listed and explained. This "Galois theory" encompassed many new theoretical and conceptual tools, such as groups, rational domains, and adjunction, and indeed went much further than Galois's original memoir on these topics. Another characteristic of Vogt's textbook was that it linked knowledge and practices derived from different mathematical readings of Galois without ever directly using Galois's original research. For instance, the results and notations of the two chapters devoted to the group concept were taken from Jordan's research, while the inspiration for the chapters dealing with rational functions came from Netto's textbook. Moreover, when he came to the algebraic solution of equations, Vogt mixed two different sets of practices to write the proofs:

> Assume ... that the equation ... has a group G which includes numerically and algebraically a function $\varphi(x_0, x_1, \ldots, x_{n-1})$, and let $\gamma_1(x_0, x_1, \ldots, x_{n-1})$ be a cyclic function whose adjunction makes the equation reducible to linear factors:
>
> $$f(x) = (x - R_0(\varphi, \gamma_1))(x - R_1(\varphi, \gamma_1)) \ldots$$
>
> Let us make the product $\Pi_\alpha [x - R_0(\varphi, \gamma_\alpha)]$ taking all the distinct algebraic values that γ_1 takes for all the substitutions of the group G. This is a function that belongs to the group of the equation and that can be rationally expressed. (Vogt 1895, 167–168)

Like Kronecker and Netto, Vogt made extensive use of algebraic calculation; the formulas were used as a visual support for the reader and as guidelines for the process of the proof. But at the same time, and in contrast to Kronecker and Netto, Vogt formulated the result with the group concept, which was needed to maintain the idea of the link between substitutions and equations.

The book by Borel and Drach was intended as a written transcription of the lectures Tannery gave at the École Normale Supérieure but was in fact very different, as Tannery explained in the preface. Moreover, according to Tannery, the algebra part, which had been entirely written by Drach, was quite unusual for a French textbook and relied almost exclusively on what Tannery qualified as the "German way" of doing mathematics: "I must admit that at first I was afraid to see [Drach] juggling with symbols that seemed devoid of all meaning to me; I persuaded myself, while thinking about it, that this fear stemmed from my own mental habits and that, in fact, it was not really justified" (Tannery, preface to Borel and Drach 1895, iv–v).

Thus, for Tannery, Drach's "German" style of thinking was, above all, an axiomatic one, in which objects were defined in a general, abstract way. Drach explained that Galois had completely renewed the theory of equations, but also that the real aim of this theory was "the study of the symbols defined by equations" (Borel and Drach 1895, 294). To understand it fully, it was therefore necessary to reorganize the whole field of algebra around Galois theory in an axiomatic way (Borel and Drach 1895, 123). For instance, the definition of a group as a set of objects on which one can do an operation satisfying given rules was quite close to the one used by Dedekind in 1856–1858 and reproduced in Weber's textbook (Borel and Drach 1895, 133). Thus, what Tannery saw as a "German way" was in fact a "Göttingen way" and remained very far from the emphasis on calculation advocated by Kronecker and Netto in Berlin at the same time.

However, Drach's contribution was not closer to the version of Galois theory found in German textbooks than the one found in French books. In fact, his section on algebra was written to emphasize the idea that Drach regarded as the very essence of Galois's memoir: the principle of the decomposition of groups by adjunction established by Galois and developed by Jordan. For Drach, the problem of solving equations was "the study of the symbols that they define," and the point was to "analyze the indeterminacy that remains in the definition of these symbols, indeterminacy which is represented by the group" (Borel and Drach 1895, 294). As a result, the Galois theory that Drach presented was completely separated from the practical issue of equation solving; it was much more abstract and general and also owed much more to Galois—and to Jordan—than the German versions of the theory. For Drach, Galois theory was neither an appendix to the classical theory of equations, as it had been for Serret, for instance, nor a new theory of equations, as it was for Vogt and Weber. It was a completely new field of mathematical knowledge, relying on the abstract group concept.

The comparison of these two textbooks thus shows that even though Vogt and Drach had been trained in the same place, they shared neither a mutual practice on Galois theory nor an understanding of it. In fact, this example puts light on a hybridization phenomenon in the making of Galois theory—both authors, in a way, mixed vocabulary, concepts, and methods coming from different sources. Moreover, this example emphasizes the plasticity of such a process—Vogt and Drach actually did not frame their book in the same way.

Galois Theory as a Collective Statement

At the beginning of the twentieth century, everybody doing mathematics would have agreed there was a mathematical object called "Galois theory," a mathematical system of questions and practices linked in some way to the theory of equations, and associated with the concepts of groups, field, and adjunction. The textbooks I have referred to here, and others, were used not only at great distances from the immediate environment in which they had been written but also beyond the context of teaching.

For instance, Weber's book was translated into French, and Netto's volume into English and Italian. Both were used for teaching the same courses in the United States, and both were considered as models for new textbooks in France. Cambridge students were advised to read both of them. Moreover, both books were very frequently quoted side by side in research papers and books published subsequently. Similarly, the works published by Vogt, Borel, and Drach were reviewed in research journals in several countries, including Germany, the United States, Portugal, and Spain, and many research papers cited them as important reference sources for Galois theory. The books were not only tools for students but also tools of communication for professional mathematicians and contributed to creating a common language about Galois theory in spite of their antagonism.

One of the reasons for this success was certainly that they met a need within the mathematical world. As the reviewer of Borel and Drach's book for the *Bulletin of the American Mathematical Society* stated: "We have now reached the stage at which the research has been carried far enough to admit of their results being collated from the journals and other scattered places of original publication, reduced to a common notation, unified and sometimes simplified, and the whole presented in an orderly systematic form" (Young 1896, 104).

However, while Galois theory was commonly defined and seen as a mathematical field of research, this did not mean that its scope and the associated way of doing mathematics were fixed, or would be fixed, in a particular local context.[19] Its boundaries were to remain vague for some time. Galois theory could be described either as the theory whose aim was "to establish the foundations of the theory of substitutions and its links with the theory of algebraic equations" (Maillet 1892, 1) or "the combination of two theories: the theory of rational functions of several variables, and the theory of substitution groups" (Pierpont 1899–1900, 113). For instance, all the reviewers explained that Vogt had written a good traditional textbook for beginners, but some of them lamented that he had not gone far enough in the conceptualization of Galois theory: they thought a real modern theory of algebraic equations should have relied much more on Galois's original research. By contrast, Drach's contribution appeared to be too difficult, because this account of Galois theory only outlined general principles. Still, some reviewers found this abstract point of view the more interesting aspect of Drach's contribution, because it indicated the road the new algebra had to take. But this bias, which we would call axiomatic today, encountered opposition within the French community and was rejected as useless "philosophical algebra" in the *Journal de mathématiques spéciales* (Longchamps 1894, 258).

Finally, when mathematicians of the beginning of the twentieth century engaged in discussions about Galois theory, the questions they asked, as well as the processes and practices they used to solve the associated problems, were not homogeneous. Mathematicians had still not developed a unified approach to Galois theory, and they certainly did not practice and apply it in similar ways.

Conclusion

Reading and using a mathematical text is not a neutral operation. Cayley's and Dedekind's readings of Galois's memoir show that, like all scientists, mathematicians have habits, ways of asking questions, of solving problems, and of understanding concepts, that come from their training and their scientific environment. However, hailing from different places did not prevent them from sharing ideas and methods. Most important, it did not prevent them from interpreting and modifying—through their own mathematical practices—the knowledge that others had found.

Transforming a corpus of research papers into a subject for teaching is not a neutral operation either. The learning questions are often taken as "context issues" that help to characterize mathematical cultures and explain their local implantation. The case of Galois theory shows that this kind of scientific production is not simply a local issue. Textbooks are made to be circulated and are often widely read outside the community they were written in. Their authors use more than the mathematics produced in their local community or research school. In the case of Galois theory, textbook authors created a common language in the 1890s that would allow mathematicians of the twentieth century to work together and to share at least some knowledge and practices, despite possible differences in the way they were doing mathematics.

Finally, this study offers an important insight into "local cultures," particularly the validity of the national context in analyzing the making of mathematical theories. More specifically, the question of scientific exchanges between France and Germany at the end of the nineteenth century cannot be seen, as some have suggested, simply as global rejection or total adoption; rather, as the present study implies, it was a desire to cooperate and to share mathematical ideas across borders that finally produced a hybridization of knowledge.

Notes

1. For an analysis of the effects of readings on a mathematical text, see Goldstein 1995.

2. This definition of a mathematical culture is taken from Chemla 2009.

3. Here I define a mathematical theory as being not only a formal compilation of definitions and theorems but also a field of activity comprising specific questions, methods, and kinds of proofs and results.

4. Solving an equation algebraically means finding a formula that obtains the solutions (i.e., "the roots") using only simple operations (addition, subtraction, multiplication, division) and radicals (square roots, cube roots, etc.).

5. The degree of an equation is the number of the power of x. For instance, the equation $x^2 - 10x + 3 = 0$ is of degree 2.

6. All translations of Galois's texts in the present chapter are from Neumann 2011.

7. The terminology is Cayley's.

8. Cayley (1860, 534).

9. The distinction between mathematical practice and mathematical content in this example accords well with the distinction between "images of knowledge" and "body of knowledge" made in Corry 1989.

10. Dedekind 1981, 64: "A set G of g substitutions is a group of order g if every arbitrary product of substitutions contained in G is still contained in G."

11. See esp. Dedekind's letter quoted at Dugac 1976, 20.

12. The book was based on the lectures given by Serret at the Paris Faculté des Sciences, where he was a professor of algebra in 1848, a position he retained for several years.

13. For details, see Ehrhardt 2012, 143–183.

14. Space is insufficient here to mention all the readings that were made of Galois's work; see Ehrhardt 2012.

15. For an illustration of Netto's use of this concept, see, e.g., Netto [1892] 1964, 266.

16. For instance, the idea of adjunction was changed into that of extension over a field, and a group that could split into smaller ones was "intransitive."

17. For a qualified study on the issue of internationalization, see Décaillot 2010.

18. Drach used Galois's ideas in the field of differential equations, whereas Vogt studied the groups of substitutions associated with a class of differential equations.

19. So this situation is quite similar to the one of "collective statement" defined by medieval historian Alain Boureau (1989).

References

Babbage, C. 1815. *An Essay towards the Calculus of Functions*. London: Bulmer.

Begehr, H. G. W., H. Koch, J. Kramer, N. Schappacher, and E.-J. Thiele, eds. 1998. *Mathematics in Berlin*. Berlin: Birkhäuser.

Belhoste, B. 1995. *Les sciences dans l'enseignement scientifique français: Textes officiels, t. I: 1789–1914*. Paris: INRP/Economica.

Belhoste, B. 2001. "The École Polytechnique and Mathematics in Nineteenth-Century France." In *Changing Images of Mathematics: From the French Revolution to the New Millennium*, ed. U. Bottazzini and A. D. Dalmedico, 15–30. London: Routledge.

Belhoste, B. 2003. *La formation d'une technocratie: L'École Polytechnique et ses élèves de la Révolution au Second Empire*. Paris: Belin.

Bertrand, J. 1845. "Sur le nombre de valeurs que peut prendre une fonction quand on y permute les lettres qu'elle renferme." *Journal de l'École Polytechnique* 18: 123–140.

Bertrand, J. 1899. "La vie d'Évariste Galois par P. Dupuy." *Journal des savants*, 389–400.

Bézout, E. 1812. *Cours de mathématiques à l'usage de la Marine et de l'Artillerie, 1e partie: Algèbre et application de l'algèbre à la géométrie*. Paris: Courcier.

Borel, É., and J. Drach. 1895. *Introduction à l'étude de la théorie des nombres et de l'algèbre supérieure*. Paris: Nony.

Boureau, A. 1989. "Propositions pour une histoire restreinte des mentalités." *Annales: Economies, sociétés, civilisations* 44(6): 1491–1504.

Cayley, A. 1854. "On the Theory of Groups, as Depending on the Symbolic Equation $\theta^n=1$—First Part." *Philosophical Magazine*, 4th ser., 7: 40–47.

Cayley, A. 1860. "Recent Terminology in Mathematics." In *English Cyclopaedia*, 5:534–542. London: Knight.

Chemla, K. 2009. "Mathématiques et culture: Une approche appuyée sur les sources chinoises les plus anciennes connues." In *Les lieux et les temps*, vol. 1 of *La mathématique*, ed. C. Bartocci and P. Odifreddi, 103–152. Paris: CNRS Editions.

Corry, L. 1989. "Linearity and Reflexivity in the Growth of Mathematical Knowledge." *Science in Context* 3(2): 409–440.

Crilly, T. 2006. *Arthur Cayley: Mathematician Laureate of the Victorian Age*. Baltimore, MD: Johns Hopkins University Press.

Dahan, A. 1980. "Les travaux de Cauchy sur les substitutions: Etude de son approche du concept de groupe." *Archive for History of Exact Sciences* 23: 279–319.

Dahan-Dalmedico, A., and J. Peiffer. 1986. *Une histoire des mathématiques: Routes et dédales*. Paris: Seuil.

Décaillot, A.-M. 2010. "Zurich 1897: Premier congrès international de mathématiciens." *Revue germanique internationale* 12: 123–137.

Dedekind, R. 1855. "Bemerkungen zu einer Aufgabe der Wahrscheinlichkeitsrechnung." *Journal fur die reine und angewandte Mathematik* 50: 268–271.

Dedekind, R. 1871. "Über die Komposition der binären quadratischen Formen: Supplement X." In *Vorlesungen über Zahlentheorie*, 2nd ed., by P. G. Lejeune Dirichlet, 423–462. Braunschweig: Vieweg.

Dedekind, R. 1981. "Eine Vorlesung über Algebra." In *Richard Dedekind 1831–1981: Eine Würdigung zu seinem Geburtstag*, ed. Wingfried Scharlau, 59–108. Brauschweig: Vieweg.

Dedekind, R., and H. Weber. 1882. "Theorie der algebraischen Functionen einer Veränderlichen." *Journal für die reine und angewandte mathematik* 92: 181–299.

de Morgan, A. 1847. *Formal Logic or the Calculus of Inference*. London: Taylor and Walton.

Dugac, P. 1976. *Richard Dedekind et les fondements des mathématiques*. Paris: Vrin.

Durand-Richard, M.-J. 1996. "L'École algébrique anglaise: Les conditions conceptuelles et institutionnelles d'un calcul symbolique comme fondement de la connaissance." In *L'Europe mathématique: Histoires, mythes, identités*, ed. C. Goldstein, J. Gray, and J. Ritter, 445–477. Paris: Editions de la MSH.

Ehrhardt, C. 2010a. "La naissance posthume d'Évariste Galois." *Revue de synthèse* 131(4): 543–568.

Ehrhardt, C. 2010b. "A Social History of the 'Galois's Affair' at the French Academy of Sciences (1831)." *Science in Context* 23(1): 91–119.

Ehrhardt, C. 2011. *Évariste Galois: La fabrication d'une icône mathématique*. Paris: Editions de l'École des Hautes Études en Sciences Sociales.

Ehrhardt, C. 2012. *Itinéraire d'un texte mathématique: Les réélaborations d'un mémoire d'Évariste Galois au XIXe siècle*. Paris: Hermann.

Ferreiros, J. 1999. *Labyrinth of Thought: A History of Set Theory and Its Role in Mathematics*. Basel: Birkhaüser.

Galois, E. 1846. "Œuvres mathématiques d'Évariste Galois." *Journal de mathématiques pures et apliquées* 11: 381–444.

Galois, E. 1997. *Ecrits et mémoires mathématiques: Edition critique intégrale des manuscrits et publications d'Évariste Galois par Robert Bourgne et Jean-Pierre Azra ; preface de Jean Dieudonné*. Paris: Gabay.

Goldstein, C. 1995. *Un théorème de Fermat et ses lecteurs*. Saint-Denis: Presses universitaires de Vincennes.

Hamilton, W. 1848. "Research Respecting Quaternions: First Series." *Transactions of the Royal Irish Academy* 21: 199–296.

Jordan, C. 1870. *Traité des substitutions et des equations algébriques*. Paris: Gauthier-Villars.

Koppelman, E. 1971. "The Calculus of Operations and the Rise of Abstract Algebra." *Archive for History of Exact Sciences* 8(3): 155–242.

Kronecker, L. 1853. "Über die algebraisch auflösbaren Gleichungen (I)." *Monatsberichte der Königlichen Preussischen Akademie der Wissenschaften zu Berlin* 4: 365–374.

Kronecker, L. 1882. "Grundzüge einer arithmetischen Theorie der algebraischen Grössen." *Journal fur die reine und angewandte mathematik* 92: 1–123.

Lacroix, S.-F. 1799. *Éléments d'algèbre à l'usage de l'École centrale des Quatre-Nations*. Paris: Duprat.

Lacroix, S.-F. 1800. *Compléments des éléments d'algèbre à l'usage de l'École centrale des Quatre-Nations*. Paris: Duprat.

Lagrange, J.-L. 1797. *Traité de la résolution des équations numériques de tous les degrés*. Paris: Duprat.

Lejeune-Dirichlet, G. 1897. "Gedächtnissrede auf Carl Gustav Jacobi: Abhandlungen der Königlich Preussischen Akademie der Wissenschaften von 1852." In *G. Lejeune Dirichlet's Werke: Herausgegeben auf Veranlassung der Königlich Preussischen Akademie der Wissenschaften von L. Kronecker Fortgesetzt von L. Fuchs* 2:225–252. Berlin: Druck und Verlag von Georg Reimer.

Liouville, J. 1843. "Réponse de M. Liouville." *Comptes rendus hebdomadaires des séances de l'Académie des Sciences* 17: 445–449.

Longchamps, G. 1894. "Bibliographie." *Journal de mathématiques spéciales*, 4th ser., 3: 257–259.

Mathieu, É. 1860. "Mémoire sur le nombre de valeurs que peut acquérir une fonction quand on y permute ses variables de toutes les manières possibles." *Journal de mathématiques pures et appliquées*, 2nd ser., 5: 9–42.

Maillet, E. 1892. *Recherches sur les substitutions et en particulier sur les groupes transitifs*. Paris: Gauthier-Villars.

Netto, E. 1882. *Substitutionentheorie und ihre Anwendung auf die Algebra*. Leipzig: Teubner.

Netto, E. [1892] 1964. *The Theory of Substitutions and Its Applications to Algebra*. Translated by F. N. Cole. New York: Chelsea.

Neumann, P. M. 2011. *The Mathematical Writings of Évariste Galois*. Zürich: European Mathematical Society.

Novy, L. 1968. "L'École algébrique anglaise." *Revue de synthèse* 84: 211–222.

Peacock, G. 1830. *Treatise on Algebra*. Cambridge: Deighton.

Pierpont, J. 1899–1900. "Galois' Theory of Algebraic Equations. Part I. Rational Resolvents." *Annals of Mathematics*, 2nd ser., 1 (1–4): 113–143.

Riemann, B. [1851] 1898. "Principes généraux pour une théorie générale des fonctions d'une grandeur variable complexe." In *Œuvres mathématiques traduites par L. Laugel*, 1–56. Paris: Gauthier-Villars.

Rudio, F. 1895. "Eine Autobiographie von Gotthold Eisenstein: Mit Ergänzenden biographischen Notizen." *Abhandlungen zur Geschichte der Mathematik* 7: 143–203.

Schmitz, M. 2004. "The Life of Gotthold Eisenstein." *Research Letters in the Information and Mathematical Science* 6: 1–13.

Schubring, G. 1996. "Changing Historical and Epistemological Views on Mathematics in Different Contexts in Nineteenth-Century Europe." In *L'Europe mathématique: Histoires, mythes, identities*, ed. C. Goldstein, J. Gray, and J. Ritter, 363–388. Paris: Editions de la MSH.

Serret, J.-A. 1849. *Cours d'algèbre supérieure*. Paris: Mallet-Bachelier.

Serret, J.-A. 1850. "Sur le nombre de valeurs que peut prendre une fonction quand on y permute les lettres qu'elle renferme." *Journal de mathématiques pures et appliquées* 15: 1–44.

Serret, J.-A. 1854. *Cours d'algèbre supérieure*. 2nd ed. Paris: Mallet-Bachelier.

Sinaceur, M. 1990. "Dedekind et le programme de Riemann." *Revue d'histoire des sciences* 43(2–3): 221–296.

Vogt, H. 1895. *Leçons sur la résolution algébrique des équations*. Paris: Nony.

Warwick, A. 2003. *Masters of Theory: Cambridge and the Rise of Mathematical Physics*. Chicago: University of Chicago Press.

Weber, H. 1893. "Die allgemeinen Grundlagen der Galois'schen Gleichungstheorie." *Mathematische Annalen* 43: 521–549.

Weber, H. 1895. *Lehrbuch der Algebra*. Braunschweig: Vieweg.

Weber, H. 1898. *Traité d'algèbre supérieure*. Translated by J. Griess. Paris: Gauthier-Villars.

Young, J. W. A. 1896. "Theory of Numbers and Equations." *Bulletin of the American Mathematical Society* 3(3): 97–105.

KARINE CHEMLA

14

Changing Mathematical Cultures, Conceptual History, and the Circulation of Knowledge
A Case Study Based on Mathematical Sources from Ancient China

Why should we attend to the various scientific cultures to which the different sources that provide evidence of scientific activity attest? The ways in which given social groups practice, and have practiced, science do not fall out of the sky—collectives of practitioners shaped them and shared them. Collectively they made them change, especially in relation to questions they were considering. In this sense, these "ways of practicing science" are an outcome of scientific activity, along with concepts, results, and theories. In the first approximation, it is to these ways of practicing science that I refer when using the term "cultures." One of the obvious reasons they should matter to historians is that they are a result of scientific activity. This is not, however, my main topic here.

In this chapter, I concentrate on another important reason that we should take scientific—or, more generally, scholarly—cultures into consideration. The thesis I propose is that the description of such cultures is an essential endeavor inasmuch as it provides historians with tools to interpret, in a more rigorous way, writings produced in the framework of these cultures.

Cultures, in this sense, grasp a kind of context for scientific activity that gets quite close to actors' practice. What do I mean by this? Understanding various aspects of scientific activity requires that we take into account distinct types of context, on different scales. My focus here is on a micro-level. I aim to concentrate on the detailed actions of actors, thus examining these collectives from a specific angle. In the ancient time period I consider, shared ways of doing science to which our sources attest are often the only means available to perceive collectives within which the documents were produced.

The difficulties attached to interpretation become striking when one deals with Chinese, Sanskrit, or Mesopotamian sources produced millennia ago. This does not mean these problems arise only for such documents. However, dealing with these sources magnifies some of the theoretical challenges posed by interpretation, thus requiring that these issues be addressed. Indeed, I am convinced that the lack of attention paid to these issues has contributed to the relative disparagement of these sources in the history of science. I argue here that a focus on mathematical cultures, through the close context it allows us to capture, helps us solve problems of interpretation, giving us clues for grasping concepts and perceiving results in our sources. What I am suggesting here is that for the history of science the description of cultures provides a tool for the practice of conceptual history.

I illustrate these claims on the basis of a single example. I concentrate on a corpus of Chinese sources, ranging from the first to the thirteenth century, that bears witness to a given tradition. Yet I argue that these sources attest to different mathematical cultures, despite significant overlap among them. I show that the description of the mathematical culture to which the earliest documents adhere allows us to identify in them a concept of quadratic equation, in a sense I shall make clear. I further argue that one can likewise perceive in that tradition, at different time periods, different yet related concepts for what we call quadratic equations.[1] We shall see that the description of mathematical cultures allows us to interpret the sources, and also to attend to the concepts to which Chinese sources attest, as well as to the ways in which practitioners worked with them.

The case study allows me to argue for several other theses regarding scholarly cultures. First, the fact that several mathematical cultures can be identified in the tradition examined between the first and the thirteenth centuries show that scholarly cultures change over time, displaying continuities with previous practices as well as breaks. The breaks partly reflect general changes in the larger environment in which mathematical practice is carried out. They partly echo actors' collective transformation of their way of working. Talking in terms of cultures thus does not imply that we deal with unchanging, static entities. I argue that historians need to take account of these changes in their practices of interpretation.

Second, this reflection on interpretation highlights another issue: my approach to this part of the history of algebraic equations shows that the concepts of equation identified at different time periods all have specificities that can be correlated with features of the mathematical culture in relation to which they were shaped. Note that this explains why the description of cultures provides tools for interpretation. There is no determinism here. We shall see that practices change, that the treatment and understanding of equations undergo transformation, and that the correlation between the two is not a deterministic one. Instead, I suggest that the correlation indicates that cultures also change partly in relation to the conceptual work done on equations as much as the concepts change in relation to how actors worked. Cultures thus prove useful not only as a tool of interpretation but also for the light they shed on the production and transformation of concepts.

The latter remark relates to an important third issue. As this case study will show, even though a correlation between concepts and cultures can be demonstrated, concepts are also not static. The range of equations covered by the successive concepts increases. The understandings of equations they reflect, as we perceive it from a present-day viewpoint, deepen. Algorithms for solving equations gain in generality. The cultural considerations I find useful to introduce thus do not lead to a static vision: in parallel with the changing cultural contexts, the concept of equation undergoes transformation. Nor do they lead to a view of cultures as bounded wholes. This issue relates to my last thesis.

It is true that the transformations in the concept of equation that can be perceived in my corpus occur within a specific tradition. All these concepts are members of a clearly defined subfamily in a larger family of equations evidenced by a corpus of sources originating from many places in the world. But nowhere else in the world do we find the way of conceiving of and handling equations to which these Chinese sources attest. The characteristic features of this subfamily of concepts can be correlated with continuities displayed in the ways of carrying out mathematical activity in the tradition of ancient China considered. Yet, this does not mean that no concept from this subfamily could be appropriated elsewhere. My conclusion mentions what appears to be a sudden occurrence of this approach to equations within Arabic sources of the twelfth century.

In brief, scholarly cultures as I see them capture specificities of ways of working in a given collective. They are by no means isolated, bounded, and unchanging cells. Moreover, concepts are not buried in cultures. Even though they display adherence to scholarly cultures in which they took shape, they circulate and can be appropriated in other cultural contexts. Through such a process, universality is constructed.

These are the theses this chapter illustrates. The argument is developed in three sections. First, I outline what I mean by "scholarly cultures" and sketch features attesting to the radical change I detect in practices of mathematics in China between the first and the thirteenth century. The second, longest section focuses on the first period, spanning the time from the first to the seventh century, and argues that we can identify a concept of quadratic equation in the earliest sources considered. I show how taking cultures into account helps us to interpret the sources and highlight the correlation between a given mathematical culture and a concept. The third section deals with a subsequent time span, ranging from the eleventh to the thirteenth century and in fact beyond. It outlines a subsequent culture and a subsequent concept of equation. Both display continuities and differences with those described in the second section. In conclusion, I return to the theses expounded at the beginning of the chapter.

Mathematical Cultures and Our Sources

One of the difficulties attached to the exercise of interpretation has already been the object of much discussion.[2] It relates to the practice that involves using modern scientific concepts to read ancient documents. I think this is a necessary step, but only a first step if we are to avoid conceptual anachronism.[3] This chapter examines another form of anachronism, which consists of approaching our sources from the viewpoint of modern textual categories and reading mathematical problems, figures, algorithms, inscriptions for computing, and so on, as their modern counterparts. My point is that the practices through which actors engaged with problems, algorithms, and more generally what was written down or inscribed in the performance of a given mathematical activity—practices that I shall designate as "elementary practices" or "elements of practice"—determine at least partly the meaning of these textual elements. Such elements of practice cannot be taken for granted a priori and need to be described. Their set constitutes essential components of the mathematical cultures I am trying to comprehend. It is at this juncture that the issue of the cultures within which our actors operated connects with that of interpretation.

Concretely, how are we to proceed, since the sources we want to interpret are, in fact, our main vehicles for carrying out this task? To start with, our sources contain many hints indicating how actors dealt with the various kinds of textual elements composing the sources—such as problems and algorithms. We can collect these hints and rely on them to describe which practices that actors shaped and shared with the textual elements leave a substantial trace in our sources. For the Chinese sources considered here, this procedure shows the elementary practices in question are specific to a given context. Moreover, in that context, these elementary practices had specific connections with one another. This nexus of practices reflects a specific organization of mathematical activity and provides a first sketch of the mathematical cultures I aim to describe.

Our sources were more generally produced within a material environment where other objects could be used to carry out mathematical activity. The material objects of the mathematical culture of the first time period under study in ancient China included an instrument for computing and blocks to practice spatial geometry. Our sources also provide evidence about some of their material features and the ways in which these features were employed. The reason is simple: because our sources were produced in the context of an activity that involved elements of text and material objects, they bear marks that derive from this contiguity and reflect the practice in general. Like above, the practices with these material objects are also components of the cultures, as I view them. The same applies to the link between these practices and the other elements of practice. Describing these practices is part of our task. In the study presented here, it proves all the more necessary to deal with these material objects that we encounter a case where inscriptions that are essential to the story and once lay outside the writings became integrated into the texts at a later date. To be able to consider a long-term conceptual history, we thus need to take into account a corpus of texts and inscriptions that present different material features at different time periods.

Let me first illustrate the abstract description just outlined, using concrete examples. The earliest extant Chinese documents attesting to a concept of quadratic equation are books dating from about the beginning of the Common Era and handed down through the written tradition. By contrast, as far as I know, none of the mathematical documents yielded by archeological excavations deal with this topic.

The earliest extant book in which we can identify a concept of quadratic equation is *The Nine Chapters on Mathematical Procedures*, whose composition I date from the first century CE. Like the other early mathematical book handed down, *The Gnomon of the Zhou*, which dates from roughly the same time period and is connected with the practice of astronomy, *The Nine Chapters* was apparently perceived as canonical shortly after its completion. Accordingly, commentaries were composed on both books. These commentaries are essential documents for describing ancient practices in mathematics. The earliest extant commentary on *The Nine Chapters* was completed by Liu Hui in 263, and in the same century Zhao Shuang authored the earliest known commentary composed on *The Gnomon of the Zhou*. Both commentaries were handed down with their respective canons.[4]

The practice of mathematics to which *The Nine Chapters* attests employs three key elements. The book is composed of problems and algorithms. The texts of the algorithms further refer to a material object, outside the text, that is, an instrument with which practitioners computed. This instrument was composed of a surface, the material features of which we can only speculate about, and also of counting rods placed on that surface and used to represent numbers according to a decimal place-value number system.[5] The canon and the commentaries provide hints on the practices with these three elements. In the absence of more substantial descriptions of these practices by the actors themselves, we can rely on these hints to show that the practices in question are quite specific and certainly different from our own practices with similar elements.

The commentators' practice of mathematics attests to a richer set of elements.[6] Commentaries systematically include proofs of the correctness of the algorithms contained in the canon, as well as second-order discussions about various facets of mathematics. Commentaries also contain references to tools of visualization, which were absent from the canons: diagrams for plane geometry and blocks for space geometry. I have argued that at the time, diagrams were, like blocks and counting rods, material objects outside the text (Chemla 2001). The commentaries give information on material features of these objects and practices with them. The same conclusion as above holds true: all these practices differ from our expectations. In brief, these sources testify to a practice of mathematics for which writings contain only discourse (problem, algorithms, proofs, discussions, etc.), whereas all the other elements (counting rods, diagrams, and blocks) are material objects used in conjunction with texts. This sketch explains why the usual appearance

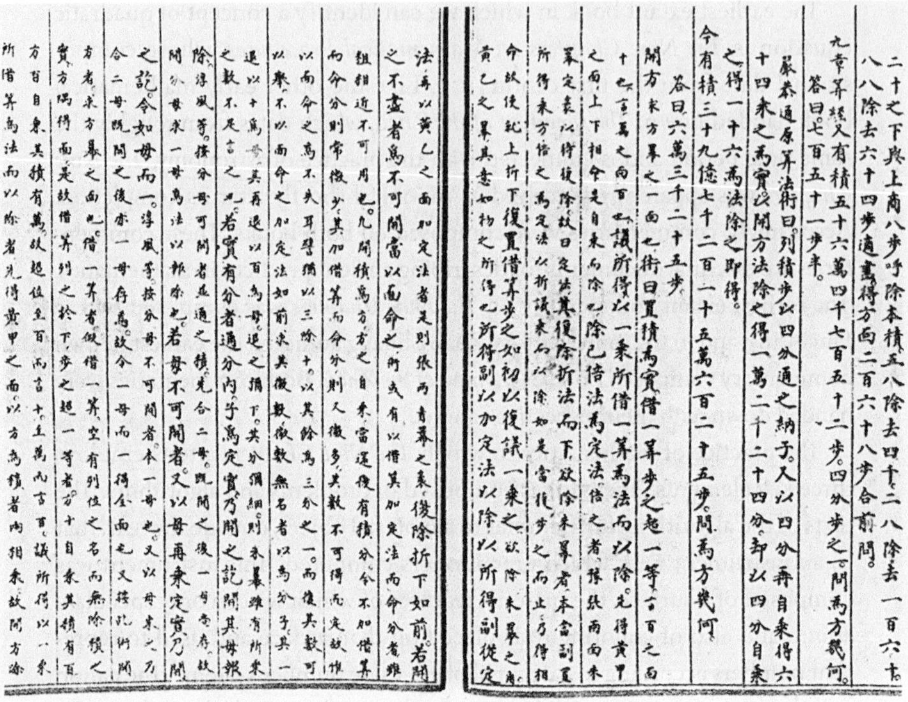

FIGURE 14.1. A page from *The Nine Chapters* with its earliest extant commentaries, as reproduced in the fifteenth-century encyclopedia *Grand Classic of the Yongle Period* (*Yongle dadian*), 1408, chapter 16344, 9b–10a.

of a page in our sources for this first time period looks like what is shown in figure 14.1. The page contains nothing but characters.

By contrast, the aspect of the sources handed down from, roughly speaking, the tenth century onward attests to a radical change in the practice of mathematics.[7] I limit myself to sources that matter for the discussion here, even though the general features described hold more broadly (Chemla 2001). These sources include remaining chapters of a thirteenth-century subcommentary on *The Nine Chapters* and Liu Hui's commentary, which Yang Hui completed in 1261 (Lam 1969; Yan 1966). The corpus also includes a book written in 1275 also by Yang Hui, *Quick Methods for Multiplication and Division for the Surfaces of the Fields and Analogous Problems* 田畝比類乘除捷法. In particular, I am interested in the last part of its final chapter, in which Yang quotes at length a book that probably dates from the eleventh century, *Discussing the Source of the Ancient (Methods)* 議古根源 by Liu Yi 劉益.[8]

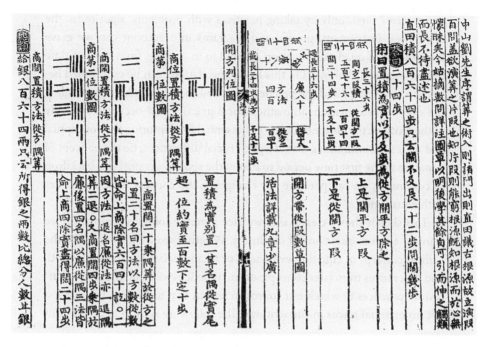

FIGURE 14.2. A page from the quotation of Liu Yi's *Discussing the Source*, in Yang Hui's *Quick Methods for Multiplication and Division for the Surfaces of the Fields and Analogous Problems*, Korean edition from 1433, reprinted in Kodama 1966 (91).

Figure 14.2 shows a page of *Discussing the Source* as quoted by Yang, which vividly illustrates the radical change evoked. The page combines discourse with geometrical figures, on the right-hand side, and illustrations of configurations of numbers represented with counting rods, on the left-hand side. The discourse still contains problems, algorithms, and proofs of their correctness. It still refers to a surface outside the text on which computations were carried out. The key point is that in the eleventh century at the latest, pages of books included new nondiscursive elements. Their inclusion within books goes together with a shift in the practices with them.[9] More generally, the mathematical culture in relation to which these writings were composed presents continuities with and differences from the earlier one mentioned above. Both the continuities and the differences are important, as I show in discussing quadratic equations.

What are the consequences of these remarks for understanding equations and the facets of their history that these Chinese sources allow us to perceive? First, only by taking practices with problems, algorithms, the computing instrument, proofs, and diagrams into account can we grasp the concept of quadratic equation attested to in writings from the first time period mentioned above. On this basis, correlations can be established between this concept and the practices attached to it. Second, understanding the deep transformations in mathematical practice that occurred probably in the tenth or the eleventh century is essential for capturing the conceptual and material continuities that, despite crucial differences, tie the concept of equation in the first time period to that of the second. Against this backdrop of continuity, we can perceive key changes in the concept of and practices with equations.

This sketch shows clearly that we face a methodological problem if we want to address questions of diachrony. Whether we want to describe the earliest concept of equation evidenced or to appreciate its similarities with and differences from later concepts of and practices with equations, we must restore practices to which our sources refer but that in the first time period left no material traces in the writings. The following section addresses this issue.

A Culture and a Concept of Quadratic Equation in the First Centuries CE

Chapter 9 of *The Nine Chapters* is devoted to the right-angled triangle. Problem 19 in that chapter is the only problem in the book to be solved by a quadratic equation. As usual in *The Nine Chapters*, this problem describes a specific setting, which I represented in figure 14.3. Note that this diagram, which I drew to help the reader follow the argument, corresponds to nothing in the sources. Nevertheless, I followed Chinese conventions that place north downward. The problem introduces a situation and particular numerical data, requiring the determination of an unknown quantity. Hints gleaned in Liu Hui's commentary show that such a problem with the procedure solving it was read as a general statement, although its formulation was not abstract (Chemla 2003). This gives a first example of the connection between the description of practices and the interpretation of a text.

Here is an outline of the problem in question. The length x of the sides of a square town, whose walls face north-south and east-west, is unknown. Someone leaves the town through its southern gate and walks a distance s. (Note that s is my notation for what in the text is a numerical value, expressed with respect to the measuring unit for length bu [step]. The same convention holds

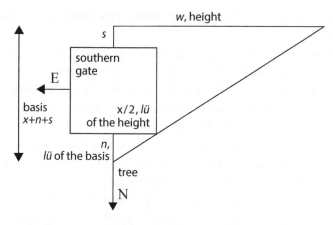

FIGURE 14.3. An illustration of problem 19 of *The Nine Chapters*

below.) At the distance *s* southward, the walker turns westward and, after walking a distance *w*, sights a tree, which is northward at the distance *n* from the northern gate. The problem asks for the length of the sides of the town.

The Nine Chapters contains a procedure for solving this problem. Its text constitutes the earliest extant piece of evidence documenting quadratic equations, and the only one dating from the first century at the latest.[10] The challenge its interpretation poses is clear, when one examines the text, which I translate as follows:

> Procedure:
> One multiplies, by the quantity of *bu* outside the northern gate, the quantity of *bu* walked westward, and one doubles this, which makes the **dividend**.
> One adds up the quantities of *bu* outside the southern gate and northern gate, which makes the **joined divisor**.
> One **divides [the dividend] by this by extraction of the square root**, which gives the side of the square town.[11]

Let us observe the main features of this text. It refers to a "procedure." The text of the procedure brings into play names of operations: multiplying, doubling, adding up, and so on. It consists of a list of operations whose execution, the text asserts, yields the unknown in question. The text thereby attests to the fact that some operations have been identified as key operations (multiplication, duplication, addition, and so on). They are the building blocks of any algorithm. Identifying key operations is one of the main theoretical goals of mathematical work in the scholarly culture observed (Chemla 2010a).

The text presupposes that the reader knows algorithms for executing these operations.

An operation bears on operands (e.g., a multiplication operates on two values). During that time period in China, specific terms were introduced only for the operands of division. I translate these terms as "dividend" and "divisor." Division has another specificity. As I show below, through various kinds of practices other key operations are shown to be analogous to it. The terminology for operations mirrors this fact, as shown by the text of the procedure quoted above. The prescription of a square root extraction reads "one *divides* ... by extraction of the square root," indicating a relationship between root extraction and division. The same feature holds true for the terms designating the operands. Similar operations have similar operands. Whereas a division operates on two operands, a dividend and a divisor, the root extraction operates on only one: the number whose root is sought.[12] That operand is called a dividend. A network of relationships is thereby established between key operations, and the terminology reflects this.

These elements of information allow me to highlight an essential feature of the text quoted above. I have marked its structure using bold characters. It prescribes the computation of two values that are taken, respectively, as "dividend" ($2nw$) and "joined divisor" ($s+n$); following this, it prescribes: "divide ... by extraction of the square root." This structure shows that a higher-level operation, related to root extraction but having two operands, concludes the procedure. It is this higher-level operation that I suggest interpreting as a kind of quadratic equation—I call it the "operation-equation."

Indeed, no modern commentator denies that the last operation of this procedure is equivalent to the quadratic equation that we would write today as[13]

$$2nw = (s+n)x + x^2.$$

Such a retrospective reading can, however, represent only the first step in the practice of conceptual history. In this chapter, I am actually interested in the problems raised by the interpretation of the text: In which sense does this operation correspond to a quadratic equation (i.e., which concept of equation do we have)? How can one argue for this interpretation? How did the ancient reader achieve this understanding? Since nothing else is added to this procedure, how did the ancient reader know how to execute the operation (i.e., to solve the "equation")? How was the operation (i.e., the equation) established? I claim that further description of the mathematical culture, in the context of which this text was written, offers clues to argue for an answer to each of these questions, and describe how practitioners worked with equations conceived in this way. The arguments outlined below aim to illustrate more generally the relationship between the issue of scholarly cultures and that of interpretation.

The Establishment of a "Quadratic Equation" and Elements of Practice with Diagrams

The procedure quoted above first prescribes the computation of two operands—$2nw$ is the "dividend" and $(s + n)$ is the "joined divisor"[14]—and then prescribes a final operation as a "square root extraction." How should we interpret the latter prescription? To begin with, let me outline, in modern terms, how the third-century commentator comments on this procedure.

Liu Hui introduces two similar right-angled triangles (see figure 14.4). He says that one of these triangles has the path described westward, w, as its "height" and the distance between the tree and the southernmost point reached by the walker as its "base." The base amounts to $n + x + s$. The second triangle has the distance northward n as its base, while half the side of the town constitutes its height $(x/2)$. Note that all quantities are related to lines identified with reference to the actual geometrical situation. This holds true for the whole commentary. By means of a rule of three, the similarity between the triangles leads to the equality $(n + x + s) \, x/2 = nw$. What is essential is that Liu Hui refers to this equality as holding between areas again by reference to the situation on the field. The value nw measures the area of the horizontal rectangle, in the lower part of figure 14.4. It is equal to $(n + x + s) \cdot x/2$, which, Liu Hui says, "occupies the half to the west." If one relies on figure 14.4, which sketches a cartographic view of the situation, it is clear that his statement refers here to the vertical white rectangle that covers the western half of the city and extends beyond to the north and the south. Liu Hui goes on: "If, further, one doubles this, one adds the eastern (part) to it, which exhausts it (the area of the rectangle) entirely." The operation of doubling yields numerically $2nw$, and geometrically it adds to the white vertical rectangle the rectangle I represent in gray on figure 14.4. Note that the commentator has thereby introduced a (vertical) rectangle, the area of which is precisely the dividend ($2nw$) yielded by the first two operations in the procedure of The Nine Chapters. In fact, the term *shi*, which I translate as "dividend," also means "area." I argue below that both meanings are active here.

Graphically, on the basis of the data (n, s, w), one has determined the area of a rectangle composed of a square, whose side is the unknown, and

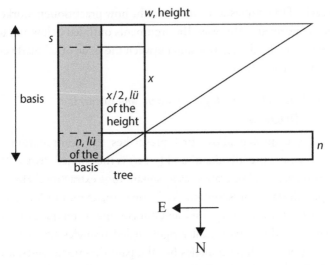

FIGURE 14.4. The geometric reasoning developed by Liu Hui to account for the correctness of the procedure solving problem 19. The text refers only to areas, not to a diagram.

two rectangles north and south of it (figure 14.5a). This is what Liu Hui recapitulates in the subsequent sentences before alluding to a reshaping of this rectangle. His commentary reads (I invite the reader to rely on figure 14.5 to understand the commentator's reasoning and to wait patiently for the elements of the passage that are still obscure):

> The area of this procedure is the area that, from east to west, is like the side of the square town and, from north to south, goes from the tree up to the end of the 14 *bu*, to the south of the town [i.e., from the northernmost to the southernmost point; see figure 14.5a]. Each of the (amounts of) *bu* north and south makes a width, and the side of the town makes the length, this is why one places the two widths side by side to make the joined divisor [see figure 14.5b]. The sum (of their areas) is taken as the area outside the corner.

This passage is extremely rich in information. A few remarks on how the text describes graphical operations will be helpful. In the tradition of writing about mathematics to which the text belongs, the designation of a length (north-south direction) and a width (east-west direction) is the usual way to point out a rectangle. Here, Liu Hui draws attention to two rectangles, one in the north and the other in the south. His last statement explains how they

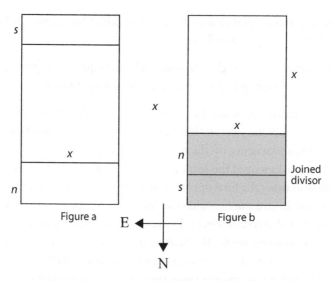

FIGURE 14.5. Liu Hui's establishment of the "operation-equation" solving problem 19 of *The Nine Chapters*: (a) the area corresponding to the dividend of the procedure. (b) the side corresponding to the joined divisor of the procedure.

are brought together graphically and how the global rectangle, whose area is known, is reshaped. The result of this operation is illustrated by figure 14.5b. Such handling of diagrams as material objects is typical of the way in which this type of visualization device was used in the mathematical culture evidenced by Liu Hui's commentary (Chemla 2010b). The absence of anything visual in the text fits with the main trends regarding the composition of writings in the context of this mathematical culture, outlined in the preceding section. These remarks constitute the first features of the practice with figures we encounter.

The statement referring to the key graphical operation is crucial. Liu Hui writes: "*this is why* one *places* the two widths *side by side to make the joined divisor*" (my emphasis). The positioning of the two small rectangles together forms a rectangle of width $n+s$. This addition is precisely the third operation prescribed by the procedure in *The Nine Chapters*, which determines the value of the joined divisor. Liu Hui's use of the particle *gu* ("this is why") at this juncture indicates he understands the graphical transformation as accounting for the meaning of the computation of the joined divisor. At this point, the commentator has established that a rectangle whose area is $2nw$ combines a square of side x and a rectangle with sides, respectively, equal to x and $n + s$. In modern terms, we could write what he has shown graphically as

the area of the rectangle (the dividend $2nw$) equals $x^2 + (n + s).x$, where he takes $(n + s)$ to be the joined divisor.

These statements conclude his commentary. The description of Liu Hui's practice of commentary shows that, in his view, he has accounted for the correctness of the procedure of *The Nine Chapters*. This interpretation implies that the establishment of the rectangle and of its structure accounts for the correctness of using the final operation of the procedure. In other terms, the rectangle and its structure provide the "meaning" of the operation in which we are interested. Accordingly, for Liu Hui, inasmuch as the rectangle writes the equality sketched above, the final operation of the procedure does correspond, as all historians interpreted it, to a quadratic equation.

The conclusion just obtained uses the commentary to establish what Liu Hui understands is the meaning of the operation-equation. We can also look at the commentary from another angle and ask what it tells us about the concept of equation as Liu Hui understands and practices it. In this commentary, Liu Hui establishes the equation in the sense that he establishes the correctness of using the operation-equation, as the procedure shapes it, to solve the problem. He does so by establishing a rectangle that brings the area $2nw$, the joined divisor $(n + s)$, and the unknown into relation. The equation is thus yielded graphically, its figure being that of a rectangle of known area a that can be broken down into a square x^2 and a rectangle bx. The term a corresponds to what *The Nine Chapters* designates as dividend, and b is the joined divisor. These facts are actually not contingent. Every time we meet an operation-equation of this type in an algorithm stated by Liu Hui or Zhao Shuang, it can be interpreted as deriving from the reading of a similar rectangle in the geometrical situation considered (Chemla 1994). The graphical writing of the operation-equation corresponds, in this context, to what we would call "equality" today. This form of inscription plays an essential part in my narrative—I return to this point later. Let me refer to this graphic formula as the first facet of the concept of equation attested to in the first time period.[15] It is all the more important to stress this facet that these diagrams do not appear explicitly in the early sources. Overcoming this difficulty was one of the methodological challenges identified in the first section. I have shown how to solve this difficulty by exploiting hints gleaned from Liu Hui's commentary. The reconstruction is also supported by later evidence. In Yang

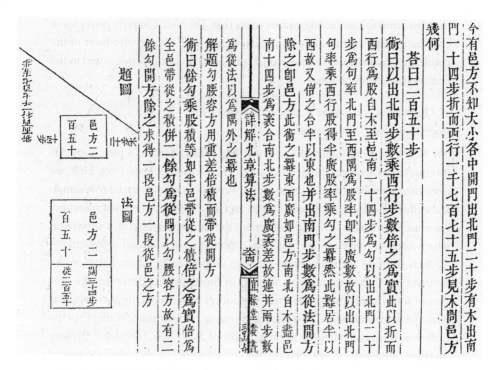

FIGURE 14.6. Yang Hui's thirteenth-century subcommentary on problem 19 illustrates the situation described and the operation-equation solving it (Yang Hui 1261, edition from the *Yijiatang congshu*, 1842, 64).

Hui's thirteenth-century subcommentary on *The Nine Chapters*, composed in the context of a mathematical culture in which writings include illustrations, a diagram illustrating the situation of problem 19 and a diagram showing the rectangle writing the equation were inserted in the pages of the book (see figure 14.6).

Liu Hui establishes the correctness of using the operation-equation but adds nothing on its execution, nor is there any explicit treatment of this question anywhere in the canon or in the commentary. And yet the equation is meant to be solved, since the outline of problem 19 is followed by an answer. Other puzzles remain. Why are *a* and *b*, respectively, designated as "dividend" and "joined divisor"? Why does the commentary refer to the rectangle *bx* as "area outside the corner"? Why is the solution of the equation prescribed using the expression "one divides . . . by extraction of the square root"? In fact, all these questions find an answer through a specific interpretation of the algorithm for extracting square roots that *The Nine Chapters* provides in chapter 4. I outline this algorithm next, which leads to a second facet of the quadratic equation and to the second methodological challenge stated in the first section of this chapter.

The Extraction of a Square Root, and Practices with Algorithms and Computations

The Nine Chapters contains the text of an algorithm to extract square roots. Its interpretation requires the description of two elementary practices. First, we need to analyze how texts for algorithms were written down in the mathematical culture to which *The Nine Chapters* and its commentaries testify. Second, we must reconstruct how the instrument for computing, that is, a surface on which numbers were represented with counting rods, was used at the time.[16]

As is the case for many other writings composed in China between the first and the thirteenth century, the text of the algorithm describes the process of root extraction by reference to another process of computation, that is, as if it were a division. This remark accounts for how the text employs the technical terms attached to the execution of a division (dividend, divisor, quotient, eliminate, move forward, move backward). It also proves essential in helping to interpret the text for square root extraction. In this way, the text further shapes an analogy between the two processes of computation of division and root extraction, showing in which respect a square root extraction is a kind of division. A statement is asserted in the way the texts display similarities and differences between the algorithms. Interpreting the text on root extraction by relying on how it refers to a process of division requires reading this analogy.

The operation corresponding in *The Nine Chapters* to a quadratic equation has two operands (a dividend and a joined divisor), whereas its execution is prescribed as a square root extraction. In the same way as the prescription of square root extraction refers to division, that of the operation-equation refers the solution of the equation to another operation, that of root extraction. Here too, it proves essential to rely on this statement of a relationship to interpret the text on which we concentrate. In the cultures considered, relationships between operations were regularly expressed using related terms to prescribe them and also describing their processes of computation in related ways. To interpret such texts and grasp the analogy they formulated, the reader apparently needed to use the reference that a text of procedure made to other texts. This practice of intertextuality is a feature characterizing the production and interpretation of texts for algorithms in the mathematical cultures under study. It echoes other features characteristic of the practice with the tool of computation. Let me now consider them.

Some of the technical terms used for division and taken up in the text of the algorithm for square root extraction designate elementary operations on the surface for computations: "moving forward" means to take a number written down with rods, using a place-value decimal system, and move it column by column leftward. Its value is thus multiplied by 10 each time it is shifted by a column. "Moving backward" refers to the reverse movement. These elementary operations exploit properties that rods lend to the representation of numbers. Numbers expressed with rods can be moved on the surface on which they are written, and their value can be modified and replaced by another value. In the text for root extraction, the operations of moving backward and moving forward are applied to numbers placed in the position of "divisors," whose shifts on the surface appear to be similar to those of a divisor in a division. More generally, the practice, outlined above, of expressing relationships between high-level operations (division, root extraction) through shaping a relationship between the texts of the algorithms that carry them out echoes a practice of expressing a relationship between operations through shaping their processes of computation on the surface in a similar way.

To shed light on this practice at the time when *The Nine Chapters* was written, we must reconstruct how the surface was used for computations, since at that time no illustrations or even generally no descriptions were included in the writings. Knowing more about the handling of computations nevertheless proves essential for answering the questions about quadratic equations raised above. This represents the second methodological challenge. Recovering ancient practices of computation also relies on hints gleaned from texts. Like "moving forward" and "moving backward," some of the terms the texts use reveal material features of the surface and its use. In this case too, results can be compared with the evidence contained in later subcommentaries on *The Nine Chapters*, composed at a time when writings included illustrations. For instance, Yang composed illustrations for his subcommentary on *The Nine Chapters*, in which he presented a cognate but different algorithm to extract square roots. His illustrations show successive moments in the computation on the surface, through arrays of numbers written with counting rods. These arrays are similar, in their material features, to what the texts themselves allow us to reconstruct about the earlier practices. Let us consider some of the features that are important here.

Computations on the surface make a crucial use of positions. The execution of a division is performed on positions arranged in three lines: the dividend in the middle row, the divisor in the row below, and in the upper row, the quotient obtained digit by digit. The fact that the text of the algorithm for root extraction uses these same three terms is correlated with the fact that the execution of the algorithm also develops fundamentally on three positions: the number whose root is sought is placed in the middle row, whereas the root is determined digit by digit, the successive digits being placed in the upper row, as in a division. As for the row below, numbers are gradually shaped throughout the process of computation so that the overall scheme and elementary operations of the root extraction can be correlated with those of a division. Accordingly, the numbers that succeed to each other in the position under that of the dividend are called "divisor."[17]

Positions of that kind are the key components of any operation considered in ancient Chinese documents. Texts for algorithms attribute names to them. The values placed in these positions change while a computation is executed. At the point of a text when a term designating a position is used, the computation involving this position picks up the value placed there at the time.

These are the first elements of a description of the practice of computation specific to the mathematical culture considered. Its key feature for the aim pursued here is that throughout the computation, the inscription of the process executing a square root extraction demonstrates that the positions named "dividend," "quotient," and "divisor" record the same events as the homonymous positions in a division. The processes of computation are thereby shown to combine the same elementary patterns of operation, which the positions allow us to grasp. In that way, a dynamic relationship is shaped between different processes of computation. Similar dynamic relationships are evidenced in various Chinese writings, composed at different time periods. One can also show that these dynamic relationships are the object of a mathematical work, since they were regularly rewritten to display the relationship in a new way.

These facts suggest that the dynamic inscription of the process on the surface was meaningful for the actors and was read as such. It is this dynamic inscription that the illustrations by Yang attempt to capture within the pages of a book. My own recovery of the practice of computation leads me in a similar way to reconstruct the successive states of the surface throughout a root extraction according to the algorithm in *The Nine Chapters*. In tables 14.1 and 14.2, I display the process as I restore it—I first use Arabic numerals instead of numerals represented with rods. Since my only intention is to provide the reader with a visual aid for the subsequent discussion, there is no need here to attempt to understand the computations.

Step 1	Step 2	Step 3	Step 4	Step 5	Step 6	(Steps)
			2	2	2	Upper: quotient
55225	55225	55225	55225	55225	15225	Middle: dividend
1	1	1	1	2	2	Lower: divisor

The related text of the procedure reads: "[Step 1] one puts the number-product as <u>dividend</u>. [steps 1 to 3] Borrowing one rod, one <u>moves it forward</u>, jumping one column. [step 4] The <u>quotient</u> being obtained, [step 5] with it, one multiplies the borrowed rod, 1, once, which makes the <u>divisor</u>; [step 6] then with this, one <u>eliminates</u>." The terms underlined are common with the process of division.

The process of extracting the square root of 55225 begins as shown in table 14.1. It suffices for our purpose to represent the subsequent steps of the root extraction symbolically. In modern mathematical notation, if A is the number whose root is sought, and if $a \cdot 10^n$ and $b \cdot 10^{n-1}$ are the first two digits of the root, the subsequent computations can be represented as shown in table 14.2. (I return below to the meaning of the computations.)

Why do these computations determine a square root, and what is the connection with understanding the nature of the quadratic equation as it occurs in *The Nine Chapters*, and how it was solved? These questions, as well as those raised above, are answered when we consider how Liu Hui fulfills the task of proving the correctness of the above-mentioned algorithm. Through sketching his reasoning, we can elucidate the meaning of the computations displayed above. Indeed, this elucidation is the crux of my argument.

The Practice of Proving the Correctness of Algorithms and a Second Facet of "Equations"

As I mentioned above, for all procedures the commentaries on *The Nine Chapters* systematically address the question of their correctness. This holds true also for the algorithm to extract square roots. In general, the commentators use a *dispositif* with which they can explicitly show "the intention" of the successive operations prescribed by the algorithm. The intention (yi 意) of an operation or a subprocedure is the meaning of its result, formulated with respect to the dispositif. At the end of the reasoning, the meaning of the final result of the algorithm is established and shown to conform to what was expected. These features partly characterize the practice of proving the correctness of algorithms in the mathematical culture examined in this whole section.

TABLE 14.2. The Subsequent Sequence of Computations of the
Square Root Extraction in Modern Terms

Step 6	Step 7	Steps 8 and 9	Step 10	Position
$a.10^n$	$a.10^n$	$a.10^n + b.10^{n-1}$	$a.10^n + b.10^{n-1}$	Quotient
$A - (a.10^n)^2$	$A - (a.10^n)^2$	$A - (a.10^n)^2$	$A - (a.10^n)^2$	Dividend
$a.10^{2n}$	$2a.10^{2n}$	$2a.10^{2n-1}$	$2a.10^{2n-1}$	Divisor
			1	Below: auxiliary

Step 11	Step 12	Step 15	Position
$a.10^n + b.10^{n-1}$	$a.10^n + b.10^{n-1}$	$a.10^n + b.10^{n-1}$	Upper row: quotient
$A - (a.10^n)^2$	$A - (a.10^n)^2$	$A - (a10^n + b10^{n-1})^2$	Middle row: dividend
$2a.10^{2n-1}$	$2a.10^{2n-1}$	$2(a.10^{2n-1} + b.10^{2(n-1)})$	Lower row: divisor
$(10^{n-1})^2$	$b.(10^{n-1})^2$		Below: auxiliary

The related text of procedure in *The Nine Chapters* reads: "[Step 6] After having <u>eliminated</u>, [step 7] one doubles the <u>divisor</u>, which gives the fixed <u>divisor</u>. [steps 8, 9] If again one <u>eliminates</u>, one reduces the <u>divisor, moving it backward</u>. [step 10] Again, one puts a borrowed rod; [step 11] one <u>moves it forward</u> as at the beginning; [step 12] with the next <u>quotient</u>, one multiplies it once." The terms underlined are common with the process of division. Steps 12 and 15, related to the digit b, set the stage for the following digit. (Steps 13 and 14 are omitted as I only reproduce the initial and the final stage.)

For square root extraction, Liu Hui establishes the meaning of each operation or group of operations by reference to a single visual device. The various hints his commentary gives enable us to restore its structural and material features, as shown in figure 14.7. Within the square of area A, Liu Hui identifies three types of areas that are essential for his proof and to which he attributes three different colors. First, he distinguishes three squares, in yellow. They represent the squares of the successive digits with the respective order of magnitude. Second, Liu Hui distinguishes two sets of rectangles, the first in vermilion and the second in blue-green. He uses these colors to refer to the tinted rectangles and to explain what the successive steps of the algorithm compute.[18] In this case too, the main structural features of the diagram conform to diagrams Yang inserts in his subcommentary. These hints reveal continuities in the practice of visual devices between the first and the second time periods considered here, beyond the break represented by the insertion of illustrations within books from about the tenth century onward.

I shall mention only some aspects of Liu Hui's proof. The part played by the diagram in the proof will turn out to have an essential role in our story. Liu Hui interprets steps 1–6 (table 14.1) as aiming to subtract from the area

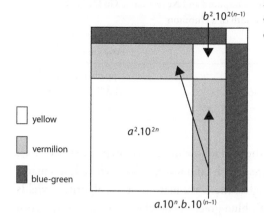

SQUARE OF AREA A

$b^2.10^{2(n-1)}$

$a^2.10^{2n}$

$a.10^n.b.10^{(n-1)}$

yellow

vermilion

blue-green

FIGURE 14.7. Restoring Liu Hui's visual device for the proof of the correctness of the root extraction.

A—which is placed in the dividend row—the area of the yellow square whose side is a .10^n. When this area is subtracted, the remaining "dividend" or "area," equal to $A - (a\ .10^n)^2$, has the shape of a gnomon. The determination of the subsequent digits of the root amounts to finding the value of the width of the gnomon. Liu Hui interprets steps 7 and 8 (table 14.2) as aiming to prepare, in the divisor row, what corresponds to the length of the two vermillion rectangles. (Note that $2a.10^{2n-1}$ corresponds precisely to twice a .10^n multiplied by 10^{n-1}, that is, the order of magnitude of the second digit. It suffices to multiply this value by b to obtain the areas of the two rectangles.) Further, Liu Hui interprets steps 10–12 as preparing, in the row below, what needs to be multiplied by b to yield the area of the second yellow square on the diagram. The sum of these two values is placed in the divisor row in step 13, and its multiplication by b yields the area of the gnomon of width b .10^{n-1}, which needs to be subtracted from the area of the overall remaining gnomon to deal with the digit b. Thereafter, the following computations (until step 15) are interpreted as determining similarly the length of the two blue-green rectangles to prepare a similar treatment for the subsequent digit.

Liu Hui's interpretation of the computations with respect to the diagram sheds light on a fact essential for our purpose. Once the first digit is dealt with and the length of the two vermilion rectangles prepared, continuing with the root extraction is equivalent to determining the width of a gnomon having an area equal to $A - (a.\ 10^n)^2$. Thus, if we forget about the first digit and concentrate on the gnomon, the part of the algorithm that starts at step 8 (table 14.3) is an operation that yields, as a result, the width of the gnomon. The

TABLE 14.3. The Steps of the Square Root Extraction in Which One
Can Forget About the Digits So Far Computed and Nevertheless Go On with
the Operation and Determine the Width of a Gnomon

Step 8	Step 16	Position
		Upper: quotient
$A - (a \cdot 10^n)^2$	$A - (a10^n + b10^{n-1})^2$	Middle: dividend
$2a \cdot 10^{2n-1}$	$2(a \cdot 10^n . 10^{n-2} + b \cdot 10^{n-1} . 10^{n-2})$	Lower: divisor

same reasoning holds true in the step following step 15 (step 16 in table 14.3): if we subtract the larger square, which contains the first two yellow squares and the two vermilion rectangles, the continuation of the algorithm yields the width of the gnomon with blue-green rectangles. Continuing the root extraction by starting at either of these moments in the execution of the algorithm, and forgetting about the first part of the root determined, amounts to operating on an array similar to that shown in table 14.3.

Several remarks will be essential for us here. The algorithm derived from the square root extraction, when one deletes the first part of the process, in fact solves a quadratic equation. For example, if we consider the shortened algorithm starting at step 8, it yields a root of the equation

$$x^2 + 2a \cdot 10^n . x = A - (a \cdot 10^n)^2.$$

If we look at the terms present in the array of numbers on the surface, whether we are at step 8 or step 16, in the middle row we have a so-called dividend, and in the lower row a so-called divisor or fixed divisor.

These terms (to which modern terminology refers, respectively, as the constant term and the coefficient of x in the quadratic equation) are precisely the two operands of the operation-quadratic equation as described in *The Nine Chapters*. In other words, in both contexts, not only do the operation-equations have only two operands (what in the modern concept of quadratic equation is the coefficient of x^2 is not identified as a term of the operation), but they also share the same two operands. Moreover, these two operands bear cognate names in both cases. It is unlikely that the correlation is fortuitous, but more is involved than these two features. It is by a kind of root extraction, that is, an algorithm derived from an execution of root extraction, that the quadratic equations read in some temporary configurations of this process are solved. This discovery echoes the prescription of the solution of the operation-equation in problem 19. Finally, in the proof of the correctness of root extraction, these quadratic equations correspond to the figure

GNOMON OF AREA $G=A-a^2.10^{2n}$

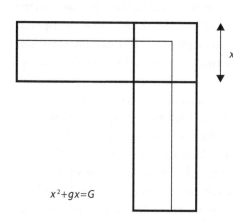

$x^2+gx=G$

of a gnomon (see figure 14.8). If one extends the gnomon, one transforms it into a rectangle similar to that obtained by Liu Hui in his commentary on problem 19. Interestingly enough, in the technical terminology at that time, gnomons were designated by a term *ju* 矩, which was also a general term for a rectangle.[19]

All these elements show that we have a complete correlation between, on the one hand, the operation-equation solving problem 19 and, on the other hand, the equation as it appears within the process of root extraction and as it can be detached from it thanks to the geometrical interpretation carried out in the context of the proof. That this relation is meaningful in the eyes of the authors of *The Nine Chapters* is clearly indicated by the prescription of the operation-equation in the procedure following problem 19 as a square root extraction.

Bringing the operation-equation of problem 19 in relation to these specific moments in the process of root extraction solves many questions I have not dealt with so far. First, the connection established between root extraction and the solution to problem 19 suggests a hypothesis for interpreting the designation of one operand of the operation-equation as a joined divisor. Indeed, in step 15 of the reconstructed process of root extraction (table 14.2), when the number in the auxiliary row below, $b. (10^{n-1})^2$, is deleted by being added to the row above it, that of the divisor, the text of the algorithm reads: "[Step 15] What has been obtained in auxiliary *joins* the fixed divisor." The commentator interprets the value computed by this operation of "joining" in the position of the divisor as representing the length of two tinted rectangles. On the one hand, it corresponds to the term in x in the equation thus constituted within the process of root extraction. On the other hand, in the correlation between root extraction and problem 19, this value corresponds precisely to what is designated as a joined divisor.[20] This fact illustrates how precise the correlation between the two contexts is. The remark even suggests that we are presented with more than a correlation here. The terminology seems to state how the operation "quadratic equation" was created on the basis of the process of root extraction. I return to this point below.

Further, we now understand how the operation-equation is solved. The execution of the operation-equation is indicated by the names of the operands and the prescription "one divides ... by extraction of the square root." This practice reminds us of how the text for root extraction was formulated with the terminology of division. The terms invite the reader to rely, in a specific way, on the algorithm given for another operation.

We also understand why Liu Hui establishes the operation-equation as he does, and how his commentary relates to the text of *The Nine Chapters* for problem 19. The connection lies in the algorithm of root extraction and the proof of its correctness, in which the two-term equation is placed in relation to the figure of the gnomon-rectangle.

These hints confirm, at the same time as they outline, the intimate relationship between the quadratic equation and the algorithm of root extraction as they occur in *The Nine Chapters*. Most important, these conclusions reveal a second facet of the concept of equation. In this framework, the equation appears to be an operation with two operands. The beginning of the second section reached this conclusion by observing the structure of the text of the procedure in *The Nine Chapters*. Now, this conclusion is confirmed through observing the process of root extraction and the procedure extracted from it to execute the operation-equation. This confirmation brings a new piece of information: we see how the two operands were probably noted on the surface. Equations fundamentally consisted of two positions, which illustrates our previous remarks about positions as key components of operations in the culture under study. I refer to such notation as "written diagrams." This fact allows me to clarify a point. Despite their close links, root extraction and quadratic equation remained distinct operations: this appears clearly when we observe that the former operation had a single operand whereas the latter had two. Similarly, the analogy shaped between division and root extraction did not mean that the two had become the same operation.

Let me summarize what I have said about this facet of the equation in *The Nine Chapters*. Its two operands are those remaining on the surface for computing at some points of the algorithm for root extraction. The execution of the operation derives from root extraction, by removing an initial part in the latter algorithm. The prescription "one divides . . . by extraction of the square root," which concludes the procedure for problem 19, indicates that the resolution of the equation was carried out by part of the algorithm of root extraction.[21]

In fact, the concepts of algebraic equation to which Chinese sources attest until the sixteenth century appear to have all been perceived as depending in the same way on root extraction. This feature characterizes a subfamily, in the larger family of concepts of equations, in which we find concepts specific to the tradition considered in this chapter. In the subsequent centuries algorithms for root extraction would undergo changes. Accordingly, the concept of equation would also experience transformations. However, throughout the centuries, the concept and treatment of equations maintain their essential link to root extraction.

The connection established between the operation-equation and the truncated square root algorithm solves another problem about the first facet of the equation (its graphical inscription as a rectangle): why the commentary on problem 19 spoke of the sum of the areas that were originally north and south of the square town as forming "the area outside the corner." The analysis above has shown that the quadratic equation deriving from square root extraction is attached to the figure of a gnomon (see figure 14.8). This gnomon of area G is composed of a square with side x and two rectangles whose total area is gx. The figure of the gnomon shows why these two rectangles can be considered as forming "the area outside the corner." In effect, the commentators use the term "corner" to designate the successive squares in the corner of the square A, whose side represents the still unknown part of the root. The fact that Liu Hui, in his commentary on the procedure for problem 19, speaks of "the area outside the corner" to designate the graphical element corresponding to the term in x makes sense in relation to his own commentary on square root extraction. Further, this indicates that Liu Hui was writing the former commentary with reference to the latter. He also related the first facet of operation-equation to root extraction.

In conclusion, every question about the solution to problem 19 or its commentary finds its explanation in the algorithm of root extraction or in the commentary Liu Hui composed on it. These various elements confirm that the quadratic equation as conceived in our first time period in China was an operation depending on square root extraction and attached to a geometrical figure deriving from the proof of its correctness. These characteristic features are so removed from those we associate with quadratic equations that some analysts fail to recognize a family resemblance. In my view it is beyond dispute. In addition, a conceptual history of science can be global only if we learn to perceive such family resemblances efficiently and appreciate the variety of forms present-day concepts have had in all conceivable pasts.

The double face of the equation explains why the two meanings ("dividend" and "area") of the term *shi* 實, which designates the "constant term" G of the equation, are both active. When the first facet of the equation is considered, the meaning of "area" refers to its geometrical figure. Its use is correlated with that of terms like "corner" to designate pieces in this extension. By contrast, for the second facet, the meaning of "dividend" comes to the fore, according with the perception of the other operand as a joined divisor. The link between concepts of equation and the figure of the gnomon will be essential in the subsequent time period.

The Connection between the Interpretation of Equations and the Description of a Culture

We have seen that the procedure for solving problem 19 brought into play an operation with two operands, the "dividend" and the "joined divisor." These terms could be interpreted as corresponding to positions remaining on the surface at specific moments of root extractions. This explained why the solution of the operation-equation was prescribed as an extraction. We have also seen that, in Liu Hui's commentary on root extraction, these two terms corresponded to elements in a gnomon. This gnomon/rectangle is precisely the geometrical figure through which Liu Hui establishes the correctness of the equation solving problem 19. The connection between procedure and commentary for this problem is exactly the same as the connection between the algorithm and its commentary at the corresponding moments of the process of square root extraction. The connection links the equation seen as an operation (first facet) and the equation written as a geometrical figure, that is, as a rectangle, opposing a "corner" to a "joined divisor" (second facet). In that way, the discussion shed light on the concept of equation at that time in China.

We can now revisit briefly how our interpretation of the final operation of the procedure given after problem 19 relied on our knowledge of the mathematical culture in the context of which these texts were written. While examining the method used, I emphasize how the concept of equation thus revealed adheres to a given way of practicing mathematics.

My interpretation of the procedure following problem 19 relied on specific features of the practice of writing down texts for algorithms. We have seen how in the sources examined, texts for algorithms made a strategic use of referring to other algorithms. This makes sense in a context where the relationships between operations are meaningful. Interpreting these texts requires that these references be taken into account. Moreover, we observed names designating key entities of an operation. Their design contributed to shaping relationships between algorithms.

These specificities in the use of terms are correlated with the fact that texts of algorithms refer to the execution of procedures on a surface on which numbers were represented in an ephemeral way and placed in "positions." Physical properties of the tool for computation, and the practice with this tool, leave their mark in specificities of the texts for algorithms recorded in writings. Most of what we know about the practice with this tool has to be recovered through clues gleaned from the texts. Restoring this practice, and in particular the process of root extraction on the surface, yielded essential information on how "quadratic equation" derived from a temporary configuration in this process. This allowed us to establish its mode of solution. It also shed light on its operands. First, the relation between the algorithm for root extraction and the operation-equation, as evidenced in *The Nine Chapters,* suggests reasons the equation in that case has only two operands. Second, it shows that operands correspond to positions, on which the procedure of solution operates: the dividend in the middle line and the joined divisor in the line under it. This kind of notation, which was not included in the pages of writings in the first time period, made its appearance in texts of the second time period. This facet of the concept of equation under consideration captures how the concept adheres to the practice with the surface used for computations. The concept of operation-equation and its solution stem from another algorithm, more precisely, from the layout and execution of this algorithm. This concept of quadratic equation derives its nature of being an operation from the surface.

We had to take into account two elementary practices of the mathematical culture under study to interpret the text: the practice of proving the correctness of algorithms and the practice with diagrams connected to it. To interpret Liu Hui's commentary on the final operation of the procedure solving problem 19, I brought into play how the practice of proof relied on diagrams to formulate an "intention" or a "meaning" for elementary operations or subprocedures in a procedure. Taking these practices into account disclosed two key facts. First, Liu Hui's interpretation of some computations within the process of root extraction (steps 7–8 and steps 15–16) introduced the gnomon as the figure with which one could make sense of the following computations in the algorithm. This figure apparently played a key part in detaching the quadratic equation from the process of root extraction and giving it autonomy. Second, conversely, this figure, with the related figure of the rectangle, might have provided tools with which equations could be established. This is how we interpreted Liu Hui's commentary on the equation for solving problem 19. However, as I have stressed, this fact holds much more generally, providing us with a key to understanding all the equations established by the third-century commentators Liu Hui and Zhao Shuang. Restoring the "intention" as formulated in the proof, and the diagram supporting it, thus helps us capture another key facet of equation in that tradition of ancient China: the equation as a standard statement. In this context, the statement is written diagrammatically, in the form of a rectangle opposing a square of unknown side and a rectangle attached to it (these pieces correspond to the colors in Liu Hui's diagram; see figure 14.7). It is extracted from the diagram and the proof in exactly the same way as the array and the procedure solving the equation were extracted from the layout and the algorithm. Moreover, both are taken out of root extraction at exactly the same point of the process.[22] Like the first facet, and for the same reasons, this second facet of the equation leaves no graphical trace in the writings before the eleventh century, yet it clearly informed the earliest sources we have.

To conclude, we can see how the interpretation I developed through taking elementary practices into account establishes the nature of the concept of equation in the tradition of ancient China under examination. I have shown that a quadratic equation had two key facets, neither of which surfaced in the sources except indirectly. However, these facets are essential to take into account, not only to grasp the nature of and work with equations at that time in China but also to understand the changes in the concept of and work with equations in the long term. Further, each of these facets adheres to distinct features of mathematical practice in relation to which our sources were written down. The surface on which computations were executed yields the equation as operation and provides the means to solve it. The interpretations in proofs of the correctness of algorithms and the diagrams on which they are based yield the equation as statement and the means to establish it. These remarks summarize in which respect we can correlate features of the mathematical culture and the concept of equation. Note that each of the two facets of the equation indicates the limitations of the concept in China at that time. The operation bore only on two operands, both of which were "positive" in modern terms. Only later will other types of quadratic equations be accommodated within this framework. This feature of the concept of equation in the first century further reinforces the conclusion of the dependence of the operation-equation on root extraction.

We now have an adequate basis to approach the concept of quadratic equation in our second time period, which seems to have begun in the eleventh century.[23]

A Subsequent Culture and a Subsequent Concept of Equation: Continuities and Differences

I take as a record of the second time period the last part of the final chapter of Yang Hui's *Quick Methods*. In fact, there Yang quotes Liu Yi's *Discussing the Source*, which probably dates from the eleventh century. It is not clear where the quotation starts, where it ends, or how exactly Yang adds to it—this is not relevant here. My only aim is to point out a new way of working with quadratic equations and a new concept of equation in the context of and in relation to another mathematical culture.[24]

As was mentioned above, the writings from the second time period are in sharp contrast with those from the first time period. The most conspicuous change is that practices of working with diagrams and with a surface for computing now leave traces in the writings as illustrations (see figure 14.2). These traces on paper attest to practices in continuity with earlier practices outlined above, but they also exhibit transformations. The transformations I use to illustrate this point are those that can be perceived in practices with diagrammatic elements. My claim is that, despite a strong continuity, by the eleventh century diagrams play a new and more fundamental role in the treatment of equations. It is essential to observe the change in the elementary practice with diagrams to interpret them adequately. As for equations, the sources testify to the fact that they are worked out within a conceptual and material framework similar to that described for the first time period. However, within this framework, the concept of equation and the algorithms solving equations show major extensions. My aim now is to describe the changes in the practice, the changes in the concepts, and the relation between the two.

To begin with, let us observe how Liu Yi's *Discussing the Source* testifies to how one worked with equations at the time. The first problem in the quotation of Liu Yi's book, problem 43 in the final chapter of Yang's *Quick Methods*, is the best introduction to this question, since this problem sets a frame of reference for the following problems. Its statement reads as follows: "The area of a rectangular field is 864 *bu*. One only says that the width fails to be equal to (that is, is less than) the length by 12 *bu*. One asks how many *bu* the width is" (Kodama 1966, 91).[25] The statement of the problem is immediately followed by the answer, "24 *bu*," and a procedure that reads: "One puts the area as dividend. One takes the *bu* (by which the width) fails to be equal to (the length) as what joins the square (or: what is appended to the square). One divides by this by extraction of the square root" (91). This procedure brings into play an operation-equation in a way similar in almost every point to the procedure solving problem 19 in *The Nine Chapters*. It describes how to derive a dividend and an operand called "what is appended to the square" (*cong fang* 從方), which corresponds to the old term "joined divisor." I return to these terms below. The procedure is then concluded by the same formulaic expression. In modern terms, the procedure corresponds to the equation $x^2 + 12x = 864$. If we set aside the change in the topic of the problem and the change in the terminology, nothing seems to have changed since the first century. In this sense at least, we perceive the continuity with what was described above. As for the changes, they appear inessential. Yet, we would be mistaken to consider these two small changes minor. I first focus on the statement of the problem.

To understand what is at stake in the change of topic, it is helpful to read a quotation of Liu Yi's preface to his book, which Yang places before problem 43 and repeats in two other places in the same book. This statement clarifies the new status of the equation and its relation to the topic of the problem: "Sir Liu [Yi] from Zhongshan said in his preface: 'As for the procedures of mathematics, from any point we start engaging [with them], one ends up with [systematically the same mathematical entity of] the rectangular field/figure [直田]'" (Kodama 1966, 91). This statement probably held true for the whole book. Whatever the situation dealt with in a problem, the quotation asserts, its solution reduces it to a "rectangular figure" and then—one might probably add—to the procedure for determining its root. If we compare the declaration to what follows, we realize that, in this new context, "rectangular figure" is a term designating the concept of quadratic equation, for which no other technical term can be found in that text. If such is the case, this remark implies an essential fact: problem 43, like those following it and similar to it, does not deal with an arbitrary situation but with the "equation" itself, as Liu Yi and Yang understood it. In this sense, the topic of the problem has undergone an important change.

In Liu Yi's text, the procedure is not followed by an equation written in modern terms but by the diagram of a rectangle (see figure 14.2, first rectangle from the right). The rectangle illustrates the situation described in the statement of the problem. It can also be interpreted as a graphic formula writing down the equation as Liu Yi and Yang conceived of it in this context.[26] Here, more precisely, the diagram writes the operation-equation, whose operands have values determined by the procedure placed after the problem.[27] The diagram shows a square, having as its side the width, to which a rectangle is appended, one side of which is the unknown and the other equals 12 *bu*. The overall rectangle is the "rectangular figure" of the problem. The captions inscribed on the diagram make this point clear. They refer to terms introduced in the procedure, including operands of the operation-equation. The caption placed under the diagram confirms this reading, since it comments: "Above, this is one piece of the square with [side] the width. Below, this is one piece of what is appended to/joins the square of [side] the width." The way in which, as we shall see, such a graphic formula of the equation is put into play shows that it was used as a base in working with the operation-equation.

It is striking that the rectangle "writing" the equation is graphically identical to the first facet of the quadratic equation in the first time period. In this respect, there is some continuity in practice, despite a change in the materiality of the diagram. There was also continuity in the way a procedure introduces the operation-equation. The terminology for its operands looked quite similar. However, the caption to the diagram seems to testify to the fact that, compared with earlier sources, the word *cong* ("join") that occurs in the designation of the term in x in both contexts, is now to be interpreted in a new way. The divisor is no longer characterized through the fact that, geometrically, it lies outside of the corner. Rather, *cong* designates the corresponding geometrical element by the fact that it is "appended to" the square.[28] A reconceptualization of the situation can be grasped in the change of terminology. It may also be what is at stake in the replacement, in this context, of the term *ju* ("gnomon/rectangle") by the term *zhitian* ("rectangular figure") to designate the rectangle.

This reconceptualization of the shape and structure of the rectangle is reflected in the two following components in the text of problem 43. First, the graphic formula that the rectangle constitutes illustrates the reason that the algorithm yielding the root of the equation is correct (see figure 14.2, second figure from the right).[29] The caption makes clear this is the function of the second diagram in the author' eyes, since it reads: "Diagram of the detail of the procedure [showing] the values of the pieces in a root extraction having [something] appended to [the square]." In that diagram, the area of the global rectangle, and consequently of its components (square and rectangle), is cut into pieces to show the meaning of all computations in the algorithm, if interpreted graphically. This diagram attests to how the writing of the equation in the form of a rectangle provides a notation used to work with the equation. This remark holds true for the following problems. The captions connect the pieces in the diagram with what happens on the surface used for computations, during the process of root extraction. This process is illustrated by the subsequent diagram, which is of a different kind. It is composed of three subdiagrams showing three configurations of numbers at different moments in the computation, much in the same way as what we reconstructed for the algorithm of *The Nine Chapters* (see figure 14.2, left page). The captions outside the subdiagrams indicate the specific moment when the configuration is extracted. They also explain what the rows contain. Below this diagram, a continuous text describes the algorithm.

The continuity with the numerical treatment of the quadratic equation of the first century, as discussed above, is manifest. However, two key transformations can be captured in the configuration of numbers. They illustrate again my thesis in this chapter: restoring material practices is an essential task, which offers tools for a more precise conceptual history.

First, in contrast to the algorithm in *The Nine Chapters*, the lower line is constantly present in the scheme of computation. It corresponds to what for us is the term in x^2 in the equation. In correlation with its presence throughout the computation, subsequent problems deal with operations-equations for which the related coefficient is different from 1 or even negative. Accordingly, modifications of the algorithm for root extraction are described. This is the first extant piece of evidence among Chinese sources that a third operand is attached to the quadratic equation.

The second key transformation can be easily described by reference to the algorithm for root extraction in *The Nine Chapters*. In the first century the operation "quadratic equation" was extracted from root extraction, after the step when the square of the first digit is subtracted from area A (step 8; see table 14.2). The operation corresponds to the shape of the gnomon, and Liu Hui accounts for the correctness of the algorithm determining the successive digits of the root by showing that each phase amounts to taking a slice out of the area of the overall gnomon, which has the shape of a thinner gnomon and whose width corresponds to the next digit determined. More precisely, for each digit of the root, the algorithm determines a divisor, which corresponds to the length of the thinner gnomon (if it is stretched), and it suffices to multiply this length by the corresponding digit of the root to obtain the area to be subtracted from the overall gnomon. Liu Yi's algorithm differs from that algorithm precisely on this point. It dissociates, and places in two separate rows, the component from the first-century divisor that, in modern terms, corresponds to the term in x in the equation (in third-century terms, this is the component of the gnomon "outside the corner"), and the component deriving from the "corner." The former component is placed in the row corresponding to "what joins/is appended to the square," whereas the latter yields the "divisor of the square."

This fact has two important consequences. The dissociated lines in the configuration of numbers now correspond each to a component in the diagram accounting for the correctness of the algorithm. The "divisor of the square" corresponds to the upper part, whereas "what is appended to the square" relates to the lower part. We thus see here the second point in which the reconceptualization of the graphic formula of the equation described above finds an echo. Here the reconceptualization is reflected in the writing of the operation-equation on the surface for computations, or, to use the expression introduced above, the written diagram. The new text is crafted in such a way that the components of the written diagram and those of the graphic formula correspond to each other in a transparent way.

The second consequence of the separation of the divisor line into two relates to the new treatment of operations-equations. The component corresponding in our terms to the term in x is being dissociated from the others, in the written diagram with which the equation is inscribed on the surface for computations. Accordingly, new operations-equations are considered, in which this operand can be negative, and new algorithms are described to deal with the various possible signs of the operands of an equation.

I have now introduced the fundamental elements that, in the context of the mathematical culture to which Yang's text attests, are involved in the new treatment of the operation-equation. They include problems, texts for procedures, graphic formulas that diagram the equation, and written diagrams that illustrate the numerical computations. We thus see how the treatment of equation presented by Liu Yi is in continuity with earlier treatments. We could not have grasped this fact had we not restored the diagrams used in the first time period.

Further, besides continuity, the text testifies to conceptual transformations in the treatment of equation. Liu Yi's text illustrates four directions in which his operation-equation differs from the previous one. First, the rectangular figure is now stated to be a universal object. Second, his writing attests to a progressive identification of a third operand—the term in x^2—in the quadratic equation. Third, we can observe in his text how negative marks on operands of the operation-equation are introduced, in relation to the widening of the range of equations considered. Fourth, Liu Yi's text testifies to the transformation of the algorithm solving the equation in a way that brings to light a higher homogeneity in the changes that execute the computation of the root. I have mentioned hints that suggest a correlation between the new way of working with equations and the transformation of the operation-equation.

I cannot describe here each feature of the changes in the mathematical culture and the conceptual approach. I examine only how the graphic formula is used in a new way, to work on new questions with the operation-equation. It is all the more important to focus on that aspect since it requires uncommon practices of interpretation. I claim that these practices of interpretation of the diagrams bring to light meanings that would otherwise remain hidden.

I have mentioned that Liu Yi's text considers new types of quadratic equations. How do the graphic formulas write down these new equations? Let us examine some cases.[30] Problem 44 deals with the same rectangle as problem 43, but now the length L is required, while the area A and the difference $L - l$ are known. The procedure given by Liu Yi involves an operation-equation corresponding, in modern terms, to

$$x^2 - 12x = 864.$$

It is written graphically as such, as is shown in figure 14.9. The caption on the left component of the diagram expresses the fact that the rectangle corresponding to $12x$ is the piece lacking in the square x^2 and the reason that the remaining area is equal to 864. In other words, the diagram connects the statement of the problem and the operation-equation for solving it. Further, the diagram writes the equation, and writes it in the standard form, expressing how a dividend/area is yielded by operations between the terms involving the unknown (divisors). The graphic formula still brings together a square and two rectangles. However, the graphical connection has changed in relation to the fact that the diagram writes a new type of equation. This is the first manifestation of a phenomenon that permeates the whole text and takes various shapes (figures 14.9 and 14.10).

Problem 46 shows a variation of the pattern just described. Its data consist of the area A of a rectangular figure with sides L and l and the sum $L + l$; the unknown sought is the width l. In modern terms, the operation-equation for solving it can be written as

$$60x - x^2 = 864.$$

Diagram of the problem

one piece of the (square) of the length when, looking for the length, it lacks the difference	width of twenty-four *bu*	The original area is eight hundred sixty-four *bu*	length of thirty-six *bu*

FIGURE 14.9. Liu Yi's graphic formula for the equation corresponding to problem 44, in Yang Hui's *Quick Methods for Multiplication and Division for the Surfaces of the Fields and Analogous Problems*, Korean edition from 1433, reprinted in (Kodama 1966, 92).

Figure 14.10 displays the graphic formula for writing the equation. Its overall rectangle has the unknown (l) and $L + l$ (which is equal to 60) as its sides. The diagram indicates how this overall rectangle combines the fundamental rectangle with area A and the square of the unknown, that is, the width. Its captions disclose interesting new elements, which are revealing of the part played by the graphic formula. The caption attached to the right designates the sum $L + l$ by the term "what is appended to/joins the square," namely, the lower square whose side is width l. This use reveals that the geometrical meaning of the expression had receded at that time and a functional meaning had come to the fore that related the geometrical element to the homonymous line, in the configuration of numbers on the surface for computations—the written diagram. This transformation in the meaning of the terms correlates with the changes in the concept of operation-equation.

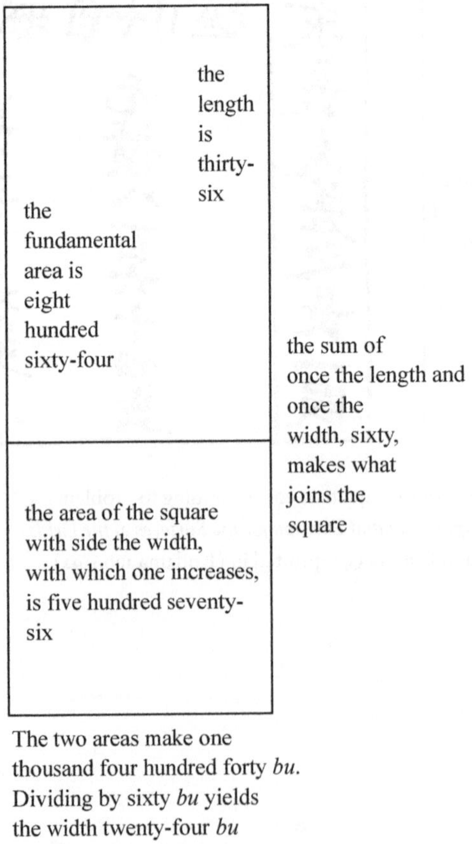

the length is thirty-six

the fundamental area is eight hundred sixty-four

the sum of once the length and once the width, sixty, makes what joins the square

the area of the square with side the width, with which one increases, is five hundred seventy-six

The two areas make one thousand four hundred forty *bu*. Dividing by sixty *bu* yields the width twenty-four *bu*

FIGURE 14.10. Translation of Liu Yi's graphic formula for the equation corresponding to the first solution of problem 46 in *Quick Methods*.

In fact, the diagram combines two functions—it writes the equation. But it also serves as a base to consider the correctness of the algorithm given to solve similar operations-equations. Thus, the notation simultaneously serves two of the purposes that we pursue with symbolic notations. However, where we today would write a sequence of formal notations, Liu Yi brings together the various uses of the notation in a single diagram.

The general conclusion, important for my main topic, is that the graphic formula writes the equation in ways that must be described systematically if we want to interpret it adequately. We can project on it neither our expectation that the equation should be written discursively nor our belief that distinct facts should be addressed by distinct statements. The kinds of statement, as well as their use by the actors, need to be attended to if we aim for a conceptual history of equations in China.

The same equation will be written in another way, for problem 47, whose outline is identical to problem 46 but whose unknown is the length. The equation is thus still

$$60x - x^2 = 864.$$

The graphic formula brings into play a new ingredient. In line with the extension of the operation-equation announced above, the term "negative" occurs in the caption of the piece corresponding to the operand "square" (in modern terms, x^2). In the following problems, another graphical element is used to denote negative operands: colors. For instance, in problem 52, the operation-equation corresponding, in modern terms, to

$$A = 312x - 8x^2$$

is written graphically. The negative operand appears in black. The mathematical culture observed in the first time period employed colors to mark diagrams, as described above, and this technique is taken up in the subsequent context. This fact displays a form of continuity between the two time periods in the practice with diagrams. However, the practice is now invested with new meanings, in relation to a major conceptual change. Equations can have negative operands, and color is used to denote them. This example illustrates how ancient features can be put into play to extend the possibilities of expression of the graphic formula. It is essential to interpret color if we want to understand how these graphic formulas fulfill two of their essential functions.

One of these functions, as we have seen, is to establish the correctness of the various algorithms put into play to find out roots, in relation to the nature of the equation solved. In this context, color is used to make a graphic formula display the key point of a proof, as illustrated by the diagram associated with the second algorithm given for problem 46.[31] The second function of the graphic formula in which color will be used is that of providing support for establishing the equation. For problem 46, the equation established derives, in the process of proof, from the equation solved. Color is also used at the beginning of the solution of a problem to establish how it can be solved using an equation.

In conclusion, if we compare this practice of diagrams with that for the first time period, we see that, although the same graphic formula of the rectangle is used, it is used in a new way, in conformity with the extension of the range of equations considered and the related change in their nature. Liu Yi's graphic formulas differ from the earlier samples, and they do more work than was previously the case. Some of the features of this new usage have ancient roots, such as colors. This continuity should not, however, hide the new meaning inscribed with these old techniques in Liu Yi's time.

Conclusion

We can now return to the theses expounded at the beginning of the chapter and see how the conceptual history of "quadratic equation" in China between the first and the eleventh century supports them. The case study I developed here shows the utility of describing how practitioners worked with various elements in their mathematical activity—here mainly problems, algorithms, the surface for computing, and diagrams—and how they connected these elements with one another. For such similar complexes of practices I suggest using the term "scholarly" or, here more specifically, "mathematical" cultures. I have argued that describing the mathematical culture in the context of which practitioners operated provides essential tools to interpret sources. This point is more enhanced in a case where sources are a challenge for the interpreter. In the case described here, this method helped me determine the nature of the concept of equation in the first century, that is, an operation-equation, and the transformations of this operation-equation in the succeeding centuries.

The descriptions of different mathematical cultures allow us to grasp continuities as well as transformations in ways of doing mathematics in China between the first and the second time period. In each of these two contexts, the concept of equation correlates with ways of working with diagrams and the tools for computations. Such an approach highlights material dimensions of conceptual history.

Despite differences, strong continuities, both material and conceptual, between the two ways of conceiving equations can be recognized. These continuities define a tradition of working with equations as operations that, to my knowledge, cannot be identified in sources other than Chinese ones. However, the existence of this tradition does not imply any kind of determinism, according to which the concept of equation, once set in a framework, could only develop within the bounded space of this framework. This fact is illustrated quite strikingly by the later history of concepts of equations in China. By the thirteenth century, Chinese sources attest to another concept of equation that presents strong continuities with earlier concepts but redefines a tradition in an entirely new way. Li Ye's *Sea Mirror of the Circle Measurements*, completed in 1248, illustrates this phenomenon. In this book, equations are noted as written diagrams, in continuity with the way the operation was inscribed on the surface on which computations were carried out. Only one of the two facets of earlier concepts of equation survives (Li 1958, chaps. 22, 23).[32] The graphic formula recedes in the background and is replaced by other (algebraic) means of establishing the operation-equation that also derive from the numerical facet of the equation. The establishment of this new way of responding to inherited tradition goes together with new developments in the understanding and treatment of equations.[33]

All the concepts of equation that can be identified through Chinese sources, however, share a common feature: they consider the equation as a numerical operation. In this chapter I have shown how this feature adheres to the practice with the surface on which to carry out computations. Despite this adherence to a stable feature of the cultures within which equations were used with in China, this approach to equations is not so typically "Chinese" that it could not circulate. In fact, a numerical approach to the solution of equations quite similar to that one, both materially and mathematically, suddenly occurs in Arabic sources in the twelfth century. In *On Equations*, by Sharaf al-Din al-Tusi,[34] the concepts and treatment of equations combine features of equations coming from the tradition established by al-Khwarizmi (first half of the ninth century) and that illustrated by Omar Khayyam (1048–1131). They further incorporate a new way of approaching the solution of equation, which presents striking similarities with the approach that had developed in China. So far, no historical evidence has been found that this was due to circulation, and yet I believe it is highly probable.[35] Whatever the case, *On Equations* testifies to the possibility of merging different concepts and treatments of equations into a single whole, thereby demonstrating that the concepts and modes of solution shaped in China could be adopted in other contexts and interact with other approaches. This conclusion suffices to establish that even though concepts and results may adhere to features of scholarly cultures, they are not condemned to remain within these boundaries and be incomprehensible for other scholarly cultures.

Notes

I am grateful to Evelyn Fox Keller and Bruno Belhoste for their comments on an earlier version of this chapter. I also thank the participants in the workshop at Les Treilles in June 2011 for their remarks, in particular my three commentators, Emmylou Haffner, Donald MacKenzie, and David Rabouin. The research presented in this chapter is part of the work that led to the European Research Council project SAW (ERC grant agreement 269804). Many thanks to Karen Margolis for sharing her thoughts with me about the formulation of this chapter. The chapter was completed while I was in Seoul, benefiting from the hospitality of the Templeton Science and Religion in East Asia project hosted by Science Culture Research Center, Seoul National University.

1. In this chapter, I take this topic only as an illustration, outlining the argument without giving any detail. The argument draws on several publications, which constitutes the core of a book I plan to write.

2. The reader can find a detailed treatment of the issue of how we can describe cultures using scientific documents in Chemla 2010a. Chemla 2009 presents an outline of the argument.

3. The symmetrical problem would be to introduce a new term for each distinct concept. Usually such historical practice is carried out in an asymmetrical fashion. This is how we end up with the idea that there was nothing in China, no philosophy, no mathematics, and so on.

4. Cullen 1996 offers a translation of *The Gnomon of the Zhou*. The commentaries still await systematic study. Chemla and Guo 2004 contains a critical edition and a translation into French of *The Nine Chapters* and Liu Hui's commentary. In the present chapter, I rely on this critical edition.

5. We can establish that, as for the number system commonly used today, the basis for the number system was 10. Moreover, a digit derived its meaning from the position in which it was put, in the same way as, when we write 123, 1 derives its meaning of "a hundred" from its position in the sequence of digits.

6. The following statements require qualification, but I must skip details (see Chemla 2003, 2009, 2010a, 2010b, and Chemla and Guo 2004).

7. For the moment, we lack evidence to date this change. It must have occurred between the eighth and the eleventh century.

8. Lam 1977 contains a full translation and discussion of *Quick Methods for Multiplication and Division for the Surfaces of the Fields and Analogous Problems*. Guo Xihan 1996 constitutes a guide to its reading. Te 1990 and Horiuchi 2000 discuss the remaining evidence on Liu Yi and in particular his treatment of quadratic equations. Here I omit

scholarly discussion on matters of date and attribution, concentrating instead on the concept of equation to which this writing bears witness and its relation to a mathematical culture.

9. I have begun to describe this shift (Chemla 2001), but further research is required.

10. All the evidence we have for the first time period shows equations having the same features as those established in this part of the chapter.

11. I use bold characters for terms on which I shall comment below. See Chemla and Guo 2004, 689–693, 732–735 for the Chinese text, its translation, and its interpretation.

12. To carry out the algorithm, other functions are introduced and terms are attached to them; see below.

13. See Li and Du 1963, 64–67, 1987, 53–55; Qian 1981, 51; Li 1990, 112–114, 404–405, 1998, 728–729; Martzloff 1997, 228–229; and Shen et al. 1999, 212–213, 507–512. I return to these authors' interpretation below.

14. I stress here that at that time the "equation" has two operands and not three. This will allow us to perceive a key change in the concept of equation later. Other historians have not noticed this change.

15. In my talk at the Stanford University–REHSEIS (Recherches en Epistémologie et Histoire des Sciences et des Institutions Scientifiques) Workshop on diagrams, organized by S. Feferman, M. Panza, and R. Netz (October 2008), to designate Liu Yi's diagrams for equations (see the third section), I borrowed the expression "graphic formulas" (*gezeichnete Formeln*) from Hilbert 1900. I also borrowed from Hilbert 1900 the expression "written diagrams" (*geschriebene Figuren*), which I use below.

16. A critical edition, an annotated translation, and references to other publications on the topic are given in Chemla and Guo 2004, 322–329, 362–369.

17. The adjustment of the value of the divisor is made possible through computations carried out in a row that the scheme of root extraction places under the fundamental three-row scheme of division. At the time of *The Nine Chapters*, this row below was considered auxiliary.

18. On the use and meaning of colors as well as unit-squared paper to make diagrams in ancient China, see Chemla 1994, 2001, and 2010b.

19. See the glossary in Chemla and Guo 2004, 943.

20. The verb "to join" occurs at the same place in the two contexts and correlates the joined divisor with the divisor of a root extraction, at the moment when it has been joined by the number in the row under it.

21. Li and Du 1963, 61–66, also interpret the quadratic equation and its solution as, respectively, a temporary configuration on the surface for computing in a root extraction and the part of the root extraction starting at this point. For these authors, too, quadratic equation derives from root extraction. However, my interpretations differ in several aspects. First, I do not restore the algorithm for root extraction in the same way. As a consequence, for Li and Du, the equation as it appears as a temporary configuration has three operands, not two—it includes a term in x^2. This interpretation does not allow them to see the transformation of the concept of equation to which later sources attest. Li Yan and Du Shiran do not refer to Liu Hui's commentary in their interpretation of the equation. They read the equation from the process of root extraction. All

these features also characterize Jean Claude Martzloff's (1997, 224–229) account for algebraic equations. As a result, Li and Du's geometrical interpretation of the process of solution, as well as the concrete numerical process they restore, is similar to that in eleventh-century sources. Moreover, they do not emphasize the geometrical facet of equation in our first time period. They adopt another view on the connection between geometry and this equation. This leads them to read some geometrical problems and algorithms as amounting to solutions of quadratic equations by radicals (Li and Du 1963, 73–76). This interpretation seems contrived. Lastly, they interpret the term "joined divisor" as "following the divisor." In their view, this refers to the fact that this divisor "has a nature comparable to that of" (Li and Du 1963, 74) the other divisor in a root extraction. Revealingly, they refer to this other divisor by the term "square divisor," which in fact surfaces only in later sources. The translation into English in Li and Du 1987, 52–55, 61–63, does not convey the meaning of the original. Qian 1981, 47–51, 58–60, offers exactly the same interpretation, which dates back to the 1950s at the latest.

22. Li Jimin (1990, 112–114) does not interpret the equation in *The Nine Chapters* as I do. For him, geometrically as well as algorithmically, the equation is a square root extraction to which an auxiliary term was "appended": a rectangle is appended to the square to write down the equation; a line is appended to a line in the root extraction to record the term in x. This is how he interprets the term *cong* that I translate as "joined." I agree that this is how the equation would be understood in the eleventh century (see below). However, in my eyes, his view mostly anachronistically projects the concept of equation that characterizes the second time period onto the first time period. The same conclusion applies to Shen 1997, 288–289, 682–683, and Shen et al. 1999, 212. Consequently, Li cannot explain why in the first century the equation has only two operands, and in fact he seems not to have grasped the importance of this feature for the later history of equations. It seems to me difficult to understand why the initial term in x would be qualified as "appended," since, when it is placed on the surface to compute, there is no other divisor to which it could be appended. Revealingly, Li has to add the following sentence at the beginning of the algorithm for root extraction: "The appended divisor makes the fixed divisor" (113). For him the algorithm solving the equation is an extension of the algorithm for root extraction and not, as I believe, a procedure deriving from it. This interpretation implies that *The Nine Chapters* does not describe how the square root extraction is modified, nor does the commentary account for the correctness of the extended algorithm. Lastly, Li does not see the general importance of the shape of the gnomon in that early phase of the history of quadratic equation. Accordingly, he does not seem to grant any part in this context to the practice of proof. The same features hold true in Li Jimin's (1998) translation of *The Nine Chapters* into modern Chinese. In the same vein as Li and Du (1963), Li (1990, 367–368) appears to offer a contrived interpretation of quadratic equation.

23. Annick Horiuchi seems to have missed the second facet of the equation in the first time period (see Horiuchi 2000, 243), and hence did not grasp the continuity in all its dimensions.

24. Since 2007 I have been preparing a critical edition and translation of this text, which I refer to as Liu Yi's writing. I refer to the edition of Yang Hui's *Quick Methods*

reprinted in Kodama 1966, 91–97, on the basis of the 1433 Korean reprint. Lam 1977, 112–133, contains a translation of the passage dealt with here. Note that in Lam 1977 the diagrams are not translated faithfully. Guo 1996, 229–279, provides elements for a critical edition and explanations.

25. The unit *bu* is used for measuring lengths as well as areas. For areas, the *bu* is a square unit having a side of 1 *bu*. The term translated as "field" acquired the more general meaning of "figure" in the third century at the latest.

26. Most historians have dealt only with the numerical facet of the concept of equation in Liu Yi's text, leaving aside the diagrams as if they were mere illustrations. We can identify two forms of anachronism in this interpretation. First, historians have projected onto the source our perception of the part played by figures in mathematics. This remark shows why recovery of the practice with diagrams is important. Second, they have read these texts through the lenses of the treatment of equations in China in the thirteenth century, as illustrated by Li Ye's *Sea Mirror of the Circle Measurements* (1248), within the context of the so-called procedure of the celestial origin. In that other context, diagrams play an entirely different role. This holds true for Li 1958, 185–188; Qian 1981, 154–157, 1966, 44–47; and Martzloff 1997, 142.

27. Horiuchi (2000) revealed the universal character of the rectangular figure and its meaning, showing the part played by the rectangle as a tool with which to work on equations. I shall explain below the two main uses of the rectangle as a support for operations that she first discussed. However, my interpretation differs from hers in that I suggest that the rectangle writes the equation and that the reader had to interpret it as such. This interpretation derives from the fact that I take scholarly cultures in their variety into account. In my view, a substantial part of Yang Hui's treatment of the equation has to be read from the diagrams and has no counterpart in the discourse. This holds true for later texts that record similar treatments. Horiuchi felt "compelled" to recognize that the text contained only a diagrammatic treatment (251), while expecting a discursive treatment. Such a conclusion calls for a critical edition of diagrams, which I am currently preparing.

28. Li Jimin and other historians have projected this reading of the equation onto texts of the first time period (see notes 21 and 22 above for exact references). I believe they have missed a subtle change in the understanding of the operation-equation.

29. In this context, operations-equations were believed to have only one root. I deal with this issue in my book in preparation.

30. The transmission of the diagrams was problematic. For what follows, I am relying here on my work on the critical edition of the text in preparation.

31. Lam 1977, 260–262, explains the algorithm, but the figure given there does not correspond to that contained in the sources. For details, see Chemla forthcoming.

32. Since previous historiography of mathematics in China has mainly emphasized the facet of the equation represented by the written diagrams, overlooking the graphic formula, the break has appeared less dramatic than it actually is. The graphic formula has offered the support to establish the equation and address the correctness of the algorithm solving it. These two activities have also been for the most part overlooked.

33. Using terms introduced in chapter 1 in this volume, the cultural history of equations in China shows two distinct forms of "path dependence."

34. Rashed 1986 provides a critical edition, a translation, and an analysis of the book. Chemla 1992 is an outline of the following argument.

35. Adolf Pavlovitch Juschkewitsch ([1961] 1964) also believed it. He was, however, relying only on the source material available at the time. The discovery of *On Equations* by Tusi does not undermine the conclusion. As Donald MacKenzie noted during the 2011 discussion at Les Treilles, this chapter "draws an analytical distinction between cultures and concepts, when in the Geertzian notion concepts are surely at the heart of culture" (see Geertz 1973). The transformations and circulations this paragraph evokes show how this distinction might help us not to "orientalize."

References

Chemla, K. 1992. "De la synthèse comme moment dans l'histoire des mathématiques." *Diogène* 160: 97–114.

Chemla, K. 1994. "De la signification mathématique de marqueurs de couleurs dans le commentaire de Liu Hui." *Cahiers de linguistique—Asie Orientale* 23: 61–76.

Chemla, K. 2001. "Variété des modes d'utilisation des *tu* dans les textes mathématiques des Song et des Yuan." Preprint given at the conference "From Image to Action: The Function of Tu-Representations in East Asian Intellectual Culture," Paris, September 3–5. http://halshs.ccsd.cnrs.fr/halshs-00000103/.

Chemla, K. 2003. "Generality above Abstraction: The General Expressed in Terms of the Paradigmatic in Mathematics in Ancient China." *Science in Context* 16(3): 413–458.

Chemla, K. 2009. "Mathématiques et culture: Une approche appuyée sur les sources chinoises les plus anciennes." In *La mathématique. 1. Les lieux et les temps*, ed. C. Bartocci and P. Odifreddi, 103–152. Paris: CNRS.

Chemla, K. 2010a. "從古代中國數學的觀點探討知識論文化" ("An Approach to Epistemological Cultures from the Vantage Point of Some Mathematics of Ancient China"). In 中國史新論．科技史分冊：科技與中國社會 (*New Views on Chinese History: Volume on the History of Science and Technology: Science, Technology, and Chinese Society*), ed. P. Chu, 祝平一, 181–270. Taipei: Lianjing 聯經.

Chemla, K. 2010b. "Changes and Continuities in the Use of Diagrams Tu in Chinese Mathematical Writings (3rd Century–14th Century) [I]." *East Asian Science, Technology and Society* 4: 303–326.

Chemla, K. Forthcoming. "How Did One, and How Could One, Approach the Diversity of Mathematical Cultures?" *Proceedings of the 7th European Congress of Mathematics*, Berlin, August 18–22, 2016, ed. Martin Skutella.

Chemla, K., and Guo, S. 2004. *Les neuf chapitres: Le classique mathématique de la Chine ancienne et ses commentaires*. Paris: Dunod.

Cullen, C. 1996. *Astronomy and Mathematics in Ancient China: The Zhou Bi Suan Jing*. Cambridge: Cambridge University Press.

Geertz, C. 1973. *The Interpretation of Cultures*. New York: Basic Books.

Guo, X. 郭熙漢. 1996. "楊輝算法"導讀 (*Guide to the Reading of "Mathematical Books by Yang Hui"*). Hankou: Hubei sheng jiaoyu chubanshe.

Hilbert, D. 1900. "Mathematische Probleme." *Göttinger Nachrichten* 3: 295; trans. M. W. Newson in *Bulletin of the American Mathematical Society* 8: 437–479, 1902.

Horiuchi, A. 2000. "La notion de *yanduan*: Quelques réflexions sur les méthodes 'algébriques' de résolution de problèmes en Chine aux Xe et XIe siècles." *Oriens-Occidens* 3: 235–258.

Juschkewitsch [Youschkevitch], A. P. [1961] 1964. *Geschichte der Mathematik im Mittelalter*. Leipzig: Teubner.

Kodama, A. 児玉明人. 1966. 十五世紀の朝鮮刊——銅活字版數學書 (*Mathematical Books Printed in Korea in the Fifteenth Century with Copper Moveable Types*). Tokyo: 無有奇奄雙私刊.

Lam, L. Y. 1969. "On the Existing Fragments of Yang Hui's *Hsiang Chieh Suan Fa*." *Archive for History of Exact Sciences* 6(1): 82–88.

Lam, L. Y. 1977. *A Critical Study of the Yang Hui Suan Fa*. Singapore: Singapore University Press.

Li, J. 李繼閔. 1990. 東方數學典籍 ——《九章算術》及其劉徽注研究 (*Research on the Oriental Mathematical Classic "The Nine Chapters on Mathematical Procedures" and on Its Commentary by Liu Hui*). Xi'an: Shaanxi renmin jiaoyu chubanshe.

Li, J. 李繼閔. 1998. 九章算術導讀與譯註 (*Guidebook and Annotated Translation of "The Nine Chapters on Mathematical Procedures"*). Xi'an: Shaanxi renmin jiaoyu chubanshe.

Li, Yan 李儼. 1958. 中國數學大綱 (*An Outline of Chinese Mathematics*), rev. ed. 2 vols. Beijing: Science Press.

Li, Yan 李儼, and S. Du 杜石然. 1963. 中國古代數學簡史 (*Concise History of Mathematics in Ancient China*). Beijing: Zhongguo qinghua chubanshe.

Li, Yan, and S. Du. 1987. *Chinese Mathematics: A Concise History*. Trans. J. N. Crossley and A. W. C. Lun. Oxford: Clarendon Press.

Li Ye 李冶. 1248. *Ce yuan haijing* 測圓海鏡 (*Sea Mirror of the Measurements of the Circle*). Tongwenguan edition, 1876.

Martzloff, J. C. 1997. *A History of Chinese Mathematics*. Trans. S. S. Wilson. Berlin: Springer.

Qian, B. 錢寶琮. 1966. 宋元數學史論文集 (*Collected Papers on the History of Mathematics during Song and Yuan Dynasties*). Beijing: Science Press.

Qian, B. 錢寶琮. 1981. 中國數學史 (*History of Mathematics in China*). Beijing: Science Press.

Rashed, R. 1986. *Sharaf Al-Din Al-Tusi: Œuvres mathématiques*. 2 vols. Paris: Les Belles Lettres.

Shen, K. 沈康身. 1997. 九章算術導讀 (*Guide for the Reading of "The Nine Chapters"*). Hankou: Hubei jiaoyu chubanshe.

Shen, K., J. N. Crossley, and A. W.-C. Lun. 1999. *The Nine Chapters on the Mathematical Art: Companion and Commentary*. Oxford and Beijing: Oxford University Press and Science Press.

Te, G. 特古斯. 1990. "劉益及其佚著《議古根源》" ("Liu Yi and His Lost Book *Discussing the Source of the Ancient (Methods)*"). 數學史研究文集 (*Collected Research Papers on the History of Mathematics*) 1: 56–63.

Yan, D. 嚴敦傑. 1966. "宋楊輝算書考" ("Examination of Mathematical Books by Yang Hui of the Song Dynasty"). In 宋元數學史論文集 (*Collected Papers on the History of Mathematics during Song and Yuan Dynasties*), ed. B. Qian 錢寶琮, 149–165. Beijing: Science Press.

Yongle dadian 永樂大典 (*Grand Classic of the Yongle Period*). 1408. The part on mathematics of this encyclopedia ranged from chapter 16336 to 16357. Today only chapters 16343 and 16344 remain and are kept at Cambridge University Library. Reprint: Beijing 北京: zhonghua shuju 中華書局, 1960.

Yang Hui 楊輝. 1261. *Xiangjie jiuzhang suanfa* 詳解九章算法 (*Detailed Explanations of The Nine Chapters on Mathematical Methods*). Edition from the *Yijiatang congshu* 宜稼堂叢書, edited by Yu Songnian 郁松年, 1842.

BRUNO BELHOSTE is Professor of History of Science at the University of Paris 1. His main research interest is the development of science in France in the eighteenth and nineteenth centuries. He is the author of *Histoire de la science moderne de la Renaissance aux Lumières* (2016) and *Paris savant. Parcours et rencontres au temps des Lumières* (2011).

KARINE CHEMLA is Senior Researcher at the French National Center for Scientific Research (CNRS), in the laboratory SPHERE (CNRS and University Paris Diderot), and the European Research Council project SAW (Mathematical sciences in the ancient world) (https://sawerc.hypotheses.org). She focuses, from a historical and anthropology viewpoint, on the relationship between mathematics and the cultural contexts in which it is practiced. Chemla published, with Guo Shuchun, *Les neuf chapitres* (2004). She edited *The History of Mathematical Proof in Ancient Traditions* (2012); *Texts, Textual Acts and the History of Science* (with J. Virbel, 2015); and *The Oxford Handbook of Generality in Mathematics and the Sciences* (with R. Chorlay and D. Rabouin, 2016).

CAROLINE EHRHARDT is Maître de conférences (associate professor) in the History of Science at the Université Paris 8 Vincennes Saint-Denis (France). Her research investigates the social and cultural history of mathematics. She has worked on the history of algebra in the nineteenth century and on the history of French secondary and higher mathematics education during the modern and contemporary periods. Her publications include *Évariste Galois, la fabrication d'une icône mathématique* (2011) and *Itinéraire d'un texte mathématique: les réélaborations d'un mémoire d'Évariste Galois au 19e siècle* (2012).

FA-TI FAN is the author of *British Naturalists in Qing China: Science, Empire, and Cultural Encounter* (2003) and numerous articles on science in twentieth-century China and on the global history of science.

KENJI ITO is Associate Professor at SOKENDAI (the Graduate University for Advanced Studies), Hayama, Japan. The main area of his research is the history of physical sciences and technology in twentieth-century Japan. Topics of his publications include the history of physics in Japan, cultural images of robots and A-bombs, and amateur videogame culture in contemporary Japan. He is currently working on a biography of Nishina Yoshio and a book on the introduction of quantum mechanics into Japan.

EVELYN FOX KELLER is Professor Emerita of History and Philosophy of Science in the Program in Science, Technology and Society at Massachusetts Institute of Technology. She received her PhD in theoretical physics at Harvard University, worked for a number of years at the interface of physics and biology, and then turned to the study of gender and science, and more generally, to the history and philosophy of science. She is the author of numerous books, the recipient of many awards and honorary degrees, a member of the American Philosophical Society, the American Academy of Arts and Sciences, a MacArthur Fellow, and recipient of the Chaire Blaise Pascal in Paris.

GUILLAUME LACHENAL is Associate Professor in the Department of History of Science at the Université Paris Diderot. His research focuses on the history and anthropology of biomedicine in Africa, from the colonial times to the contemporary global health era. He has recently published *Le médicament qui devait sauver l'Afrique* (2014; English translation is forthcoming) and coedited the catalogue *Traces of the Future*, an archeology of medical science in Africa (2016).

DONALD MACKENZIE is Professor of Sociology at the University of Edinburgh. He is a sociologist and historian of science and technology. His current research is on the sociology of financial markets, in particular the development of automated high-frequency trading and of the electronic markets that make it possible, with a special focus on how trading algorithms predict the future. His books include *Inventing Accuracy: A Historical Sociology of Nuclear Missile Guidance* (1990); *An Engine, Not a Camera: How Financial Models Shape Markets* (2006); and *Material Markets: How Economic Agents Are Constructed* (2009).

MARY S. MORGAN is the Albert O. Hirschman Professor of History and Philosophy of Economics at the London School of Economics and holds a visiting fellowship at University of Pennsylvania. She is an elected Fellow of the British Academy, and an Overseas Fellow of the Royal Dutch Academy of Arts and Sciences. Her research interests cover questions about models, measurements, observation, experiments—the practical side of economics. Her two most recent books are *The World in the Model* (2012) and *How Well Do Facts Travel?* (2011). She is currently working on poverty measurement, the performativity of economics, and narrative explanation in the sciences.

NANCY J. NERSESSIAN is Regents' Professor (emerita), Georgia Institute of Technology. She currently is Research Associate in the Department of Psychology at Harvard University. Her research focuses on the creative research practices of scientists and engineers, especially how modeling practices lead to fundamentally new ways of understanding the world. This research has been funded by the National Science Foundation and National Endowment for the Humanities. She is a Fellow of American Association for the Advancement of Science and of the Cognitive Science Society as well as a Foreign Member of the Royal Netherlands Academy of Arts and Sciences. Her numerous publications include *Creating Scientific Concepts* (2008; Patrick Suppes Prize in Philosophy of Science, 2011) and *Science as Psychology: Sense-Making and Identity in Science Practice* (with L. Osbeck, K. Malone, and W. Newstetter, 2011; American Psychological Association William James Book Prize, 2012).

DAVID RABOUIN is Senior Research Fellow (CR1) at the French National Center for Scientific Research (CNRS) in the research group SPHERE (CNRS and Université Paris Diderot). His interest is in the history of philosophy and mathematics in early modern times, with a special focus on Descartes and Leibniz. He also works in contemporary French philosophy. He is the author of *Mathesis universalis: L'idée de « mathématique universelle » d'Aristote à Descartes* (2009). With Norma B. Goethe and Philip Beeley, he edited the collection titled *G.W. Leibniz, Interrelations between Mathematics and Philosophy* (2015).

HANS-JÖRG RHEINBERGER is Director Emeritus at the Max Planck Institute for the History of Science in Berlin and a molecular biologist and historian of science. His current research interests include the history and epistemology of experimentation, the history of the life sciences, and the relation between the sciences and the arts. Among his recent publications are *On Historicizing Epistemology* (2010); *An Epistemology of the Concrete* (2010); *A Cultural History of Heredity* (2012, with Staffan Müller-Wille); "Culture and Nature in the Prism of Knowledge" (*History of the Humanities* 1, no. 1, 2016).

CLAUDE ROSENTAL is Research Professor of Sociology at Centre National de la Recherche Scientifique, Director of the Center for the Study of Social Movements at the École des Hautes Études en Sciences Sociales, and head of the Science and Technology Group at Institut Marcel Mauss in Paris. His publications include works on the sociology of science, logic, and public demonstrations. He is the author of *Weaving Self-Evidence: A Sociology of Logic* (2008), *Les capitalistes de la science* (2007), and *La cognition au prisme des sciences sociales* (with B. Lahire, 2008).

KOEN VERMEIR is Senior Researcher at the CNRS laboratory SPHERE (University of Paris-Diderot). He is a Global Young Academy Fellow and co-directs the project Cultures of Baroque Spectacle. As a historian and philosopher specializing in early modern science, religion, and technology, he is now working on the long-term history of the concept of culture. Before coming to Paris, Vermeir has held positions in the United States, the United Kingdom, Belgium, Germany, and Switzerland. He is on the editorial boards of *Journal of Early Modern Studies, Society and Politics, Artefact, Studies in History and Philosophy of Science,* and *International Archives of the History of Ideas.*

Duméril, Constant, 255
Du Shiran 杜石然, 393n18, 393n21, 394nn21–22

Education, 60, 71, 228–41, 245n45, 247, 303, 305, 311, 333, 339
Ehrhardt, Caroline, 19–20, 23, 327–51, 399
Einstein-Podolsky-Rosen thought experiment, 60–61, 66, 66n4
Eisenstein, Gotthold, 335, 350–51
Elementary practices, or elements of practice, 355–56, 363, 368–70, 379–81
Elisabeth, Princess, 211, 220n30
Elite, 80, 82, 238, 302, 314–15
Engineering sciences, 118–88
Engineering values, 126, 133, 139–40
English Algebraic School, 333
Epistemic cultures, 2, 10, 12, 15, 21, 30, 47, 118–19, 228, 241, 289
Epistemic object, 11, 15
Epistemological cultures, 2, 10–12, 21, 99, 107–11, 118–19, 133, 139, 142, 228, 250–54, 271, 274
Epistemological factor, 2
Epistemological relativism, 105
Epistemological universalism, 105
Epistemological values, 10, 15, 18, 108, 110, 251–53, 267, 328
Epistemology, 82, 107, 109, 163–64, 197, 202–3, 207, 210, 251, 267, 294, 310, 401; historical, 278, 290–91
Equation(s), 19, 206, 212, 221, 327–34, 338–46, 348n18, 354–59, 374–96; algebraic, 210–11, 328, 332, 337, 346, 347nn4–5, 350–54, 377, 394n21; quadratic, 20–21, 353, 355–74, 377–86, 390, 392n8, 394nn21–22
Essence, 44, 51, 55, 114, 245, 333, 335; essentialism, 4, 7, 29, 99, 102–9; essentializing, 228, 239, 242n6, 315
Ethnographic studies, 2, 117
Etymology, 229–30
Euler, Leonhard, 212, 222–23
European Commission, 170, 189n5, 191–94
Exchange, 170, 173, 178, 181–82, 184, 193, 347
Experience, 35, 97, 100, 105, 148, 150–64, 167n27, 172, 186, 284, 304, 310–11, 321–22
Experiential knowledge, 162–63
Experimental culture, 17, 126, 278–79, 287–88, 290, 292

Experimental system, 16–17, 126, 248, 278, 286–87, 290, 292
Expertise, 40, 79, 90, 133, 161–63, 167n26, 186, 190, 287, 310–13, 320

Facts, 160; fact-value interdependence, 164; middle-level facts, 146; supply-side theory, 161; travelling facts, 169
Family resemblance, 105, 378
Fan, Fa-ti, 17–18, 188, 296–323, 399
Faujas de Saint-Fond, Barthélémy, 254, 257, 262–64, 277
Febvre, Lucien, 6, 242, 246
Feferman, Solomon, 214, 222, 393
Feminism, 11, 101–3, 312
Feminist establishment, 159
Feminist movement, 103, 154, 157, 162
Feminist theory, 7, 99, 101–3, 114
Fermat, Pierre de, 202, 212, 219, 222–23, 349
Ferreiros, José, 213–14, 222, 335, 349
Feynman diagrams, 50, 59, 61–65, 67
Figures, 32, 130–31, 149, 158, 259–60, 264, 269, 304, 358–59, 361, 364–65, 367, 373, 375, 387–88
Fitch, 34
Fleck, Ludwik, 1–2, 24, 190n32, 191, 220n28, 222
Forms of communication between cultures, 12, 17, 184
Foucault, Michel, 252, 277
Fox Keller, Evelyn. See Keller, Evelyn Fox
Fraenkel, Abraham, 214
Francesconi, Marco, 155–56, 167
Frank, Jeff, 155–56, 167
Fraser, Craig, 212, 222
Fréchet, Maurice, 199–200, 219n16, 221
Fukushima Nuclear Accident Independent Investigation Commission (NAIIC), 65, 67
Furner, Mary, 162, 167

Galois, Évariste, 19, 327–47, 347n6, 348n14, 348n18, 349–51
Galois theory, 19, 327–28, 334–36, 338–47
Gaussian copula, 36, 42, 45nn12–13, 47
Gayon, Jean, 199, 203, 217nn1–2, 222
Geertz, Clifford, 3, 25, 106, 204, 222, 279, 293, 396n35
Gender, 3, 7, 101–4, 108, 110, 152, 157, 167–69
Gene, 124, 284, 288, 293
Geoffroy Saint-Hilaire, Etienne, 254, 256

Printed and bound by CPI Group (UK) Ltd, Croydon, CR0 4YY

27/10/2024

14580226-0005